AGING AND TECHNOLOGICAL ADVANCES

NATO CONFERENCE SERIES

I   Ecology
II   Systems Science
III   Human Factors
IV   Marine Sciences
V   Air–Sea Interactions
VI   Materials Science

III HUMAN FACTORS

Recent volumes in this series

AGING AND TECHNOLOGICAL ADVANCES

Edited by

Pauline K. Robinson
Judy Livingston
and
James E. Birren

Ethel Percy Andrus Gerontology Center
University of Southern California
Los Angeles, California

Associate Editors
Victor A. Regnier
Arnold M. Small
Harvey L. Sterns

Published in cooperation with NATO Scientific Affairs Division

PLENUM PRESS · NEW YORK AND LONDON

Library of Congress Cataloging in Publication Data

Symposium on Aging and Technological Advances (1983: University of Southern
  California)

  Aging and technological advances.

  (NATO conference series. III, Human factors; v. 24)
  "Proceedings of a Symposium on Aging and Technological Advances, sponsored by
the NATO Special Program Panel on Human Factors, held August 22–26, 1983, at the
University of Southern California, Los Angeles, California"—Verso of t.p.
    "Published in cooperation with NATO Scientific Affairs Division."
    Includes bibliographical references and index.
    1. Aged—Employment—Congresses. 2. Technological innovations—Social aspects
—Congresses. 3. Aged—Services for—Technological innovations—Congresses. 4.
Gerontology—Congresses. I. Robinson, Pauline K. II. Livingston, Judy. III. Birren,
James E. IV. Nato Special Program Panel on Human Factors. V. North Atlantic Treaty
Organization. Scientific Affairs Division. VI. Title. VII. Series.
  HD6279.S96 1983   331.3'98   84-17819

  ISBN-13: 978-1-4612-9464-1   e-ISBN-13: 978-1-4613-2401-0
  DOI: 10.1007/978-1-4613-2401-0

Proceedings of a Symposium on Aging and Technological Advances, sponsored by
the NATO Special Program Panel on Human Factors, held August 22–26, 1983,
at the University of Southern California, Los Angeles, California

©1984 Plenum Press, New York
Softcover reprint of the hardcover 1st edition 1984

A Division of Plenum Publishing Corporation
233 Spring Street, New York, N.Y. 10013

PREFACE

The chapters and reports in this publication have been selected from presentations at a Symposium on "Aging and Technological Advances" held in August, 1983 at the Ethel Percy Andrus Gerontology Center of the University of Southern California. The Symposium was made possible by a grant from the NATO Special Programme Panel on Human Factors, and the support of this program is gratefully acknowledged.

Members of the Symposium Advisory Board were James E. Birren, Judy Livingston, Erhard Olbrich, Victor Regnier, Pauline Robinson, Thomas Singleton, Arnold Small, Harvey Sterns, and Alvar Svanborg. Professor Lambros Houssiadas also provided invaluable encouragement.

Appreciation is also extended to the Andrew Norman Institute for Advanced Study in Gerontology and Geriatrics for support of planning activities leading up to the Symposium and for support of events surrounding the Symposium itself.

A generous gift from The UPS Foundation to the Ethel Percy Andrus Gerontology Center made possible the compilation, editing and preparation of this manuscript and helped to support Symposium activities.

We thank David Bergstone and Mary Margaret Ragan who together carefully and skillfully organized and carried out the typing of the manuscript.

Pauline K. Robinson
Judy Livingston
James E. Birren

CONTENTS

SECTION II: HEALTH AND STRESS

AGING AND TECHNOLOGICAL ADVANCES:

INTRODUCTION

Pauline K. Robinson and James E. Birren

Andrus Gerontology Center
University of Southern California

The Symposium on Aging and Technological Advances, sponsored by the NATO Special Programme Panel on Human Factors, was convened at the Ethel Percy Andrus Gerontology Center at the University of Southern California in August of 1983. Over 100 participants from 15 countries (both NATO and non-NATO countries) and from a number of scientific disciplines participated in the five-day symposium. The chapters and brief research reports included in this volume were selected from the presentations at the meeting; many were revised following the exchange that took place during the sessions. The diversity of opinion of the authors was one of the most interesting aspects of the symposium and it is reflected in the papers in this volume.

The organization of the book follows the format of the symposium, in which participants met both in full plenary sessions and also in four separate sections. The plenary session papers appear first in this volume, followed by the four major divisions of the book which reflect the topics of the symposium sections, namely: Labor Force Participation, Health and Stress, Human Factors, and Home and Community. Reports on the discussion in the four symposium sections are at the end of each section of the book. Reflecting the focus of the Human Factors Panel of NATO, the symposium deliberately did not include topics related to advances in medical technology.

The conveners, James E. Birren and Pauline K. Robinson, presented several challenges to the participants at the opening of the symposium. The first charge was to recognize from the outset that this scientific gathering was faced with a heavily value-laden topic. We would be examining both the promises and the hazards of

technological advances for the aged, and underlying our discussions
would be value assumptions about the importance for the aged of
independence, integration in society, interpersonal contact,
opportunities to contribute productively to society, control over
one's environment, self esteem, and the quality as well as the
length of life.

The theme of "good" and "bad" consequences of technological
advances, from the perspective of the impact on the aged, emerged
continually during the symposium and is reflected in the papers in
this volume. Some of the favorable outcomes of technological
advances examined were the reduction of physical work load, the
facilitating of communication, compensation for infirmities, and the
increased safety of individuals living alone in their homes. On the
other hand, some of the undesirable outcomes discussed were the
stress of the faster pace of work, increased obsolescence of older
workers, displacement of older workers, the reduction of
face-to-face interaction, and the weakening of family and other
societal ties.

A second charge to the participants was to recognize the
diversity of the older populations under study. The impact of
technological advances on the "aged" varies between the more and
less developed regions, among nations, and between urban and rural
areas within nations. Furthermore, discussions of technology and
aging cover a wide age span, from workers in their forties and
fifties to the very old. Within this age span are individuals who
are quite frail and dependent. The very definition of when old age
begins varies from the less developed regions, where life expectancy
is lower and old age begins earlier, to the more developed regions
of the world.

Third, symposium participants were reminded of some of the
viewpoints that have been gained from past research in gerontology.
One of these is that age differences are due not only to the effects
of aging per se, but also to cohort effects--that is, the effects of
growing up and growing older in different periods of time. In
discussions of technology and its effects on the future aged, it is
therefore important to keep in mind that the aged of tomorrow will
be unlike the aged of today in many important respects. For
instance, they will have gained more years of formal education and
more direct experience with technological innovations such as the
computer.

Another viewpoint to be gained from gerontological research is
that the composition of populations of the aged will be different in
the future. The aged of the world of the future will increase, and
will include a greater representation from the less developed
regions and a greater proportion of the very aged. The growth rate
of the 60+ population of the world between 1980 and 2000 exceeds

that of every other broad age segment; the number of 60+ will double, or nearly double, in countries such as the Philippines, Korea, India, and Thailand during this period. In Europe the 60+ segment will be 20% of the total population by 2000. The number of persons 60 years old and older from the less developed regions will increase from less than half to over half of the world's population of 60+ by the year 2000.[1] There will be a world-wide increase in the size and proportion of the very aged population, since the number of 80 year old and older persons will increase faster than the number of 60-79 year olds. This increase of the very old will be particularly acute in the less developed regions, where the number of 80+ will double by the year 2000 and then will double again by the year 2020.[2]

There are a number of frustrations in the attempt to study technology and aging. The direction of technological innovations in the future is almost by definition difficult to predict; and even with technological trends, their impact is difficult to measure. Technological change sometimes seems to take on a life of its own, largely out of our control. Sometimes the very concept of "aging" in the context of the symposium topic is elusive, since chronological age is often less significant to the discussion than is presence or lack of technological skills, disability, infirmity, and dependence--at any age. Many important empirical and policy questions remain unresolved regarding the impact of technology on the aged.

Nevertheless, the importance of technological change for the well-being of the aging population is evident in the papers and research reports in this volume. The growing numbers of older persons are especially vulnerable to technological change at the same time that their quality of life can be enhanced by technology. This volume deals with the intersection of two major changes in our world, the accelerating pace of technological advances and the accelerating growth of older populations.

NOTES

[1] The United Nations population projections designate the More Developed Regions as North America, Japan, Europe, Australia, New Zealand, and the USSR; the Less Developed Regions are designated as as Africa, Latin America, other East Asian countries (excluding Japan), South Asia, Melanesia, and Micronesia-Polynesia.

[2] The population data in this introduction are from Jacob S. Siegel, "Demographic Aspects of the Health of the Elderly to the Year 2000 and Beyond," World Health Organization, 1983.

AGING IN A TECHNOLOGICAL SOCIETY:

DESIDERATA AND DILEMMAS IN DECISION-MAKING

Ida Hoos

Space Sciences Laboratory
University of California
Berkeley, California

INTRODUCTION

Every society has its ways of dealing with its aging and
elderly members. To say that these are a reflection of the
society's mores, its ethos, its religion, and also its economic and
political condition may sound like a truism. But it is an
undeniable social fact of life that this milieu determines, in large
part, the place of and prospects for all sectors, and certainly the
elderly. It is, therefore, to the priorities and values prevailing
in the final decades of the twentieth century that we must look for
the shape of the future. Only a flip of the calendar away from
1984, we already live within the Orwellian shadow. In the rush into
tomorrow's world, science and technology play a powerful role.
Historiographers offer theories as to the nature of social change.
Spengler, Toynbee and Sorokin had differing views except that for
all of them technology was the major driving force. Sorokin[1]
put forward the notion of the Zeitgeist, the spirit of the times,
pervading the particular period; he said that every era has its own
character, expressed in its system of arts, truth, ethics,
government, law, institutions, etc. The names reveal the nature and
values of past historical periods. There are the Dark Ages, the
Renaissance, and so on. Ours having been called the Space Age, the
Computer Era, the Scientific Society, we can be expected to envince
a high regard for and receptivity toward technology in its various
manifestations. We find ourselves in the Sensate phase of Sorokin's
cyclical dynamic, its salient characteristics being this-worldly,
pragmatic, utilitarian, and hedonistic. Other features of the
Sensate phase: quantity instead of quality; know-how instead of
knowledge.

5

Some twenty years ago, Ralph Lapp[2] wrote a book called The
New Priesthood, which, as the name implies, foretold the advent of
technology as a theology. He also introduced a useful concept--the
technological imperative, about which he wrote:

> Technological possibilities are irresistible to man.
> Once an invention or discovery is made, it takes on an
> imperative of its own and will be used. Moreover,
> besides creating its own further demand, it carries
> within it a self-justifying mechanism that rationalizes
> that usage. (Emphasis added)

There may be some persons among us who remember when the adage,
"Necessity is the mother of invention," had some validity. Now, in
these days of the technical fix and determined technology transfer,
it is apparent, as Lapp suggests, that invention has become the
mother of necessity.

Increasing complexity and interrelatedness of phenomena have
encouraged and lent credibility to the notion that there are
techniques that can bring order to the chaos, that there is a
methodology that is "logical," "rational," and "scientific" to solve
the tangled problems of modern society. Such a techno-logical model
for decision-making finds a hospitable environment in our
Technological Era and has been brought to bear in every facet of our
lives...health, welfare, education, energy, war, and peace. The
tools are derived from management science and are known under a
growing glossary of names. Basically, they have the same
genealogical roots, contain the same elements, and perpetuate the
family characteristics.[3] Born as operations research in the
Battle of Britain, spawned in the Pentagon and used in the War on
the Potomac as systems analysis, nurtured in government agencies as
program budgeting, applied assiduously in environmental impact
assessments, and emerging as the core of risk analysis and
technology assessment, the methodology is our dominant
paradigm.[4] The methodology has thoroughly saturated
decision-making processes.

It has certain features of special interest in this context:
(a) it is intolerant of and even antagonistic to unquantifiable
variables; (b) it harbors a bias against social and human factors;
(c) it has a strong now orientation; (d) its ethic is that of the
marketplace; (e) it is generally tinged with technical optimism.
Called by McNamara "the biggest bang for the buck," when he applied
the principle to weapons development, cost/benefit analysis is still
the linch pin of the "rational" decision-making techniques. This
ratio, which must, of course, always appear favorable in
quantitative terms in order to justify any program or course of
action, has come to constitute the iron law of budgeting. Not
surprisingly, the entire nation performs this kind of St. Midas

dance to satisfy the all-powerful Office of Management and Budget,
which evaluates proposals. In all branches of government, and in
business, too, the canons of cost/effectiveness prevail, no matter
what the issue may be--air pollution, highway safety, public
education policy, nuclear hazards, deregulation of nursing homes.
The logic is as simple as it is simplistic and misleading: decisions
involve trade-offs and these must be made on the "rational" basis of
comparative payoff.

     We must remind ourselves, however, that cost/benefit analysis
is a subjective exercise implicit in almost any choice and quite
jejune at that. Only through its re-attribution in consort with
systems analysis did this common-sense rule of thumb exercise become
imbued with the mystique permeating our times and accredited as part
of the arsenal of "powerful tools of technology." By supplying the
weapon for executing economic ideology-turned-theology,
considerations of cost/effectiveness became the fulcrum of the
budgetary process--the numbers game played to "rationalize"
regulation policies and to condone reliance on the marketplace to
take care of the public's health, safety, and environment. With
trade-off the epicenter of public management, long-term costs to the
populace-at-large rate low priority when vested interests are at
stake. Social costs can never be calculated numerically and even
dollar cost/benefit comparison involves an arbitrary assignment of
weights. Who is the Peter being robbed and the Paul being paid is a
value judgment. Cost/benefit analysis is a teeter-totter that tilts
as it is loaded. If, as is my contention, the cost/benefit
stricture in public planning supports and reflects a doctrine of
immediate relevance and pay-off, then herein lies discrimination,
ratified "rationally," against certain types of programs and certain
sectors of the population. Caught in the fabric of disadvantage are
poor people, young and old. Candidly stated, the fact of the matter
is--they are not cost/effective. How insidious and antisocial is
the bias can perhaps be appreciated as we ponder the special
case--that of the elderly, and especially in relation to some of the
areas designated for study during this symposium.

TECHNOLOGY AND WORK LIFE

     Common sense would prescribe that before embarking on a
discussion of how advancing technology affects the work- and
after-work-life of persons advancing in years, we should establish
some reference points. Just who, for example, are the elderly? How
many of them are there? Where are they? How are they deployed in
the labor force? Are they employed? Are they unemployed? Are they
employable? Are their chances for work enhanced or are they
vulnerable to technological changes in the workplace? It is indeed
paradoxical, in an era when information technology generates so much
data, that we still lack definitive answers to many of these

questions. There are, to be sure, statistics, but looking to them
for enlightenment is much like Plato's savages' scrutiny of the
shadows on the wall of their caves for clues to reality.

Phillipe Aries[5] reminds us that aging, like childhood, is a
social construct, until recently in European society more a matter
of physiology than of chronology. Loss of teeth and eyesight,
diminished mental and physical capacity were the prime indicators.
Medical technology has increased longevity and counteracted some of
these disabilities but chronology has remained the benchmark in the
United States since 1935, with enactment of the Social Security Act,
although the number is strictly arbitrary. There have been attempts
at definitional refinement but these appear to be more euphemistic
niceties than serious taxonomy. If we reconcile ourselves to
acknowledging that, in view of medical advances, it is idle pedantry
to pursue any answer beyond the 65 figure, then we will have to
content ourselves with Browne and Olson's "ballpark" statement[6]
about their number, viz.:

> Twenty-three million older Americans 65 and over now
> belong to this group (the elderly), which increased
> from approximately 10 to 12 percent of the population
> between 1970 and 1980. Within thirty-five years, if
> trends continue, 15 percent of the population will be
> over 65 years of age.

According to Myers,[7] there are 24,928,000 persons aged 65
and older in the United States in 1980. He says that in 2000 there
will be 31,822,000 persons in that group, with the number of
extremely old persons, 80 years and over, increasing 56.1 percent
and thus constituting 24.4 percent of the aged. Such dramatic
demography condones all manner of generalization. It is also a
temptation for just such stereotypes and cliches as Binstock[8]
includes on his list of "axioms of public rhetoric." He recommends
that, if we are to achieve an accurate understanding about the
elderly, we must disaggregate the data. This, however, could lead
to several undesirable consequences. However we break out
categories for special scrutiny, the sources, quality, and
reliability of the statistics upon which we draw are subject to
question.

It is clear that we lack robust data on both numbers of elderly
persons and their employment status. The infirmity and infidelity
of figures being recognized, small wonder that data are always being
doctored! No cure is in sight. In fact, the condition is bound to
deteriorate, thanks to the zeal of the Office of Management and
Budget for cost/effectiveness within its operations. Construing
cost-cutting as a benefit per se, that agency in 1982 eliminated its
Statistical Policy Branch, the responsibiliy of which was to assure
coherence and integrity of activities and standards among various

federal statistics programs. Affected was the Labor Department's Employment and Training in Administration, which might have been a useful source of information in matters which concern us here.

Some caveats are in order, moreover, when we look to research with a special focus, such as women, minorities, elderly poor, older persons living in rural areas, or such. Data gathered for a purpose may lack objectivity and may cause a certain amount of distortion. We know all too well Kaplan's[9] story of the drunkard's search. The inclination to search where the light is may be natural, innocent and free of guile, but not necessarily so. As gerontologists are aware, many disparate and divergent issues have crystallized vis-a-vis aging policy. Programs purported to serve the needs of the elderly are interlaced with agenda sometimes hidden and interfaced with other policy objectives. According to Browne and Olson[10], they "have been superimposed on ongoing market structures and shaped by powerful and self-serving political and economic interests." Margaret E. Kuhn[11] of the Gray Panthers contends, as does Carroll L. Estes,[12] that old people have become the bailiwick of an array of specialists, whose prescriptions might advance their own needs and interests rather than those of the aged. The service providers, i.e., the gerontological professionals, stand to gain from the "self-perpetuating service delivery network."

If, consistent with the dictates of a technological era, we look to conventional technology assessments for clues as to what is happening to work and jobs, we find ourselves in the position of the six blind men "who went to see the elephant, though all of them were blind."[13] Like them, each of us may be partly in the right; on the other hand we may be all wrong. And certainly we are "prone to prate about an elephant that none of us has seen." So fast are changes occurring and across so broad a front that we hardly known what they are, to say nothing of what their impacts will be. We do, however, know why there are blinders:

> Some of the developments are already further advanced
> than many realize--indeed, many companies are loath to
> trumpet their new systems for fear of encouraging
> hostile trade unions reaction.[14]

For clues as to how technology is affecting the labor market, historical reference to the Industrial Revolution or even to more recent innovation is irrelevant. In the past, new fields replaced those affected by automation. And, in periods of expansion, farm workers could move into industry; there was room in offices for some of the factory workers. There was even occasional evidence that some upgrading could occur as workers moved from heavy manual labor to tasks requiring more than muscles. In earlier times, especially during periods of economic expansion, firms could pursue a policy of

humane retrenchment as automation took over jobs. There were
incentives for early retirement, attrition rather than lay-off,
relocation, and the like.

Automation in today's terms is defined[15] as "the use of
micro-electronic and other technologies that either reduce the need
for people, enable people to perform more work, or perform functions
that people can not." Information technology is progressing rapidly
on all fronts, in manufacturing, in processing, in all branches of
office work, and in many kinds of service work at a time when the
world economy is in a state of decline. Usilaner cites predictions
pointing to potential loss of millions of jobs in the manufacturing
sector alone because of the use of robotics.

Usilaner cites two instances of what he calls "short-run
displacement": automated typesetting equipment, use of which has
led to the lay-off of many well-paid and highly-skilled typesetters
and robotics in the automobile manufacturing industries, where
machines are doing the jobs which once provided men with a living
wage. His "short-run" label is a misnomer; those jobs are gone
forever and with some of them has gone a whole avenue of access to
employment--from apprentice to journeyman. Perhaps it is a
manifestation of the commonly-held compensation theory, a
whistling-in-the-dark notion that losses in one sector will be
offset by growth in another. There is no reason to believe that
this technological revolution, which is taking place across a broad
front and affecting office, farm, and factory, will bring jobs.
Quite the contrary. It is being introduced and fostered because of
its labor-saving potential. Even the "high tech" industries are
highly automated and, hence, not a promising source of jobs.

A commonly-proposed solution--one which should probably be
attributed charitably to technical optimism--is of the
let-'em-eat-cake variety. Imbued with nostalgia for a past that
never existed, proponents of this prescription advise workers
threatened with unemployment to get retrained and upgrade their
skills. But this counsel is based on two fallacious assumptions:
(1) that there are jobs waiting to be filled and (2) that these jobs
are at higher skill levels than those made obsolete. As to the
first, we have already seen the displacement-without- replacement
impact of the new technologies--a matter which will be discussed
further in the context of office automation. For the moment, it is
necessary to consider the oversupply of labor and ponder its
implications for older workers. There are already some twelve
million Americans counted as unemployed. This number does not
include 20 percent of the workers holding only part-time jobs and
the untold thousands of "discouraged workers" who have simply given
up the job hunt. Length of periods of joblessness is growing; many
workers have voluntarily downgraded themselves in the desperate hope
of getting a job. Economists who promise pie-in-the-sky, cyclical

upturns, and Kondratiev long waves choose to overlook the
job-gobbling capability of the technology to which they look for
recovery. The prospects on the global scale are equally
gloomy.[16]

> To absorb the net increase in the labor force in the
> 1980's, 11 million jobs will need to be created in the
> United States on the assumption that the decade begins
> with an unemployment level of six million. In the
> United Kingdom as many additional jobs will need to be
> created as are already needed for the number of people
> currently unemployed; and France, starting the decade
> with an unemployment figure of about one million, will
> have to at least double the number of jobs that were
> created in the past 20 years. The Federal Republic of
> Germany entered the 1980's with about one million
> unemployed, and nearly two million jobs will need to be
> created during the decade to absorb the labor force
> increase.

There simply are not enough jobs to go around. This year's college
graduates joined the ranks of somewhere between two and three
million* who are either unemployed or in jobs for which a college
degree is not necessary. The fact that young persons with college
degrees willingly take jobs which high school graduates could do
speaks volumes about opportunities for older persons. Their life
span extended and their capacities prolonged by medical
technology,their expectations for continuing an active and
productive life have been raised. Indeed, the aging are exhorted by
a barrage of propaganda from the media, counselors, and gerontology
professionals to keep working, acquire new skills, and so on. The
problem is that, with the exception of "showcase" instances,
opportunity is severely limited. In a tight labor market,
competition is keen and age a handicap.

As to whether retraining is a solution for any sector of the
labor force, regardless of age, we return to the second of the
assumptions that we earlier labelled as fallacious, viz. that jobs
associated with new technologies are at higher skill levels than
those made obsolete. Evidence is mounting to substantiate the
contention that the "scientific management" introduced by Taylor to
"rationalize" production by breaking down tasks into standardized,
repetitious units first brought the assembly line, then automation,
and now electronic robotization into being. Cost/effective
operation in the factory called for first the deskilling and then
the elimination of a number of jobs.

Since, in earlier times, ascent up the occupational ladder went
from blue- to white-collar work, the office has been traditionally
regarded as a haven of opportunity. This is no longer so, thanks in

large part to microelectronics. The full impact on jobs is
temporarily masked by current expansion in the banking and insurance
industries, but already the rate of job growth in office work is
showing an annual decline while volume of transactions is climbing.[17]
Several officals in the French Ministry of Industry foresee a
30 percent reduction of clerical workers in the finance industries
by 1990. In offices, as was the case in factories, a continuous
routinization of previously skilled jobs seems to be preparing the
way for total elimination through the next phase of automation.[18]
According to two Stanford economists, the very proliferation
of high-technology industries is "far more likely to reduce the
skill requirements for jobs in the U.S. economy than to upgrade
them." They cite studies showing that the new technologies further
simplify and routinize tasks and reduce opportunities for worker
individuality and judgment. What should, of course, concern us as
we see the takeover by technology is the pressure on payrolls needed
as a cushion, such as Social Security, to ameliorate the impacts of
joblessness. The burdens of unemployment have always been
concentrated on certain sectors of the population; automation is
making them heavier.

We began this paper with special emphasis on jobs because work
has always occupied a place of central importance in our lives. Our
place in society, in our families, and in our own self-esteem
relates directly to what we do or have done. Whether a financial
necessity or not, work is considered so vital to well-being as to be
prescribed as therapy for the many maladies attributed to
retirement. But technology creates a double dilemma here: medical
advances are prolonging life expectancy and automation is gobbling
up the jobs. The average life expectancy for males is 73 and for
females 81, according to figures cited at the annual meeting of the
American Association for the Advancement of Science in May,
1983.[19] That this increase in life expectancy should have caused
several economists from the Office of Management and Budget to issue
"warnings" about the "ominous growth potential" in costs of Social
Security and could be interpreted by various experts as "a threat to
the nation's budget" is a prime and horrible example of the
scape-goating so cogently described by Carroll L. Estes.[20] As
she so correctly points out:

> The attempt to link the problems of the economy with
> the issue of aging is, many view, one of the most
> serious threats to a viable public policy for the
> elderly to appear in many years.

Professor Estes attributes the red herring of the "crisis" to
current political ideology, which is sanctioned by economic theory
and promulgated by powerful interests. It is also, as we have
pointed out, ratified by the "Space Age" techniques of
decision-making that are quantitative in their orientation,

antithetical to human factors, and slavishly subservient to the
canons of cost/effectiveness. According to the calculus of "fiscal
economists" from the Office of Management and Budget, the older old
people grow, the greater the burden they impose on society. These
"experts" calculate that if the federal government had not been
mercifully rescued by the premature death from cancer, heart
disease, and violence, it would, in 1978, have "suffered a net loss
of $15 billion." These "experts" are quoted as urging consideration
of such options as extending the retirement age or reimbursing the
government from the estates of those who die after long lives! One
cannot but wonder how much its "suffering" would be alleviated if
the government could designate premature death as a desirable
objective in budgetary management and promote programs to achieve
that end! Perhaps we are failing to grasp the meaning of the
handwriting on the wall and this is already occurring in the
concerted push toward relaxing environmental protection, tolerating
the use of deadly toxins, allowing the movement of hazardous
materials, deregulating and lowering safety standards at work and in
construction of vehicles and buildings.

   The "option" of extending retirement age, as proposed by OMB's
fiscal economists, is a nice bit of bureaucratic logomachy. Option,
by Webster's definition, means right of choice, a luxury not
accessible to every older worker. Persons required to stay in the
workforce may find that they have worn out their welcome. Their
skills obsolesced, they may, even on the job, have suffered the
ignominy of having outlived their usefulness. That the
long-heralded recovery will not resuscitate the job market is
assured by the assiduous application of the principles of
cost/effectiveness inherent in the dictates of productivity, with
continued substitution of electronics for manpower. Not only is
this occurring in industry, but also in retailing and in the service
sector generally. Gone are the elevator-operator jobs of
yesteryear, and the night watchman has long since been replaced by a
sophisticated alarm system.

   Not all workers remain strong and healthy. In fact, the more
we learn about the long-term effects of exposure to cotton dust,
silicon, asbestos, radiation, and other toxic substances in the work
environment, the less sanguine can be our expectation for good
health in the later years. Current pressures for the lowering of
safety standards and for deregulation bode ill, the more so because
such policies are rationalized by the Space Age decision-making
apparatus called risk analysis. This technique, which embodies the
cost/benefit ratio, is a manifestation of the Administration's
predilection for (a) disallowing regulations costly to industry and
hence, presumably, disadvantageous to the state of the economy and
(b) allowing industry to police itself and thereby save the
government the expense! Human costs, too numerous to recount, do
not count in these equations. That ultimately the victims, their

families, and society at large must pay heavy costs simply does not
enter into the calculations.

In this connection, it would be well to note Browne and Katz's
observation[21] that many of the problems of the elderly did not
begin when they reached age 65. Besides the infirmities that are
often associated with aging, there are, especially for persons of
low socio-economic status, the disadvantages of inadequate
education, health care, housing, and so on. These manifest
themselves during the working years, a social fact that emerged
poignantly in studies on retraining programs for welfare recipients
and others considered "hard core unemployed."[22] The poverty
syndrome has a cumulative effect: a factor in workers' occupational
history, it shapes their occupational future. Difficulties
encountered in earlier years become more pronounced as age advances
and opportunity shrinks. That older persons be kept in the labor
force longer as a solution to the fiscal fiascos wrought by
jingoistic economic theories is yet another example of the current
cavalier disregard for the human side of the cost/benefit ratio.

A combined demographic twist and technological thrust may cause
the onus of this ill-conceived "option" to fall on women.
Statistics show that not only do they live longer than men but that
their chances for employment are, at least for now and the near
future, better. If, as is generally conceded, the rosier prospects
are due to an increase in office work, then it is important that we
consider the technological changes taking place there. We pointed
out long ago[23] that automation was making a factory out of the
office. Just as "rationalization" broke down manufacturing
processes to their smallest and simplest repetitious components and
imposed rigid controls over every step, so too automation introduced
the same conditions into office work. With each successive
generation of equipment, the once-valued perquisites of white-collar
work faded. Pace of work, work intensity, possibilities for human
contact, freedom from close supervision, physical work environment,
health and safety are the traditional points on which quality of a
job are rated. Developments over the past two decades have caused
deterioration on all counts. And microelectronics accelerates the
process and exacerbates the problems.

The advent of the word-processor is a case in point. It has
increased the intensity of the typing job, reduced the little
variety it once had, and so downgraded the skills of the operator
that she does not even need to know how to spell. The machine is
the know-it-all, clocking her every movement and counting her every
finger stroke. The word-processor has invaded the once-sacrosanct
bailiwick of the private secretary. Nor are executives immune, many
of them turning out their own reports on word-processors. The
effects of technological changes on office work are mirrored in the

results of a study which revealed that female office workers developed coronary problems at nearly twice the rate of other employed women.[24]

The video display terminal (VDT) is central to electronic information systems and to it has been attributed a formidable and frightening array of physical effects on operators. Unexplained clusters of miscarriages and birth defects, eye disease, cataracts, and other forms of visual damage, skin rashes, and a variety of strains and pains have been linked to VDT use. A recent compendium of news items[25] about incidents, hearings, research and reports having to do with health hazards indicates the scope and complexity of the problem, which is international. Because long-term effects cannot be assessed, many of the "official" positions taken reflect a what-you-don't-know-won't-hurt-you philosophy. The National Academy of Sciences, for example, concluded a two-year study with the statement that there was "no scientific evidence that eye disease, cataracts, and other forms of visual damage result from working with VDT's." This may be the truth, but it is not the whole truth, which perhaps will make itself known later. Workers' problems stemming from non-ionizing radiation are being addressed by "technical experts" and government agencies in exactly the same way as were those of workers in the nuclear industry and of soldiers exposed to radiation from A-bombs testing.[26] Radiation effects on the eyes, on the reproductive functions, and on neurobehavioral mechanisms do not make themselves known immediately. The same experts who so confidently perform technology assessments proving that there is "negligible risk" are not required to produce data establishing that links between exposure and cancer do not exist. That there is an inverse relationship between time and truth becomes apparent when we review experience with other toxicities. The less we knew about them, the more sure were we of their safety. A fool's paradise of ignorance, as well as some cost/effectively calculated deception, kept us confident when we should have been skeptical. We now know that ailments once dismissed as attributable to chance or a statistical quirk could take on ominous significance over the years.

Technological progress, which brought microelectronics into factory, office, and home, is not, as we have pointed out, without its dysbenefits. Prime among them is the marked increase in stress, a condition long recognized as detrimental to health and well-being. The full consequences are just becoming known, as experiments establish a link between stress and cancer because of weakening by stress of a body's immune system. It is interesting to note that the medical information system known as MEDLARS (Medical Literature Analysis and Retrieval System) carries over 500 entries under the heading of stress, with nearly 90,000 technical articles appearing in the past 30 years.

The problems show up in statistics on alcoholism and drug abuse
at every level of the workforce. Not unexpectedly, special
attention has been paid to computer and data-processng personnel.
Dr. Stephen C. Duvall, long associated with Control Data
Corporation's Employee Assistance Program, provides these
insights:[27]

> Some people use drugs to alleviate boredom. Many of
> the jobs are structured in ways that are repetitive and
> are not particularly creative or challenging...
> Keypunchers basically are just sitting there checking
> in data eight hours a day. Programmers and systems
> analysts do work that can be aggravating or
> exhausting. It takes forever to find the flaws in the
> programs and the frustration factor is severe and
> real. Sometimes drink or drug use is a real way to
> alleviate tension, or to relax, or to get away from it
> for a bit--or just "space out" intentionally.

A source of stress not anticipated by enthusiasts who prate
about the office without walls is the blurring of the distinction
between place of business and place of residence. Already, many men
and women conduct their work at the computer terminal at home.
While this may have advantages, such as elimination of the nuisances
of road congestion and parking, it exacerbates other annoyances, not
least among which is the constant interaction--the
"togetherness"--which tries and sometimes finds wanting many a
marital partnership. That having the business as part of the
household will perturb and put a strain on relations with children
as well as upset other facets of family life is to be expected. We
are still accustomed to having the wage-earner go off to work and,
if truth be told, this provides a compartmentalization and
preservation of much-prized "private space."

There is a danger, in bringing VDT's into the home, of adding
to the already numerous but numberless sources of radiation with
which modern technology is bombarding us. Taken as individual
increments, emissions from microwave ovens, smoke detection devices,
alarm systems, television sets, VDT's, or any other cathode ray
tubes may be dismissed as insignificant; the cumulative and
long-term dosage may be another matter. Thresholds for and
bio-medical effects of ionizing and non-ionizing radiation are
controversial issues and, as in so many similar instances, what we
do not know can hurt us.

Lest the bizarre brand of logic applied by budgetary experts
still beguile us into making the older population pay for the
extravagances of military spending and general mismanagement in
public enterprise, we might remember that even those persons who do

not suffer retirement gladly would not opt for a return to work.
Much of the nostalgia for the good old days on the job is akin to
the camaraderie and good-old-buddy memories of war years. Time and
circumstance have swathed them in a rosy aura that reality cannot
fade. There are many reasons for workers to dwell on the
discontents of retirement. Just because they miss the routine or
long for companionship should not, however, be interpreted as
evidence that they are ready, willing, or able to continue working.

When we come to ask how medical technology is affecting the
health and medical care of the elderly, we run into the yin and yang
of the technological era. On the one hand, we prize medical
advance; we make a great brouhaha over "breakthroughs." We even
value prolongation of life. But we have a marked antipathy toward
aging and are downright apathetic about the care and comfort of old
people. The benefit side of the Faustian bargain dealt by
technology is well-known. Advertisements for Monsanto, Ma Bell, and
Amtrak have made it abundantly clear. Optimists are brimming with
ideas about electronic networks that will serve as extended
families, telecommunications that transmit one's heart beat to the
doctor's monitor screen, and a panoply of gadgetry to compensate for
infirmity.

What needs concern us here are the ways in which the technology
of decision-making have penetrated and taken over the medical
field. The macro-approach was touted as the macro-solution and an
army of economists, econometricians, operations researchers, systems
analysts, and information experts was deployed. Not unexpectedly,
cost/benefit analysis was applied to every aspect of health
services, with results sometimes ludicrous but generally detrimental
to human values.[28] During the past decade, cost/effectiveness
became the benchmark for government spending and the concept of
health maintenance organization (HMO) emerged. The objective,
according to Joseph A. Califano, Jr., then Secretary of Health,
Education and Welfare, was to improve access to care by injecting
competition, presumably to lower costs already boosted by the
third-party reimbursement system. The model looked fine but it
failed to work as anticipated. Califano's [29] ex post facto
comment is eloquent:

    In our rush to provide access, the Great Society had
    let the health industry set the prices, and had
    acquiesced in its reimbursed systems. Over the
    intervening decade, the industry had used America's
    quest for broad access to quality health to protect
    and enhance its financial interest and solidify its
    legislative and regulatory position. The health
    industry was seated confortably at a groaning table set
    by taxpayers. (Emphasis added)

What are the implications for the health and well-being of society when policy is dictated by the new "medical-industrial complex," which Dr. Arnold S. Relman[30] describes as the network of proprietary hospitals and nursing homes, diagnostic laboratories, home-care and emergency-room services, hemodialysis, and a wide variety of other services? Investor-owned businesses supplying health services for profit apply the market-place model, which has "McDonaldized" health care, as evidenced in the franchising of facilities all over the country. The coalition of corporations orchestrate national health policy in accord with their own profit-maximization objective and have already lobbied successfully for relaxation of regulations, lowering of requirements, and other measures giving them "more flexibility."

When we attempt a summation of our views on technology and the ways in which it is affecting the quality of life for the elderly, we are caught between the Scylla of trivializing the effects by dwelling on petty personal irritants and the Charybdis of polemicizing about the Faustian bargain with benefits extolled and costs untold. It has not been our intent to engage in tedious debate about technology's autonomy nor its culpability. We do not see the technological imperative as a deus ex machina but, for all its revolutionary impact, part of an evolutionary process. Thus, automation in office, field, and factory has been rationalized by its potential to reduce costs, increase efficiency, and raise productivity and profits. In every instance, a worthy end could be conjured up to justify the technological means.

The mismatch between hyperbolic scenario and the reality seems to have gone unnoticed, so great has become our tolerance for incongruity. Computers introduced to make jobs easier have taken them away altogether. Electronic systems devised to handle information have overloaded us to such an extent that we can no longer distinguish treasure from trash. Technology purported to extend our capabilities has dulled our capacities: the calculator does our arithmetic, the word-processor, our editing. Computers are programmed to digest the avalanche of output from busy-working computers. In education, it is vale to Mr. Chips and ave to microchips. In medical care it is all VDT and no TLC. A society that has a no answer to "Am I my brother's keeper?" has, through interactive systems, made Big Brother a permanent and live-in member of every family with a television set. Lest anyone be perturbed by these developments, technology has come to the rescue. Already on the market are video-cassettes designed to soothe the psyche and 'induce relaxation'. For anywhere from $35 to $50, the listener can simulate tranquility at the push of a button and the twiddle of a dial.

NOTE

*The actual number is not known because of vagaries in the Bureau of Labor Statistics' data.

REFERENCES

1. Pitirim A. Sorokin, "Social and Cultural Dynamics," Porter Sargent, Boston, Massachusetts (1957).
2. Ralph Lapp, "The New Priesthood," Harper & Row, New York (1965).
3. Ida R. Hoos, "Systems Analysis in Public Policy," University of California Press, Revised Edition, Berkeley, California (1983).
4. Thomas J. Kuhn, "The Structure of Scientific Revolutions," University of Chicago Press, Second Edition enlarged, Chicago, Illinois (1970), p. 175.
5. Phillip Aries, "Centuries of Childhood," Knopf, New York (1962).
6. William P. Browne and Laura Katz Olson, "Aging and Public Policy," Greenwood Press, Westport, Connecticut (1983), p. 3.
7. George C. Myers, The Aging of Populations, in: "International Perspectives on Aging: Population and Policy Challenges," Robert H. Binstock, Wing-Sun Chow, and James H. Schultz, eds., United Nations, New York (1982), p. 18.
8. Robert H. Binstock, The Elderly in America: Their Economic Resources, Income Status, and Costs, in: "Aging and Public Policy," William P. Browne and Laura Katz Olson, eds., op. cit., pp. 19-20.
9. Abraham Kaplan, "The Conduct of Inquiry," Chandler, San Francisco (1964), p. 11. Kaplan tells the story of the drunkard on his knees under a lamppost. He is looking for keys, dropped some distance away. When asked why he does not search there, he replies, "It's lighter here."
10. Browne and Olson, op. cit., p. 14.
11. Margaret E. Kuhn, "Open Letter," Gerontologist, 18, (10/78):423.
12. Carroll L. Estes, "The Aging Enterprise," Jossey-Bass, San Francisco, California (1979), p. 224.
13. John Godfrey Saxe, "The Blind Men and the Elephant."
14. Tom Forester, ed., "The Microelectronics Revolution," The MIT Press, Cambridge, Massachusetts (1981), p. xvii.
15. Brian L. Usilaner, Associate Director, Accounting and Financial Management Division, National Productivity Group, U.S. General Accounting Office, Statement, Hearing before the Subcommittee on Labor Standards of the Committee on Education and Labor, House of Representatives, "New Technology in the American Workplace," Ninety-seventh Congress Session, Washington, D.C., (6/23/82), p. 6.

16. J. Raga, "The Impact of Microelectronics," International Labor
     Office, Geneva, Switzerland (1980), pp. 76-77.
17. Simon Nora and Alain Minc, "The Computerization of Society,"
     The MIT Press, Cambridge, Massachusetts (1980).
18. Henry M. Levin and Russell W. Rumberger, "High-Tech Requires
     Few Brains," The Washington Post, (1/30/83).
19. Philip M. Boffey, "Longer Lives Seen as a Threat to Nation's
     Budget," The New York Times, (5/31/83).
20. Carroll L. Estes, "The Elderly: Scapegoat for the Economic
     'Crisis'," Grey Panther Network, (7/8/83), p. 4.
21. Browne and Katz, op. cit., p. 14.
22. Ida R. Hoos, "Retraining the Workforce," University of
     California Press, Berkeley, California (1974).
23. Ida R. Hoos, "Automation in the Office," Public Affairs Press,
     Washington, D.C. (1961).
24. S.G. Haynes and Manning Feinleib, "Women, Work, and Coronary
     Heart Disease: Prospective Findings from the Framingham Heart
     Study," American Journal of Public Health, Vol. 70, No. 2,
     (2/80).
25. Louis Slesin and Martha Zybko, "VDTs--Video Display Terminals:
     Health and Safety," Excerpts from Microwave News, New York
     (1983).
26. Ida R. Hoos, Risk Assessment in Social Perspective,
     "Proceedings," National Council on Radiation Protection and
     Measurements," Washington, D.C., (3/14/79).
27. Marvin Grosswirth, Stoned at the Offices, in: Datamation,
     (2/83), pp. 30-36.
28. Ida R. Hoos, "Systems Analysis in Public Policy," op. cit.,
     262.
29. Joseph A. Califano Jr., "Governing America," Simon and Shuster,
     New York (1980), p. 142.
30. Arnold S. Relman, "The New Medical-Industrial Complex," The New
     England Journal of Medicine, (10/23/80), pp. 963-970.

THE NEXT 25 YEARS: IMPACT EXPECTATIONS FROM TECHNOLOGY ON AGING

Earl C. Joseph

Anticipatory Sciences, Inc.
Minneapolis, Minnesota

INTRODUCTION

In this paper, future perspectives relative to forecastable new
revolutions in technology and computers for advancing health care
for the aged are presented. Included are forecasts of "expert
systems" for amplifying individuals, doctors, nurses and other
health care professionals and technicians. It is about future eras
and the many trend paths we are on taking us through the decade of
the 1980's and toward first quarter 21st Century technology. It
discusses both short-term and long-term futures of interest,
especially for assisting the aged and for reducing stress.

For example, "expert systems"--artificially intelligent
computer systems and a set of AI programs that use a stored
knowledge base and inference procedures--are becoming available to
assist humans. Artificial Intelligence (AI) research is a subfield
of computer science that investigates imitating human processes
(within computer systems), like learning, symbolic reasoning,
inductive discovery, deductive analysis, intuition, problem solving
and other human intelligence processes including machine
representation of knowledge for use in inference tasks. Imagine
expert systems for assisting the aged--and how the societal
equations could be tipped allowing older persons to participate as
if they were younger!

INFORMATION AGE FUTURES

Massive forces are building for drastically altering which set
of alternative futures will become most likely for the remainder of

this century--and beyond. Rapid and accelerating advances are
occurring in science and technology--especially in computers,
communications, artificial intelligence, genetics, agriculture,
microbiology, space and chemistry--for altering our future way of
life. Economic, political and social forces are also building to
cause step function changes.

The emergence of a new information age societal framework,
based upon the expanded use of and need for information, supplied by
electronic computers and communications, is altering the way in
which social, business, economic, educational and political
exchanges are conducted. Information age technology is rapidly
thundering-in on most jobs and into many homes. Further, since
information age technology affects, impacts and alters the way
knowledge is created, stored, retrieved and applied, the character
and type of jobs we work at, and the networks in which people are
linked are changing the infrastructure of society. These will in
turn impact and change the tools we use in the information age, how
we use them and for what purposes they will be employed. This is a
revolution for society--and in the architecture, organization,
processing and dissemination of information/knowledge--for which
future computers will perform a central role (but in decentralized
embodiments), and will speed the societal transformation underway.

To get a handle on what futures are being grown and what they
will be all about, one first needs to understand the information age
environment being spawned. It has many characteristics and
dimensions including an:

> Information Ecology, Environment & Sociology
> Information Economy and Capitalism
> Information Technology, Tools & Systems
> Information Resources
> Information Conferencing and Dialoguing
> Information Politics

Information ecology includes a new age sociology of many
environmental dimensions--a few of which are: information
interactions, real-time knowledge access and amplification,
information/ knowledge engineering, information management and
politics, new job creation and old job displacements, and new
individual freedoms and protection options and impacts. In an
information age, an information economy transforms capitalism in
many forms. Information becomes a basic need and the major social
capital as well as societal power source. In an information
economy, information further substitutes knowledge for capital,
energy, jobs, materials and travel. It de-industrializes jobs by
substituting information flows to automations and robotic tools for
performing work. An information age thus radically alters societal

roles, values, jobs and needs by redesigning the infrastructure of the economy and society.

Information technology marries computers and communications into many other things, as well as into our daily lives, and in the process makes "things" smart and intelligent. Part of this information technology tool kit consists of: people amplifier appliances (computers, calculators, expert systems, etc.), micro-technology/computers/communicators information appliances (e.g., word processors, "paperless book" systems, computer mail systems, etc.), information systems, information networks, information utilities, information software, data bases and information services. Information technology is part of the information resource environment which additionally includes a rapidly expanding information industry, laws, controls, standards, knowledge bases, media, telecommunication and data communication systems, and computer-aided-systems.

In this new information age, people increasingly "discourse" and "dialog" directly with information via computer terminals, tele- and computer-conferencing and the like, instead of just with people. Information conferencing allows us to conquer distance via electronic computer/communication networks with information appliances backed-up with information services, data bases, knowledge bases, information retrieval and information management software systems. Thus, in such an information environment, policy making, management decision making, people-to-people, people-to-machine, machine-to-machine interactions, and machine operation increasingly are performed remotely via computer networks using CRT video screens, voice channels, keyboards and data bases--instead of face-to-face in conference rooms, back-rooms, offices or in factories.

Information politics includes individual privacy and freedom considerations relative to "data-basing people," jobs and employment, societal power structuring changes, transborder data flows, taxes on information, and the like. In the process, the information economy now growing increasingly moves a major portion of the GNP from being supplied by the sale of products to sales of information services; i.e., sales of information, information systems, information networking, and information assistance.

Taken in total such a future information environment, now birthing, points to a long list of information related new computer applications--many of which have their precursors in the primitive (current) data base management system, management information systems, decision support systems and computer-aided-design/ instruction systems now in place. Thus, for the computer world and the movement into this neo-modern information age, which we are leaping into, evolutionary advance is the name of the game rather than revolution--even though for society it is a revolution relative to the magnitude of change occurring and the impacts expected and forecastable.

For example, the current computer revolution differs from the early 1950's, when computer and automation were introduced, in several critical aspects. First, the computer application arena is changing 1) from large centralized and costly computer hardware and software to distributed, personal and low-cost systems, 2) from use of computers by a few large firms, the military and Federal Government, to use by individuals at work and at home, as well as being embedded in a wide variety of machines, 3) from displacing "low level" jobs to displacing "high-level" jobs, 4) from creating more jobs than displaced toward the real possibility of displacing jobs faster than society can create new ones, 5) from automating industrial age systems to automating post-industrial information age systems and jobs, and 6) from computer systems that required considerable education and software development in order to apply them to easy to use systems with a vast array of ready to use packaged software.

But the major change occurring is that total computer system and usage costs (hardware, software, terminals, memory, communications, etc.) are at the turning point wherein their cost is diving ever lower at a faster clip than the inflation rate--and at the same time their functionality and applicability are increasing and widening. Thus, computers have passed the threshold of affordability and are entering an era of being involved in our daily lives, almost world-wide, with billions of people already entering such a computer age.

FUTURE ARTIFICIAL INTELLIGENCE FOR ASSISTING THE AGED

AI assumes heuristic knowledge to be of equal importance with "factual" knowledge--in fact, for AI purposes, heuristics is assumed to be the process defined as "expertise." Heuristics goes beyond the use of logical procedural-oriented strings of instructions operating on streams of data, or on data bases--as that which occurs in standard computer program execution.

AI heuristics include logical inference procedures which allow semantic access (closely related to natural language) of knowledge bases which use AI processes for making "expert" judgements. AI expert systems require capturing and storage of the known expertise of a field, like medical diagnoses, and the translation of such knowledge, via AI programs and hardware. AI offers intelligent assistance to a practitioner in a field (i.e., for amplifying a doctor, with its stored knowledge, and AI heuristics for interpreting such knowledge)--or for assisting any person by

providing expert knowledge that can be applied in the real-time of
that person's need.

Thus, an expert system uses AI inference coupled with a
knowledge base for assisting as a machine "consultant" in solving
problems, planning, making decisions, assessments, diagnosis, and
judgements or for creating (discovering or inventing)
opportunities. Expert systems allow the tackling of problems that
are difficult enough to require solutions which go beyond simple
arithmetic or logic, and that require heuristics of significant
power for approaching what heretofore required human experts for
their solution. The knowledge and AI heuristic processes necessary
to perform at such an expert level, plus the AI inference algorithms
used, can be viewed in the AI expert system as a model of the
expertise of the best human expert practitioners in that field.
Knowledge, once captured in such a fashion in an AI Expert system,
could allow a future non-expert human to apply such knowledge and
heuristics to nearly match and often exceed the average unaided
human expert in that field--but to also vastly amplify human
experts. Further, AI expert systems can be constantly updated as
society gains new knowledge. Obviously, such "convivial" and
"congenial" intelligent amplifiers will greatly assist the aged to
participate as up-to-date and knowledgeable members of modern
society--and will help them with their health maintenance--and thus
reduce stress.

In forecasting the future of AI expert systems there are a
number of obvious and expanding application areas. Perhaps at the
top of the list for the course of future events for the 1980's is AI
advice-giving and consulting systems. Included are expert systems
for medical diagnosis, pharmacy, medical lab analysis, intensive
care nursing, chemistry, hospital architectural design, molecular
generic design, programming, office mangement decisioning,
management, home advice (e.g., financial, medical, lawn, repairs),
and much more including advice to patients and the aged.

However, perhaps for the shorter term future, the biggest
market could be for home entertainment. Games have been used in AI
research from their beginning (dating back to the 1950's) for
testing AI features and programs. Therefore, AI games are the
natural outcome of expert systems research. Further, intelligent,
knowledge-based games are expected to spur on the home entertainment
market as well as for education and training of medical/health care
professionals. The money-making potential of this market could make
microcomputer TV-based expert system home entertainment the dominant
market for early-on expert systems. Further, for the longer-range
future, expert systems would allow home entertainment systems to
quickly evolve to include consultation and advice about a broad
range of subjects of interest to the aged, including medical
advice--and later on for working and learning at home. That is,
expert knowledge-based systems will go far beyond traditional

"how-to" books by allowing real-time expert (interaction) assistance
tailored heuristically for the task at hand--learned from home
entertainment systems.

Additionally, besides consulting and advice, AI expert systems
also can assist in the creative and invention arts, as well as to
give diagnostic and prescriptive advice, and to dialogue giving
recommendations for the tasks at hand. Such dialogues involve the
AI expert system threading itself through its knowledge base via
"IF-THEN-AND-ELSE" heuristic (logic) rules together with the human
that it is advising. They will also help with planning and design.

But what is a knowledge base? The process of building a
knowledge base for use with an AI expert system requires the
compilation of an extremely "factual" taxonomy of the (each)
specialized field--and the heuristics for its application. Such
knowledge base taxonomies turn out to be far more accurate,
comprehensible, and reliable, and therefore more useful, than
today's manuals and textbooks. "Knowledge engineering", requiring
knowledge engineers, is a rapidly growing new profession. The task
basically is the creation of "knowledge bases" for use in "expert
systems". In the future, most professionals from all fields will
become, in one form or another, knowledge professionals. Today
expert systems are computer programs employing artificial
intelligence operations using knowledge bases for advising people
(in an expert fashion) in the "real-time" of the process of doing
something--like assisting a doctor in diagnosing what ails an ill
person, or cooperatively designing a new building structure or for
cooperative management decision making.

Future expert systems, in the form of a person amplifier,
present factual data or information advice or give "opinions" based
upon the "knowledge" contained in their knowledge bases. Further,
and importantly, an expert system can backtrack to tell the logical
and heuristic process and information that it used to arrive at its
"expert" advice or opinion.

The knowledge base of an expert system consists of "facts"and
heuristics. Heuristics in expert systems include inductive and
deductive inference reasoning, learning, thinking, (etc.) type human
imitated processes. The "facts" constitute a body, or taxonomy of
knowledge (information) that is similar to the information that a
human expert would use for whatever expert task such an expert would
be performing. But herein lies the stumbling block - what does an
expert (human) do? That is, it is no easy task to create a
knowledge base which contains expert information and knowledge which
is generally agreed upon by experts in a field. Further, not all
expert knowledge is a set of "black and white" facts--much expert
knowledge is codifiable only as alternatives, possibles, guesses and
opinions (i.e., as heuristics). Heuristics, thus, consist of rules

of good judgement, rules of plausible reasoning, as well as hard and
fast logical reasoning, rules of good guessing, and the like, that
are characteristic of expert-level decision making. Therefore, the
performance level of an AI expert system is primarily a function of
the size and capacity of the knowledge base, the quality of its
contained expert information, the completeness of its taxonomy, and
the number and characteristics of its stored or programmed
artificial intelligent heuristics (inference rules and procedures).

Thus, there should be little doubt for the future that as AI
expert systems evolve to become ever higher level "experts", we will
possess very powerful amplifying tools to assist us for almost any
task we tackle. In fact, because a knowledge base arranges
knowledge in a somewhat procedural fashion, like a computer program,
it must be more complete, correct and comprehensible than the
typical text book. Therefore, experience with current expert
systems shows that when compared with traditional sources of
knowledge (books, tapes, classrooms, etc.), present and future
knowledge based systems are 10 to 1,000 times more complete, correct
and comprehensible. Additionally, AI expert systems allow knowledge
application in the real-time of decision making.

COMPUTER TECHNOLOGY BEYOND SILICON CHIPS

Today, and for the relatively long range future, silicon chip
like hardware semi-conductor technology is expected to continue to
grow in usage and expand into a wider range of applications. But
what comes after silicon chip technology farther out in the future?
Answering this question, which is being asked more often, takes us
along many avenues which are opening for the future. They include:
gallium-arsenide, wafers (instead of chips) and bio-physical
technology.

The first is a non-silicon technology, which portends higher
circuit densities and higher performances (in speed and/or
reliablility), thus allowing continuing evolutionary technological
progress. The latter two suggest that for the future, however,
step-function revolutionary advances are also possible--and likely.

Today, on a single silicon chip it is becoming possible to
integrate together over a hundred thousand electronic circuits--and
soon more than a half million using submicron geometries. Each
circuit consists of approximately two and a half transistors, (micro
logic switches), resistors and capacitors - all made in the silicon
by imbedding (doping) other elements (impurities) together with
layers of evaporated metal and resistant materials. Such complexity
of circuits per silicon chip allows us to design and make complete
machines on a single chip as a single component. And as
evolutionary advances occur we are on a track into the future to

integrate larger and larger (in capability) machines onto the silicon. Such "chip machines" or "component machines" become building blocks for bigger machines. For example, the advent in late 1971 of the "microprocessor" "calculator chip" made possible the modern hand-held calculator--which grew and evolved since then into the more complex models now in common use. Today, component machines are being imbedded in wheelchairs, home appliances, office and factory machines, heart pacemakers, and the like, to make them smarter and more capable--expecially useful to assist the aged.

The small silicon chips are made by breaking up large silicon wafers three to six inches in diameter. Wafers are sliced from pure silicon crystals grown (manufactured) from sand in silicon foundries. It is expected/forcasted that before the end of this decade we will/could be using the full wafer as the component--instead of breaking it up into little chip pieces. Today, this step is necessary in order to separate and to toss out the bad chips (bad areas of the wafer). As our semiconductor manufacturing technology further matures, allowing purer silicon with fewer processing imperfections, it will become increasingly unnecessary to break the wafer into chips.

As we begin to use major portions of wafers, and later total wafers, as larger components we will be able to integrate many millions of circuits - perhaps billions of circuits. When we reach such a period, in the late 1980s or early 1990s on the agenda for the future, then looms the ability to construct component institutions which can be used to assist the aged in the real-time of their thinking, discourse and actions. Some possibilities are:

* Component health care machines/systems
* Component libraries
* Offices-on-a-wafer/component offices
* Component schools
* Component "hospitals".

That is, we will be able to design, construct and manufacture "component institutions" as basic building blocks to integrate with systems (or to imbed within) for making bigger and more capable and more intelligent machines and/or as people amplifier devices.

But what comes after wafer technology? One possible next step in miniaturization, beyond submicron geometries, could be a VSD (Very Small Device) with near nano-meter features. That's three orders of magnitude, or 1000 times in each dimension, smaller than current silicon circuit element sizes and interconnection line widths.

There are few materials capable of reaching these ultra-small circuit geometries. One candidate now in reaserch is bio-physical molecular switch technology. Such bio-molecular switches can act as conductors and semiconductrs, and therefore can be used like computer logic circuits. One way these bio-switches operate is via the use of electron transporting enzymes in electro-active polymers.

Since enzymes are living things, controlled by DNA, they require genetic engineering techniques for the subassembly process to build "bio-circuits" and "bio-chips". That is, recombinant-DNA cutting, splicing, editing and transcription process technology would be used to assemble the circuits. The assembly of the switchable organic molecules is accomplished via genetically engineering computer DNA amino acid sequences to produce specifiable proteins as templates for the self-construction of larger molecules--e.g., to grow future computers and ohter medical machines for use in future health care systems. That is, one likely longer-range future involves "growing" our future computers as living systems--as well as for constructing bio-genetic parts for repairing people and bio-genetic adjuncts for amplifying people.

CONCLUSION

Taken as a whole, these technology trends allow for both the long and short-term future reality of going beyond science fiction--allowing "Star Trek"-like computers and communications systems--and next steps toward (artificial intelligence) "inference engine"-type computers. Rapid, technological change always has been the norm in the computer field and, recently, in the bio-genetic and communications fields. In the past, technology-driven change has forced an increasing diversity; and from the foregoing, we now see that the same technological-advance trends are forcing the merger of some of this diversity. However, most future technology watchers see this merger as a way for making way for a new form of diversity. The most likely form that such future splintering, now forcastable, will take is along "smart"/"intelligent" vs "dumb" technology lines, and along application areas.

But, whichever multiple directions advances in technology take into the future, there should be little doubt that these developments will allow technological systems to penetrate deeper into society. Thus providing and making new opportunities and causing considerable change and impact--for providing opportunities for individuals, institutions and society, and most importantly for the aged.

ATTITUDES AND PERCEPTIONS OF OLDER PEOPLE

TOWARD TECHNOLOGY

Cyril F. Brickfield

American Association of Retired Persons
Washington, D.C.

Technological advances have been changing our society and our attitudes to the world around us with a startling swiftness. We have seen men walking on the surface of the moon and heard their voices coming to us through the vastness of space. We have plumbed the depths of the ocean and sent a space vehicle on a voyage millions of miles beyond our planetary system. Through an unparalleled explosion in communication techniques, we now live in a sort of global village. We can sit comfortably in our living rooms and view events--tragic or otherwise--taking place halfway around the earth.

The computers, the microchips, the satellite systems are not going to disappear. We could not, if we desired, reverse the onward rush of technological innovation. Our primary aim is not to resist but to harness its forces for the benefit of mankind.

As Executive Director of the American Association of Retired Persons, a non-profit, non-partisan organization with a current membership of more than 15 million Americans who are 50 or older, I have a twofold interest: To what extent do older Americans welcome or oppose technological advances? To what extent can older people help guide technological developments that may impinge on their daily lives?

Today's older Americans have lived through more technological change than any previous generation in history. They are not alien to technological concepts. In fact, they accept much of today's technology as a normal part of their lives. Like younger people, they accept without serious misgivings the changes in life style brought about by automobiles, airplanes, telephones, and

31

television. These technological developments are acceptable because
they have become familiar.

Some other less obtrusive products of technology, such as
computers and robotic devices, still require some getting used to.
Based on rather sketchy evidence that we have, older people appear
to be more reluctant than younger people in acknowledging the value
of newer, less familiar forms of technology. For example, some
older people tend to view computers as having a dehumanizing effect
on society--of being an electronic substitute for person-to-person
relationships.

Relatively few studies have been undertaken to determine the
attitudes of older people toward technological developments. The
association I serve has undertaken one survey, and we shall conduct
a second survey. One of the difficulties we have encountered is to
phrase our questions in a language that is precise and readily
understandable to older people.

In our first survey, undertaken in June 1981, we conducted
telephone interviews with a representative sample of 750 adults 45
years or older who live in various parts of the nation. The survey
revealed that older people have less positive views toward a variety
of technologies than do younger people. They are also less likely
to use the new products of technology. The higher the age of the
person interviewed, the less likely it was that he or she had used
such products as an electronic calculator, cable television, a
computer, or a video recorder. Table 1 indicates by age groups the
percentage of those interviewed who have used six products resulting
from recent technological developments.

Table 1. Use of Technology by Age

Technology	Age		
	45-54	55-64	65+
Calculator	76%	67%	41%
Cable TV	39%	39%	37%
Computer	40%	30%	19%
Video recorder	12%	6%	6%
Automatic teller machine	19%	14%	9%
Video games	28%	13%	12%

Employment status also has an effect on one's use of the products of technology. Persons employed in white collar jobs were more likely to have used these products than persons in blue collar jobs. In addition, those in higher level positions within white and blue collar occupations were more likely to make use of technology than those in lower positions. The amount of one's annual income is also directly related to the use of products developed by technology as is indicated in Table 3.

The educational level attained is also an important factor in determining one's acceptance and use of products developed by technology (Table 4).

Table 2. Use of Technology by Employment Status

Technology	White Collar		Blue Collar	
	Upper	Lower	Upper	Lower
Calculator	82%	71%	54%	43%
Cable TV	45%	37%	40%	32%
Computer	51%	39%	24%	14%
Video recorder	16%	12%	5%	2%
Automatic teller machine	24%	17%	12%	8%
Video games	28%	18%	20%	11%

Table 3. Use of Technology by Income Groups

Technology	4K	4-12K	12-18K	18-30K	30+K
Calculator	22%	38%	61%	70%	80%
Cable TV	29%	33%	42%	36%	43%
Computer	5%	14%	32%	35%	44%
Video recorder	2%	3%	7%	6%	17%
Automatic teller machine	2%	7%	11%	19%	28%
Video games	2%	13%	23%	19%	31%

Table 4. Use of Technology by Education Level

| Technology | Education Level | | | |
	High School	High School Graduate	Some College	College Graduate
Calculator	29%	60%	69%	87%
Cable TV	27%	38%	43%	41%
Computer	7%	25%	42%	44%
Video recorder	3%	8%	9%	15%
Automatic teller machine	7%	11%	20%	21%
Video games	8%	19%	23%	25%

In addition to determining the extent of use by older Americans of technological products, our survey included a number of more generalized questions with the intent of determining attitudes toward technology. On the basis of responses we received, a pattern of positive and negative attitudes emerges.

Men are more likely than women to have a positive reaction to technological advances. When we asked for a reaction to the statement, "Life is going to be better when people are able to do their banking, shopping and a lot of their other business from home using telephones or cable television," 60 percent disagreed and only 34 percent agreed. And in response to the statement, "Machines make life too impersonal," 64 percent agreed and 21 percent disagreed.

It should be pointed out, however, that attitudes are likely to change with the passage of time. Future generations of elderly will be better educated and will have greater opportunity to familiarize themselves with new products and services created by technology. It is not unreasonable to assume that their reactions to technological change may be more positive.

As we look to the future, in what directions should we seek to develop new technologies that will benefit the elderly? First, we need to develop new technologies in health care that will enable more people--both young and old--to receive medical treatment in their own homes rather than being cared for in hospitals, nursing homes, or other institutions.

Older people are most interested in products that will meet their basic personal needs. They are less likely than younger people to be swayed by fads of the moment.

Older people are not attracted to products for which they are singled out as consumers because of their age. For example, they resist purchasing foods advertised as specially suited for elderly.

Older people are attracted to products that enhance their capability for independent living, both in their communities and in their homes. Thus older people are more likely to accept new technologies which give them a greater sense of physical and emotional well-being. Those whom we interviewed gave higher priority to the development of medical and emergency alarm systems.

Older people are likely to reject any technological advance that decreases their opportunity to socialize and that tends to isolate them. Those who maintain reasonable mobility would prefer to venture out to shop or to share a meal with others than remain secluded in their homes.

I recognize, of course, that some major technological advances have already been made. The New York Times ran an article about a man who had undergone surgery for stomach cancer. After the surgery, he developed a narrowing of the esophagus, and his weight dwindled from 176 to 110 pounds. To keep him from starving to death, doctors implanted a catheter in a vein near his heart so he could be fed intravenously. When the man insisted that he wanted to leave the hospital, a company named Home Health Care of America set him up with an intravenous pump at his home and taught him how to use it. He hooks himself to the pump before going to bed, then unhooks himself in the morning. In two months, he had gained 20 pounds and was able to return to his job as service manager at an auto repair shop.

Another development which has brought mobility to seriously handicapped persons is the "smart wheelchair" which will respond to spoken commands and even to movements of the handicapped person's head. It has a microcomputer system that provides automated guidance roughly equivalent to an aircraft autopilot. As one might expect, this wheelchair is quite expensive. Those are two examples of technological advances that permit a number of seriously disabled people to lead nearly normal lives. But many other home health-care aids can be developed through technology.

Second, we need to develop better medical alert systems. We can monitor the pulse of an astronaut thousands of miles away but cannot do so simply or cheaply for an older person living alone or in a nursing home. Some medical alert systems have already been devised. One of these, called Med-E-Lert, was designed by an

electronics engineer living in Florida. It consists of a miniature
radio which, when gently squeezed, sends an emergency signal to a
control unit plugged into a telephone jack in the user's home. When
activated, the control unit sends a specially coded signal to a
24-hour emergency monitoring center. The designer of Med-E-Lert
cites a number of instances whereby older people have received
emergency help. For example, an 84-year-old woman in Minnesota's
icy winter slipped on the stoop outside her house and was unable to
regain her footing. She squeezed her Med-E-Lert device and within
10 minutes had been carried inside her home.

Third, we need further refinements of voice-activated warning
systems. I am thinking of devices that will warn an older person
that the stove is still turned on or that the oven temperature has
reached the broiling point.

Fourth, we can help older people by providing memory aids for
them. It is possible to manufacture a portable hand held calculator
which would serve this purpose. It can remind someone to take
certain medications or keep an appointment; it will store the
telephone numbers of friends and relatives; and it will provide
other information to help one get through the day in orderly
fashion.

Most older Americans are reasonably self-sufficient. In fact,
only about five percent of them are receiving institutional care.
So I ask: How can technology best serve the great majority of older
Americans who are mobile and able to take care of themselves?

We can make better use of audio-visual techniques and of the
media in helping people prepare themselves financially and
emotionally for retirement. Ideally, this preparation should begin
at least 10 years before retirement. And one of the prime
objectives of the association I serve is to help older Americans
live their retirement years with a sense of dignity, purpose, and
independence.

Technology has increased the opportunities to develop new
skills and interests in retirement. For example, with equipment
that is now available, one can become a fairly decent carpenter,
cabinet maker, or molder of stained glass. Heretofore a long period
of apprenticeship would have been required. Although technology to
some degree has tended to dehumanize the workplace, particularly for
those who do repetitive work on an assembly line, technology also
permits the development of new cottage industries in the home. Many
older people want to continue work, at least part time, and the
tools that are now available make it easier for them to do so.

Technology, if used with imagination, can also serve as a
catalyst to bring older people together--to give them an opportunity

to feel integrated with their neighbors and the world around them. In some retirement communities and congregate housing sites, older residents have gathered to watch video cassettes of stimulating television programs, such as the NOVA series. These programs invariably have stimulated a spirited discussion.

Older people appear to be more ambivalent than younger people in their acceptance of new technologies, but this does not mean that they resist all technological development. For example, a home for the aged in suburban Washington won high favor with its residents by introducing computer games. Since most of the residents were over 80 and infirm, the games were adapted for poor vision and reduced manual dexterity. In playing the games, the older people felt they were part of the modern age, and, perhaps of greater importance, they had a common interest to share with children and grandchildren.

Now I come to my next question: What contributions can older people themselves make in the development of technology? I suggest that scientists, engineers, and manufacturers could profit by having older people serve as consumer advisers. These older people could identify needs that are susceptible to a technological solution, and they could help designers and manufacturers avoid the costly error of producing products that would have little acceptance among the elderly. Older people could be used to test new products and to suggest ways these products might be improved before they are marketed. Older people could also be encouraged to instruct their age peers about new technologies and their uses. The use of older persons as instructors would help diminish whatever fears or skepticism other people may have.

And, finally, retired scientists and engineers--as well as retired plumbers, electricians, carpenters, and others with journeyman skills--can contribute to technological advances. As an example, in the Washington, D.C. area, a number of older professional people have formed a group known as Senior Scientists and Engineers. Among their objectives, they intend to use their years of experience in improving the quality of instruction in science and mathematics and in serving as advisers to organizations such as AARP that represent older Americans. This group holds regular meetings, and is seeking to recruit other older scientists and engineers to support them. They also invite persons who have a wide knowledge of technology to address their meetings.

At a recent meeting, an engineer from the National Aeronautical and Space Administration explained how technology gained from the space program could be adapted to the needs of the elderly and the handicapped. He told how a sensor can be implanted in the body to detect heart fibrillation and then deliver an electric shock to restore the heart to a normal rhythm. He also described a

prosthetic urinary sphincter consisting of a simple reliable valve
which is implanted to restore urinary control.

This quest by older Americans for new applications of
technology to serve the elderly could be replicated in many
communities. Certainly it has the enthusiastic support of our
association. In an assessment of technology, we must admit that the
Industrial Revolution has tended to have some negative aspects. In
western industrialized societies many older persons have felt
isolated from the mainstream of life. They have often been treated
with less respect and veneration than are older people in more
traditional, less developed societies. But I firmly believe that
these negative opinions about the elderly are changing. Moreover, I
think we all recognize that technology in most respects has been of
immeasurable benefit to older people.

Medical technology has done much to eliminate ancient scourges
and to enhance life expectancy. In fact, we have reached the point
where there is serious debate concerning extraordinary measures by
doctors and hospitals to sustain the life of patients who obviously
have no chance of recovery.

Technology can continue to serve the elderly, but to do so
effectively the scientific, engineering, and industrial communities
must work hand-in-hand with the elderly and with those who are their
advocates. There is still an urgent need to identify areas in which
technology can assist older people in a cost-effective way--and to
develop new products that older people will actually use to their
benefit.

Technology is only a means to an end. It is a tool that can be
used well or abused. In all of our technological developments, we
must never forget the human element. We should strive for a
technology that brings people together in a greater sense of
brotherhood, not a technology that tends to isolate people or to
intrude in interpersonal relationships.

I am reminded of two quotations that I consider relevant. John
F. Kennedy said: "Man is still the most extraordinary computer of
all." And Walter Lippman said: "You cannot endow even the best
machine with initiative; the jolliest steamroller will not plant
flowers."

NOTE

An earlier and somewhat different version of this paper appeared in
Vital Speeches, September 15, 1983.

AGING AND LABOR FORCE PARTICIPATION

Robert L. Clark

Department of Economics and Business
North Carolina State University
Raleigh, North Carolina

INTRODUCTION

Population aging has been a highly significant economic and social force in NATO and other developed countries during the past three decades. Moreover, should fertility decline, as expected, in the less developed countries, these nations will also experience the aging of their populations. Accordingly, it is important to assess the economic and social implications of population aging. The interaction between the changing age structure of the population and the composition and size of the labor force is the primary concern of this paper. In addition, the economic consequences of these changes are examined in conjunction with potential technological advances.

POPULATION PROJECTIONS

The age structure of a population is determined by its fertility, mortality and net immigration rates, with national fertility experience exerting the predominant influence on the country's age composition. In the NATO and other developed countries, fertility rates have approximately stabilized at levels which if continued imply near zero population growth. As a result, the percent of the population aged 65 and over in these countries rose from 7.6 percent in 1950 to 10.5 percent in 1975 while the proportion of the populations aged 0-14 years declined from 27.9 to 25.0 percent during the same period. The apparent reduction in the fertility rates of the less developed nations has not as yet produced significant population aging primarily because of the accompanying increase in infant survival rates. The proportion of the population 65 years and older in

39

these countries is estimated to have declined from 4.4 percent in 1950 to 3.8 percent in 1975, while the proportion aged 0-14 increased from 38.7 to 40.4 percent (United Nations, 1979a, p. 43). These divergent age structure changes are further illustrated by the rise in median age in the more developed regions--from 28 to 30 years--in contrast to a slight decline in the less developed countries. However, over the period 1965-75, the old age population in the less developed countries was growing at a higher rate than other population age groups in these nations (United Nations, 1979b, p. 123-7).

The continuation of low fertility in the developed countries will generate further population aging as these nations approach stable stationary populations; indeed, some countries actually face the prospects of declining population. The expected decline in births in the less developed regions will give rise to future population aging. Table 1 shows the medium variant population projections of the United Nations for developing and developed nations. Table 2 indicates similiar projections for the NATO regions. These data illustrate the effect of anticipated declines in fertility and mortality rates over the next century by showing the increase in older populations of these regions. In the NATO regions it is expected that there will be a gradual increase in the proportion of the population that is 65 and older over the remainder of the twentieth century before the growth of the older population accelerates during the first quarter of the next century.

DEPENDENCY RATIOS: EFFECTS AND IMPORTANCE

Population and economic dependency ratios, which are measures of a population's age structure, have been used as indicators of the potential well-being of a nation. These ratios are derived by dividing the number of youths and/or elderly by the size of the economically active population, usually assumed to be everyone between 15-64 years of age. The principal argument is that a lower dependency ratio implies relatively more workers and that fewer resources need to be diverted to the dependent populations. Within this framework, fertility declines and the concomitant population aging would be associated with increased economic potential of a society because the proportion of the population 15-64 years of age increases as the rate of population growth falls. Expected changes in the dependency ratios for various regions of the world are shown in Tables 1 and 2.

The apparent favorable effect of population aging implied by the simple population dependency ratio concept must be further examined because other variables affect labor force size. First, labor force participation rates vary greatly by sex and age and are influenced by the level of fertility and stage of development. For example, the fertility declines that increase the proportion of the

Table 1. Age Structure of Future Population in More Developed
Regions and Less Developed Regions, 1950-2100 for
Medium Variant

	More Developed Regions				Less Developed Regions			
	All ages	0-14	15-64	65+	All ages	0-14	15-64	65+
	Population (million)							
1950	832	231	537	64	1,681	641	976	64
1980	1,131	260	742	129	3,284	1,286	1,869	129
2000	1,272	274	831	167	4,927	1,684	1,014	229
2050	1,381	268	869	243	8,394	1,768	5,583	1,043
2100	1,390	269	871	250	9,135	1,764	5,707	1,664
	Percentage Distribution							
1950	100	28	65	8	100	38	58	4
1980	100	23	66	11	100	39	57	4
2000	100	22	65	13	100	34	61	5
2050	100	19	63	18	100	21	67	12
2100	100	19	63	18	100	19	63	18
	Annual Rate of Increase (%)							
1950-1980	1.0	0.4	1.1	2.4	2.2	2.3	2.2	2.3
1980-2000	0.6	0.3	0.6	1.3	2.0	1.4	2.4	2.9
2000-2050	0.2	-0.0	0.1	0.8	0.8	0.1	1.2	3.0
2050-2100	0.0	0.0	0.0	0.1	0.2	0.0	0.0	0.9

Dependency Ratio

	Total	Child	Old Age	Total	Child	Old Age
1950	55	43	12	72	66	7
1980	52	35	17	76	69	7
2000	53	33	20	63	56	8
2050	59	31	28	50	32	19
2100	60	31	29	60	31	29

Source: Population Division, United Nations, Long-Range Global
Population Projections, New York, August 1981, p. 19.

Table 2. Age Structure of Future Population in North America and Europe 1975-2000 for Medium Variant

	Population (millions)				Percent Distribution			Dependency Ratio		
	Total	0-14	15-64	65 and over	0-14	15-64	65 and over	Total	Child	Old Age
North America[a]										
1975	236.4	59.7	152.4	24.4	25.3	64.5	10.3	55.2	39.2	16.0
2000	289.5	62.9	191.6	35.0	21.7	66.2	12.1	51.1	32.8	28.3
2025	315.0	62.0	199.3	53.7	19.7	63.3	17.0	58.1	31.1	26.9
2050	317.3	61.8	199.9	55.6	19.5	63.0	17.5	58.7	30.9	27.8
Western Europe[b]										
1975	152.3	35.0	96.6	20.8	23.0	63.4	13.7	57.8	36.2	21.5
2000	157.6	31.1	103.4	23.1	19.7	65.6	14.7	52.4	30.1	22.3
2025	158.7	29.9	99.7	29.1	18.8	62.8	18.3	59.2	30.0	29.2
2050	154.3	29.7	97.2	27.4	19.2	63.0	17.8	58.8	30.6	28.2
Northern Europe[c]										
1975	81.6	19.0	51.4	11.2	23.3	63.0	13.7	58.8	37.0	21.8
2000	84.5	17.2	55.0	12.3	20.4	65.1	14.6	53.6	31.3	22.4
2025	87.0	16.7	55.3	15.0	19.2	63.6	17.2	57.3	30.2	27.1
2050	86.0	16.6	54.2	15.2	19.3	63.0	17.7	58.7	30.6	28.0

[a] Canada and the United States.
[b] Austria, Belgium, France, Federal Republic of Germany, Luxemburg, Netherlands, and Switzerland.
[c] Denmark, Finland, Iceland, Ireland, Norway, Sweden, and United Kingdom.
Source: Population Division, United Nations, Long-Range Population Projections, New York, August 1981, Table 4A, pp. 49, 60, and 62.

population aged 15-64 also tend to be associated with increases in
the participation rate of females. The rising proportion of females
in the labor force is not captured by the population dependency
ratio and this indicates the importance of considering the
proportion of the population that is actually in the labor force.

The development process also affects labor force participation
rates of youths and older persons, and as a result would alter the
appropriate ages of the economically active population. As economic
development proceeds, entry into the labor force is delayed in order
for younger workers to obtain increased schooling and technical
skills, and retirement at early ages becomes increasingly
prevalent. Labor force statistics indicate that the participation
rate of older men declines as economic development proceeds. For
example, Duran (1975) finds that labor force participation rates of
males aged 65 and over range from 63.1 to a level of 29.5 percent in
the most developed nations. Hence, the age limits of dependency are
a function of development and economic growth. When the age limits
of dependency are varied jointly with the rate of population growth,
i.e., slowing population growth is paired with lowering the age of
labor force withdrawal and raising the age of entry, the favorable
effect of slowing population growth on dependency is moderated.

Taking expected labor force participation changes into account,
Table 3 illustrates a measure of old age economic dependency. These
data indicate the number of persons in the labor force under age 65
for each person aged 65 and over in some of the developed countries
for the last half of the twentieth century. Combining expected
labor force and population changes, the number of workers per older
persons is not projected to decline in most of these nations during
the next twenty years.

Governmental policies clearly influence the labor supply
decisions of married women, youths, and the elderly. The economic
dependency burden can be reduced by policies encouraging labor force
participation. In general, population aging should be consistent
with later retirement. The trend toward early retirement in the
more developed countries could probably be reversed by altering
public policies toward work of the elderly. Mandatory retirement
policies could be eliminated. These and other government retirement
policies could be used to encourage continued work.

Developing countries may be able to avoid sharp declines in the
labor force participation of their elderly by avoiding policies that
encourage early retirement. Specifically, they could resist the
temptation to lower the age of eligibility for their maturing social
security systems. However, many developing countries have relatively
low ages for old age benefits and these countries may need to
consider raising these ages as the life expectancy improves. By
maintaining relatively high rates of market work by persons aged 60

Table 3. Number of Persons in Labor Force under Age 65 for Each Person Age 65 and over, Selected Countries 1950-2000

Year	Austria	Belgium	Canada	France	Germany, Federal Republic	Japan	Netherlands	Sweden	United Kingdom	United States
1950	4.5	3.6	4.7	3.9	4.8	8.5	4.9	4.1	4.2	5.0
1955	4.1	3.4	4.6	3.7	4.5	8.2	4.3	3.8	4.0	4.5
1960	3.8	3.2	4.8	3.5	4.2	7.9	3.9	3.5	3.8	4.2
1965	3.3	2.9	4.7	3.3	3.7	7.6	3.8	3.4	3.8	4.1
1970	2.9	2.8	4.9	3.1	3.3	6.9	3.6	3.2	3.5	4.2
1975	2.8	2.7	4.9	3.1	3.1	6.3	3.5	2.9	3.3	4.1
1980	2.8	2.7	4.9	3.1	3.1	5.7	3.4	2.7	3.2	4.1
1985	3.3	2.9	4.6	3.6	3.6	5.4	3.5	2.7	3.3	4.0
1990	3.3	2.8	4.4	3.4	3.5	4.9	3.3	2.7	3.3	4.0
2000	3.4	2.7	4.4	3.2	3.1	3.8	3.2	3.1	3.6	4.2

Source: Illene Zeitzer, "Social Security Trends and Developments in Industrialized Countries," Social Security Bulletin, March 1983, p. 56. Table was derived from data in International Labor Office, Labor Force Estimates and Projections, second edition, 1977.

and over, countries can substantially reduce the burden of income
transfers for the elderly, which generally grow with population
aging.

In assessing the usefulness of dependency ratios and the
seemingly favorable effects of population aging, a second issue that
should be considered is the relative cost of maintaining young and
old dependents. In most developed countries, the income and health
care of the elderly are increasingly provided through social
security systems. The primary source of transfers to children
remains the family, although schooling is usually provided by the
state. The extent of public intergenerational transfers varies
across countries, with the family remaining the principal provider
for young and old dependents in some societies.

While slowing population growth lowers the total dependency
ratio, it increases the old age dependency ratio. As a result, the
relative cost of young and old dependents will determine the effects
of population aging on total dependency costs. In the United
States, the public cost for each elderly person has been estimated
to be approximately three times the per-youth expenditure (Clark and
Spengler, 1980). Mueller (1976) estimates that in the less
developed countries consumption requirements of those 65 years old
and over exceed those for youths 0-9 and are less than the needs of
youths 10-14 in medium consumption profiles. Clearly, the costs of
supporting the elderly are a direct function of public policy.
Population aging will have a more significant cost effect, the
higher the quality of care and the greater level of income support
for the elderly. Thus, improvements in these support programs
should be made only after assessing their long-run costs.

Another problem with the dependency rate concept arises from
the fact that there also are significant changes in the composition
of each dependent group and those of working ages associated with
slowing population growth. As the rate of population growth slows,
an increasing percentage of the elderly are in the higher age
groups, e.g., over 75 years, and also a larger proportion of youths
are nearer the age of entry into the labor force. These
compositional changes will tend to increase the cost of supporting
the elderly, since a greater proportion of them are likely to
require additional public services, while an increasing number of
youths can partially support themselves. The costs of supporting
the very old are likely to increase with further declines in
mortality rates at higher ages and with the advancement of medical
care that is increasingly expensive.

A final area of concern that may be affected by the changing
age structure of the population is the average productivity of
workers. The labor productivity of an economy can be affected by
population aging because the age composition of the labor force is

altered. Evaluation of this effect requires that the productivity
or ability profile over one's worklife be known. In general, people
enter the labor force relatively untrained, gradually acquiring job
skills and knowledge during their worklives. At older ages some of
these skills may begin to deteriorate, but the importance of this
decline depends on the physical and mental demands of the job
(Kreps, 1977).

Understanding of this life-cycle pattern of human capital
acquisition, combined with analysis of the age structure changes in
the labor force, provide the criteria for assessing the effects of
population aging on labor productivity. In the more developed
countries, technological advances have made job requirements less
and less physically demanding. Reduction in these requirements and
the greater experience of older workers explain the research
findings that older workers frequently are as productive as younger
workers. On balance, increasing numbers of older workers should not
adversely affect labor productivity in the developed countries.

A decline in the relative number of younger, less experienced
workers should raise the average level of productivity. These
workers are more likely to be unemployed as they enter the labor
force and search for satisfactory employment. Population aging will
reduce the proportion of the labor force composed of inexperienced
workers and should also have a favorable effect on the nation's
unemployment rate.

Summarizing, the dependency ratio analysis indicates that in
general slowing population growth and its concomitant population
aging raises the proportion of the population in the economically
active years and lowers the dependency ratio. But declines in the
age of retirement, higher costs for older dependents and the rise in
the proportion of the very old may offset this favorable effect of
population aging. Government policies can be used to encourage
delayed retirement, which would moderate these potentially adverse
effects. As populations age, it is increasingly important to
consider the long-run costs of improvements in their retirement
plans.

DECLINES IN MALE LABOR FORCE PARTICIPATION IN THE U.S.

The preceding discussion indicated that the effect of population aging on dependency ratios and economic support systems depends
to a large degree on future changes in labor supply of older persons.
This section outlines the continued decline in labor force participation of older men in the United States, while the following section
attempts to explain these trends.

Throughout this century, the proportion of men 65 years and older who are in the labor force has been declining. For example, the LFP rate of men 65 and older declined from 71.3 percent in 1890 to 41.5 percent in 1940, or almost 6.0 percentage points per decade (Long, 1958). During the past three decades, the proportion of older men in the labor force fell from 45.8 percent in 1950 to 17.8 percent in 1982 (see Table 4).

Within this general decline, there were substantial fluctuations in the rate of decline throughout this 32-year period. Dividing the period into 10-year intervals shows that the average annual rate of decline was 1.3 percentage points per year from 1950 to 1960. The annual decline slowed to 0.6 percentage points in the 1960s and accelerated to 0.8 percentage points per year from 1970 to 1980 (see Table 5). The percent decline in the LFP rate shows a greater acceleration during the 1970s due to the lower LFP rate in 1970 compared to that of 1960. Thus, the decline in participation rate was 29.1 percent in the 1970s compared to 19.0 percent in the 1960s.

The acceleration of the decline during the 1970s in the LFP rate of men 65 and over may come as a surprise to some who thought that the participation rate of older men had stabilized. Part of the explanation lies in the volatility of this decline during the 1970s. For example, the LFP rate of men 65 and over fell from 26.8 percent in 1970 to 20.2 percent in 1976, or an average annual decline of 1.1 percentage points. By contrast, the rate actually increased to 20.4 percent in 1978. However, since 1978 the rate has fallen to 17.8 percent, or an average annual decline of 0.65 percentage points.

In the past twenty years, the decline in the LFP rate of men has been extended into the 55-64 year old age group. The participation rate for these men was virtually the same in 1950 and 1960; however, it declined by 3.8 percentage points from 86.8 to 83.0 percent between 1960 and 1970. During the next decade, the decline was 10.9 percentage points so that the 1980 rate was 72.1 percent (see Tables 4 and 5). The pattern of decline during the 1970s paralleled that of the 65 year old and older group. In the early 1970s, the LFP rate declined by 1.5 percentage points per year. The decline slowed to 0.5 points per year between 1976 and 1978 before accelerating to 0.8 percentage points per year between 1978 and 1982. During the early 1970s, the participation rate of men 45-54 years of age also declined sharply. Prior to 1968 the participation for men of this age exceeded 95 percent; however, the proportion of these men in the labor force fell from 95.2 percent in 1967, to 91.1 percent in 1977. This participation rate has remained relatively stable during the past five years and was 91.4 in 1981.

Table 4.   Civilian Labor Force Participation Rates, United States

Year	Aged 45-54	Males Aged 55-64	65 and over
1950	95.8	86.9	45.8
1955	96.5	87.9	39.6
1960	95.7	86.8	33.1
1965	95.6	84.6	27.9
1970	94.3	83.0	26.8
1971	93.9	82.1	25.5
1972	93.2	80.4	24.3
1973	93.0	78.2	22.7
1974	92.2	77.3	22.4
1975	92.1	75.6	21.6
1976	91.6	74.3	20.2
1977	91.1	73.8	20.0
1978	91.3	73.3	20.4
1979	91.4	72.8	19.9
1980	91.2	72.1	19.0
1981	91.4	70.6	18.4
1982	N/A	70.1	17.8

Source: Employment and Training Report of the President 1982,
Washington, D.C.:   USGPO, p. 155 and unpublished data from the
Bureau of Labor Statistics for 1982.

DETERMINANTS OF LABOR FORCE PARTICIPATION

Retirement from the labor force is a decision made by most
workers within an economic, political and personal framework. For
the most part, this is a voluntary choice of older persons as they
allocate their available resources over their remaining lifetime.
The desirability of various work/not-work choices depends on current
economic conditions, the value of one's time, availability of
retirement benefits, wealth, and health. This section examines
factors influencing the retirement decision and attempts to apply
them to the time series data to see if they provide an adequate
basis for explaining the trend toward early retirement.

Market Value of Time

Most studies of the labor supply of older persons indicate that
higher wage rates increase the probability that a person will be in
the labor force. Despite these findings, it is generally agreed
that long-run economic growth which raises the lifetime wealth of
persons leads to a reduction in a lifetime work effort.   This

Table 5.   Declines in Labor Force Participation Rates of Men,
           United States

Age Group	Interval			
	1950-60	1960-70	1970-80	1980-82
Men, 65 years and older				
Percentage point change in LFP during the interval	-12.7	-6.3	-7.8	-1.2
Average annual percentage point change in LFP	-1.27	-0.63	-0.78	-0.6
Percent change in LFP during the interval	-27.7	-19.0	-29.1	-6.3[a]
Men, Aged 55-64				
Percentage point change in LFP during the interval	-0.1	-3.8	-10.9	-2.0
Average annual percentage point change in LFP	-0.01	-0.38	-1.09	-1.0
Percent change in LFP during the interval	-0.10	-4.4	-13.1	-2.8[b]

[a]If this rate remains constant throughout the decade, the 1990 LFP would be 13.7 and the change during the 1980s would be -27.9 percent.

[b]If this rate remains constant throughout the decade, the 1990 LFP would be 62.6 and the change during the 1980s would be -11.3.

Source: Table 4.

seeming contradiction occurs because wage differences at the end of life for members of a cohort are primarily the result of life-cycle investment decisions and do not represent exogenous increases in wages. Anticipated economic growth raises the expected wage profile which life-cycle theories predict will have an income effect, thus lowering lifetime labor supply. For a variety of reasons, much of this reduction in work time comes late in life in the form of earlier retirement. It seems likely that much of the decline in participation rates of older men during the past century is due to real economic growth.

In the past decade, real economic growth has slowed and real wage rates have actually fallen. During this time, the decline in labor force participation rates has continued. In contrast to increases in wages over the entire worklife, temporary fluctuations in wage rates should have stronger substitution effects as people respond to changes in the price of their time. Between 1973 and 1981 real wages fell for many workers in the United States. For older workers, these changes will have only minor effects on the real value of their entire lifetime earnings. However, the value of work time in their last working years has fallen, and as a result they may shift to an increased consumption of home time by retiring earlier. Thus, a slow-down in the growth of real wages may have been a factor in acceleration of early retirement in the last few years, but the continuation of low growth rates should encourage delayed retirement in the future as people reallocate their time to this changed economic environment.

Social Security

The social security program is defined by a complex set of rules that have been changed periodically since 1940. The growth of benefits and the influence of social security rules on individual budget constraints has altered the resource allocation decisions of older men. Various aspects of the social security program are described below along with their effects on labor supply.

Significant improvements in social security benefits since 1940 have given past cohorts of older Americans unanticipated windfall gains in their lifetime wealth. These wealth increases have encouraged older persons to withdraw from the labor force. An example of the effect of large increases in benefits may be the sharp decline in participation rates following the 1972 benefit increases of 20 percent. Thus, one possible explanation for the pattern of change in the participation during the 1970s is that this substantial improvement in social security benefits resulted in relatively large drops in the LFP rate in the early 1970s. The relative stability of the participation in the mid 1970s allowed the rate of decline to move toward its long-run average, which was reestablished during the last part of the decade. The decline from 1978 to 1982 is approximately the same as the average annual decline for the entire decade.

The automatic indexation of social security benefits to equal the year-to-year change in the consumer price index (CPI) has insured the real value of this source of income. In the past, changes in the CPI have tended to overstate the change in prices and thus this indexation has tended to overcompensate retirees during the past inflationary decade. The indexation of benefits during a time when real wages were falling may also explain some of the decline in participation during the 1970s.

Another aspect of the social security program that may affect the timing of retirement is the age of eligibility for receipt of benefits. Many studies find that current eligibility for benefits reduces the probability of being in the labor force. Initially only men 65 years and older were eligible to receive benefits; however, in 1962 early retirement benefits were made available to men aged 62-64. This extension of benefits may help to explain the subsequent fall in LFP of males less than 65. For example, the participation rate of men aged 55-64 was 87.3 percent in 1961. Between 1950 and 1961, this rate ranged without trend from 88.7 to 86.8. After the extension of early retirement benefits to men, the LFP rate for these men dropped to 86.2 percent and continued to decline slowly through the 1960s before the decline accelerated in the 1970s.

Until a few years ago, economists were in general agreement that the earnings test for social security benefits was a significant factor in reducing labor force participation of older persons. Recent studies have incorporated the actuarial adjustment for delayed acceptance and the automatic benefit recalculation along with the earnings test into an assessment of the effect of social security on the compensation from continued employment. The combined influence of the three effects does not reduce the net wage for most persons prior to age 65. The few studies that have attempted to introduce the change in the wealth value of social security benefits into a retirement equation have found that the greater this value is, the lower the likelihood of withdrawal from the labor force.

If these effects are understood, then the change in the value of social security benefits with continued work should not encourage withdrawal from the labor force prior to age 65. In fact, it has been argued that labor supply prior to age 62 should be increased due to social security. After age 65, the actuarial adjustments decline sharply and the incentive for delaying benefit acceptance falls. Thus, the change in benefits with continued work and delayed acceptance is not a cause of the fall in the LFP rates prior to age 65 but may help to explain the drop at age 65 in the probability of being in the labor force.

Employer Pensions

The expansion of coverage of employer pensions during the post-World War II period and the grandfathering of existing workers conveyed unanticipated wealth gains to older workers. Real benefits were raised by increasing the generosity of pension benefits and reducing salary averaging periods. Such unanticipated gains in the wealth value of pension benefits have encouraged earlier withdrawal from the labor force. Pensions have also encouraged early retirement by lowering the eligiblity age, reducing early retirement penalties, and sometimes offering bonuses to those who take early retirement. In addition, after a certain age workers may not

receive any further credit in calculation of pension benefits from
extra years of service or an increased salary. All of these factors
affect the change in pension benefits with continued work. If this
value declines with age, perhaps becoming negative, then older
workers will be encouraged to retire.

The effect of pensions on the trend toward early retirement has
probably declined in the last 15 years as the growth of pension
coverage has slowed and existing systems have matured. Counter-
acting this to some degree may be the increased use of early
retirement options in response to the decline in economic growth
during the 1970s.

Mandatory Retirement

Prior to 1978, firms could enforce mandatory retirement at age
65, and there was a popular belief that these provisions were widespread and a significant cause of retirement. Amendments to the Age
Discrimination in Employment Act in 1978 raised the minimum age for
mandatory retirement to 70 in the private sector and eliminated its
use in most federal jobs. Several economic studies conducted
shortly after the passage of the amendments predicted that their
aggregate effects on the number of older persons in the labor force
would be small. This conclusion was reached because the evidence
suggested that, even in the presence of mandatory retirement
policies, most people retired voluntarily. Certainly, the decline
in LFP rates for men between 62 and 68 years of age since 1978 does
not provide support for the view that previously large numbers of
older workers were forced from their jobs.

Health

All studies of retirement conclude that health status is an
important determinant of early withdrawal from the labor force.
Older persons in poorer health are more likely to leave their jobs,
and poor health is often reported as the major factor leading to
early retirement. However, it is only a change in average health
status of older persons that could affect the trend in
participation. Data on the health status of older persons over time
are sparse. If rising life expectancy is used as an indicator of
improving health, then recent declines in mortality rates suggest
improving health and this should have encouraged continued work.
One shortcoming of this conclusion is that reduced mortality may
mean that more people in poor health survive to older ages and thus
tend to reduce labor force participation rates of the elderly.

Aggregate Economic Conditions

Aggregate economic conditions also influence retirement decisions. High unemployment rates make it more difficult for older

persons to find a job and more likely that they will stop searching
for employment and start drawing retirement benefits. In addition,
declines in demand for workers make firms more likely to offer early
retirement incentives to older workers. Estimates from studies on
discouraged workers confirm that the LFP rate of older persons is
sensitive to changes in unemployment. Since unemployment rates
fluctuate with business cycles, it is difficult to ascribe a
downward trend in participation rates to levels of unemployment.
The declines in LFP rates of older men have come in years when
unemployment was rising and in years when it was falling. To
confirm this, one need only compare changes in participation rates
to changes in unemployment rates during the 1970s.

Inflation may also influence retirement decisions by altering
real wealth and real wages. It is commonly believed that inflation
harms the elderly and thus might encourage older persons to remain
in or return to the labor force. Available evidence indicates that
older persons have not suffered a loss in real income because of inflation during the 1970s. Certainly, the pattern of LFP decline in
the presence of high rates of inflation during the 1970s does not
support a conclusion that inflation encourages continued market work.

During the past three years, the fact that price increases have
exceeded wage increases indicates that real wages have fallen.
During the same period, social security benefits and federal pensions have been increased by the same rate as inflation. Thus many
retirees have had their benefits raised by a rate that exceeded the
potential increase in wages. This pattern of income changes will
tend to encourage retirement and is one factor in explaining the
decline in LFP rates since 1978.

SUMMARY

Without significant increases in the current fertility rate,
the aging of the populations in the developed nations will continue
in the coming decades. Current projections indicate a gradual
increase in the fraction of the population 65 and over during the
next 20 years followed by a more rapid increase in the following 25
years. The impact of population aging on the economy depends, in
part, on possible changes in the labor force participation of
younger persons, primarily women. In addition, the proportion of
older persons who remain in the labor force instead of retiring will
have an important effect on the size of income transfers to the
elderly.

Using data from the United States, this paper has illustrated
the decline in the incidence of market work among older males. Such
declines have occurred throughout the world in conjunction with economic development. The cost to society of maintaining economically

dependent older persons is magnified by population aging. Thus, the pattern of early retirement and its determinants becomes a concern of public policy. An examination of the retirement decision suggests several economic variables that influence the choice of older workers to leave the labor force. The future pattern of withdrawal from the labor force will be governed by changes in these variables. While some of the determinants are not subject to direct governmental control, others such as pension policies are.

It seems likely that changes in social security systems will play a major role in influencing the trend toward early retirement. Reductions in the age of eligibility for benefits that some European countries have adopted or are considering have substantial long-run cost implications. Population aging and increasing life expectancy are inconsistent with public policies that promote early retirement. The gradual raising of the normal retirement age and the adoption of policies to facilitate employment of older persons could significantly reduce the increased cost of government transfers associated with population aging.

REFERENCES

Clark, R. and J. Spengler,"Dependency Ratios: Their Use in Economic
        Analysis," in Julian Simon and Julie DeVanzo (Eds.). Research
        in Population Economics, Vol. 2, Greenwich, Conn.: JAI Press,
        1980, p. 63-67.
Duran, J., The Labor Force in Economic Development, Princeton:
        Princeton University Press, 1975.
Kreps, J., "Age, Work, and Income," Southern Economic Journal,
        April, 1977, pp. 1423-37.
Long, C., The Labor Force Under Changing Income and Employment,
        Princeton, N.J.: Princeton University Press, 1958.
Mueller, E., "The Economic Value of Children in Peasant
        Agriculture," in R. Ridker (Ed.). Population and Development,
        Baltimore: Johns Hopkins University Press, 1976.
United Nations, The World Population Situation in 1977, Population
        Studies No. 63, New York 1979a.
United Nations, World Population Trends and Policies: 1977
        Monitoring Report, Vol. 1, Population Studies No. 62, New York
        1979b.

DEMOGRAPHIC TRENDS AFFECTING THE AGE STRUCTURE OF

THE LABOR FORCE: 1950 TO 2000

Howard N. Fullerton, Jr.

Office of Economic Growth and Employment Projections
Bureau of Labor Statistics
Washington, D.C.

INTRODUCTION

The structure of the labor force depends on the structure of
the underlying population and on labor force participation
(activity) rates. These two factors are not independent, but
synergistic. This paper will present an analysis of the demographic
structure of the United States labor force in the last half of the
twentieth century. "Technology," as defined by the Office of
Technology Assessment, includes "soft" technology, such as law,
regulation, and indeed social custom, as well as the more
conventional aspects of "hard" technology. Although the main thrust
of his paper is descriptive, we will touch on some aspects of
changes in technology on the composition of the future labor force.

The structure of the population is determined by past trends in
births, deaths, and migration. The effects of the current pattern
of births will be felt well into the next century, while the current
structure of the labor force and population was determined by the
changing pattern of births since 1920. Death and migration have
played a lessor role in that period.

Labor force activity has changed even more than fertility
rates, resulting in a labor force that has changed significantly
since 1950: women's participation has increased greatly, while the
participation of men has dropped. Thus, there were changes in the
composition of the labor force by sex, which was not a significant
factor for the overall population. In addition to the changing age
composition of the population, there were important changes in the
composition by marital status and educational attainment. Changes

55

in fertility also affect the parental status of the population and
thus the labor force.

POPULATION

    The structure of the population is one of the important factors
affecting the composition of the labor force. There are three
elements affecting the structure of the population: births, deaths,
and migration. Of these, by far the most important for the
structure of the population is births; changes in fertility will be
translated into a changing population composition.

Fertility

    Births have fluctuated in long cycles over the past century.
(See Figure 1.) During the 1920's, fertility dropped. The birth
rate began rising in 1933 and continued to rise until 1958. There
was a sharp increase in births with the end of World War II, but the
highest level of births occurred in the 1950's. For a period of 11
years, from 1954 through 1964, births exceeded 4 million. Between
the late fifties and the middle seventies, births dropped. Since
then, births have been rising and are expected to peak in 1988 at
3.9 million. After that, as the babyboom generation moves from the
peak childbearing years, the number of births would drop, even as
the rate is projected to rise slightly through the end of the
century. Fluctuations in births generate different sizes of
generations (cohorts), so that the age composition of the population
will vary in size.

    Looking at births in terms of decades simplifies matters,
though of course changes in demographic trends do not occur on
decade boundaries (Figure 2). Births in the forties were about the
level of the twenties though with a greater population and lower
birth rates. The level of births in the fifties, sixties, and
eighties was about the same. Again rising population levels
occasioned by the births meant that the rates were lower in the
later decades. The total fertility rates displayed in Figure 1 are
"period" rates, relating to a synthetic cohort of women, that is, a
group of women assumed to have the various age-specific fertility
rates prevailing in that year. No actual cohort is likely to
experience the period rate. In general, the period rates overstate
the cohort experience, being too low during baby busts and too high
during the boom (Campbell, 1975, 1978). Finally, changes in the
technology affecting births now will not affect the level of those
65 and older until about the middle of the next century.

    The number of births were also affected by trends in marital
status, labor force participation, and technology controlling the
number and spacing of births. In the immediate post-World War II
period, the age of first marriage dropped, a high proportion of the

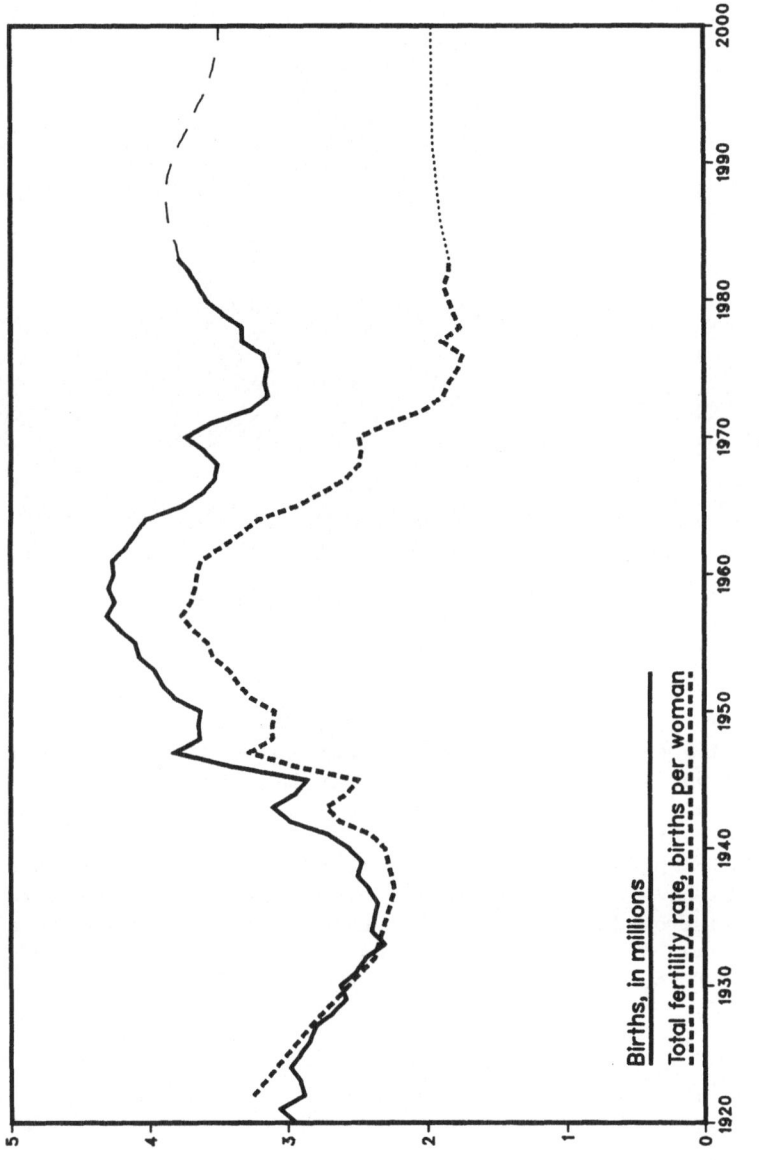

Figure 1. Births and the total fetility rate, 1920 to 2000.

Millions

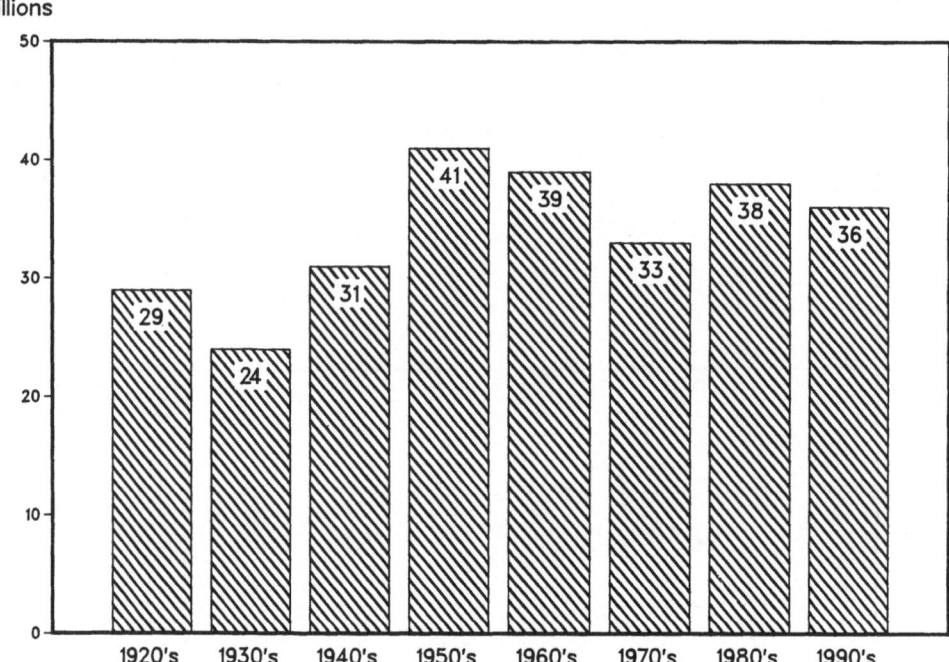

Figure 2. Births by decade, 1950-2000.

population married, and there were lots of babies. More recently,
the baby boom generation has been postponing marriage (or forgoing
it entirely) and the birth rate has been lower. At the same time,
labor force participation by young women has risen sharply;
participation is regarded as competing with childbearing. The most
important effect of technology on fertility was to allow individuals
to plan their childbearing, both in terms of the number of births a
couple would have and in the timing of births. Births now are more
concentrated around the mean age of childbearing. There are also
fewer unplanned births. Both of these trends have reinforced the
propensity of young women to be in the labor force.

Mortality

     Death rates dropped steadily over the 1920-58 period, then
remained steady over the next decade before resuming their fall
(Crimmins, 1981). Thus, life expectancy increased by 6 years over
the twenties (comparing the figure for the death registration area
in 1920 with the rate for the country in 1930), by 3 years over the
thirties, then 5 years over the forties, but by only 6 years over
the 1950 to 1980 period. (See Figure 3.) Life expectancy for women
is projected to rise from 78.3 in 1981 to 81.3 in 2005; for men,

Years

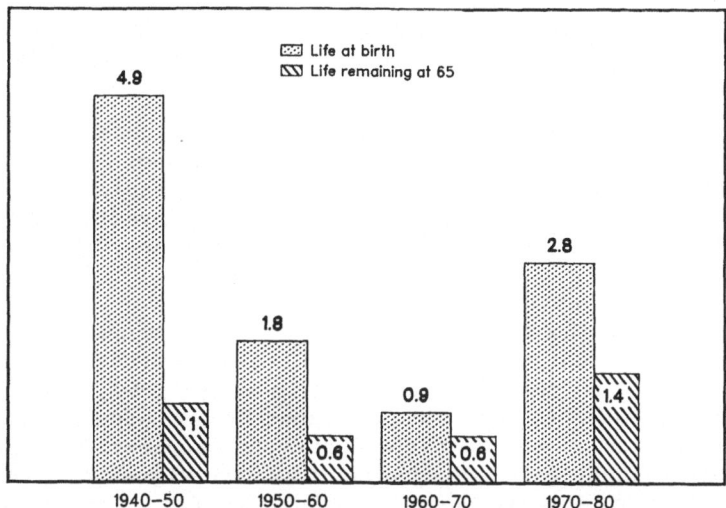

Figure 3.   Improvements in life expectancy at birth and at 65, by
decade, 1940-1980.

life expectancy would rise from 70.7 in 1982 to 73.3 in 2005. That
is, life expectancy is projected to increase by little more than a
year a decade. Changes in life expectancy do not affect the working
age population as much as either younger or older ages. The recent
changes in mortality have mainly affected the number and sex
composition of the older population.

Life expectancy at age 65 increased by 2 years from the turn of
the century until 1950 or 0.4 years a decade. Over the fifties,
expected life remaining at this age also increased by 0.4 years.
Between 1960 and 1970, though overall mortality improved only
slightly, life expectancy at 65 increased by 0.9 years. During the
seventies life expectancy increased by 1.2 years, to 16.4 years by
1980 (Figure 3). Thus, life expectancy at the age of entitlement to
Social Security has not just increasd, the amount of the increase
has been increasing.

When contemplating the effect of longer life expectancy on the
length of worklfe, the question rises as to whether the increased
years of life are of the quality that permits a person to continue
working. If so, then do older persons <u>want</u> to continue working? If
they do not, then what combination of incentives and sanctions will
induce them to work? From the view of either "soft" or "hard"
technology, those three questions appear to be vital questions for
those desiring to maintain or expand the length of worklife. For
discussions of the relationship between mortality and morbidity

trends, see Verbrugge (1983) and Fries (1980).

Migration

Aside from births, the only way for a population to grow is by
immigration. Generally, migrants are of working age and are
interested in work. Migration (either external or internal) is
difficult to monitor. The United States has accepted the greatest
number of international migrants for years. Migration also affects
the sex-age composition of the population, and thus the labor
force.

Immigration was fairly high during the twenties, then fell
during the thirties. After World War II, migration returned to a
level of more than two million a decade; since 1960, immigration has
accounted for more than a tenth of inner-decade population change.
When immigration accounted for a quarter of population change, the
laws (soft technology) were changed to limit migration and to
restrict immigrations to specific ethnic groups. After World War
II, immigration laws were changed to limit immigration and to remove
restrictions on the ethnic groups which could immigrate to the U.S.
The characteristics of migrants have changed as a consequence of the
changes in the law. Despite the limitations imposed by successive
laws, immigration has grown during the seventies; as a consequence,
immigration was 20 percent of the overall population change. (See
Table 1.) The Census Bureau projects a net migration of more than
four million a decade. Again, the range of migration levels
projected by the Census Bureau do not affect the size of the
projected labor force greatly, even though immigrants are more
likely to be in the labor force than the general population is.

Table 1. Population and Migration, by Decade, 1920-1980

Year	Population	Immigration	
		over the decade	as a percent of population
	(Millions)	(Millions)	change
1920	106	4.3	25
1930	123	0.7	8
1940	132	0.9	5
1950	151	2.5	8
1960	181	3.2	13
1970	205	4.3	20
1980	227	---	--

SOURCE: Historical Statistics of the United States, Statistical
Abstracts.

Immigrants can affect dependency ratios by increasing the number of workers; they can also offset low birth rates. Thus, one technical response to the greater proportion of older persons in the next century would be to increase international migration. Generally, use of international migration to manage national human resource shortages has not been an unqualified success, since it has proved difficult to have the migrants leave when the receiving nation perceives the migrants are no longer needed. Even if all the net migrants projected to enter the country were to be in the labor force, they would be less than four percent of the labor force. Of course, they would be a more substantial component of the increment to the labor force, about a quarter.

Population Structure

The consequence of rising and falling births is that the age structure of the population and the labor force fluctuates. As Table 2 and the population profile charts (Figure 4) indicate, there has been fluctuation in the age composition of the population. (We follow a specific birth cohort in Table 2 by moving diagonally down as we move to the right.) In Figure 4, the birth dearth of the thirties, represented in 1950 at ages 10 to 19, is always smaller than its surrounding cohorts. The baby boom of the fifties is always larger than its surrounding cohorts. However, the proportion of older people, specifically those 60 and older is growing over the entire half century.

We can also see the effect of recent lower birth rates; birth during the eighties will remain high, as almost a third of the

Table 2. Age Distribution of the Population, 1950 to 2000

Age	1950	1960	1970	1980	1990	2000
0 to 9	19.5	21.8	18.1	14.7	15.1	13.5
10 to 19	14.4	16.8	19.6	17.3	13.5	14.3
20 to 29	15.8	12.1	15.0	18.1	16.0	12.8
30 to 39	15.1	13.7	11.1	14.0	16.8	15.2
40 to 49	12.8	12.6	11.8	10.0	12.8	15.6
50 to 59	10.2	10.0	10.4	10.3	8.8	7.4
60 to 69	7.3	7.4	7.7	8.3	8.3	7.4
70 to 79	3.6	4.2	4.6	5.1	5.7	5.9
80 and up	1.3	1.6	1.9	2.3	3.1	3.8

SOURCE: U.S. Bureau of the Census, for 1950 and 1960, Statistical Abstract, 1962, Table 18; for 1970 and 1980, Current Population Reports, P-25, No. 917, 1982; for 1990 and 2000, Current Population Reports, P-25, No. 922, 1982.

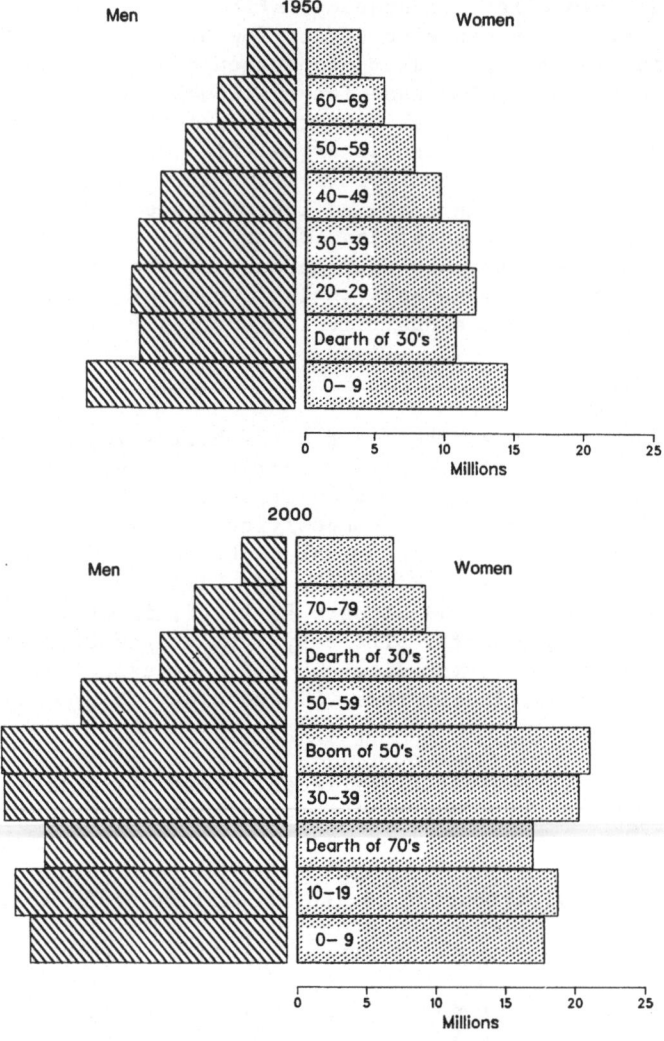

Figure 4. Age profile of the population, 1950 and 2000.

population will be between the ages of 20 and 40. Thus, the
proportion of those under the age of 10 should increase between 1980
and 1990. After that, births will drop--by 2000, the proportion
under the age of 10 will be the lowest recorded. Easterlin (1968,
1980) identified the relation between relative cohort size and
fertility. Larger cohorts experience greater competition for jobs
and feel worse off relative to their status growing up. They adjust
to this by deferring family formation; young women would participate
more intensely in the labor force. Thus, if this hypothesis is
true, we would find long fluctuations in births. We would also

expect to find that women have high participation rates currently.
We would also expect to find the birth rate climbing soon and young
women's participation remaining stable.

For those aspects of technology which are affected by the level
of a population component (such as the need for school rooms), the
effect of comparable levels of births in different decades should be
the same. Taking the example of school children, in the fifties,
the trend of higher births had to be discovered and confirmed--in
the sixties, the appropriate action of building schools and hiring
teachers had to be done. Since the trend of higher births was
projected to continue indefinitely, in the seventies, corrections
had to be made. Thus, even with comparable levels of births, the
effect of a specific population level was different in different
decades. Further, the effects occured at different years for
different levels of schooling. Because of internal migration these
effects were more intense in some parts of the country than in
others. We would expect comparable consequences of changes upon the
labor force.

Because different levels of government in the United States
support different age groups of dependents, the changing age
structure affects the relationship between different levels of
government. When there are many young dependents, the sub-national
government bodies must provide schooling and welfare. On the other
hand, the aged population is one in which the responsibility for
providing support shifts from the state and local to the national
levél.

LABOR FORCE

Labor force growth is influenced by the population composition
and by the proportion in the labor force (participation or activity
rates). The overall labor force will change as the composition of
the population changes; these changes can be offset or emphasized by
changes in participation. In addition to the changes in the age
composition of the population, the composition of the labor force is
affected by changes in marital status, parental status, and
educational attainment. Further, the labor force sex structure is
different from that of the population because participation rates
vary by sex.

Participation by Age and Sex

Figure 5 indicates the striking differences between participation patterns of men and women, and the great changes in
participation that have occurred. Male rates are higher than women's
rates--at all ages. Both sexes increase participation rapidly
during the teens and early twenties, but women's participation peaks

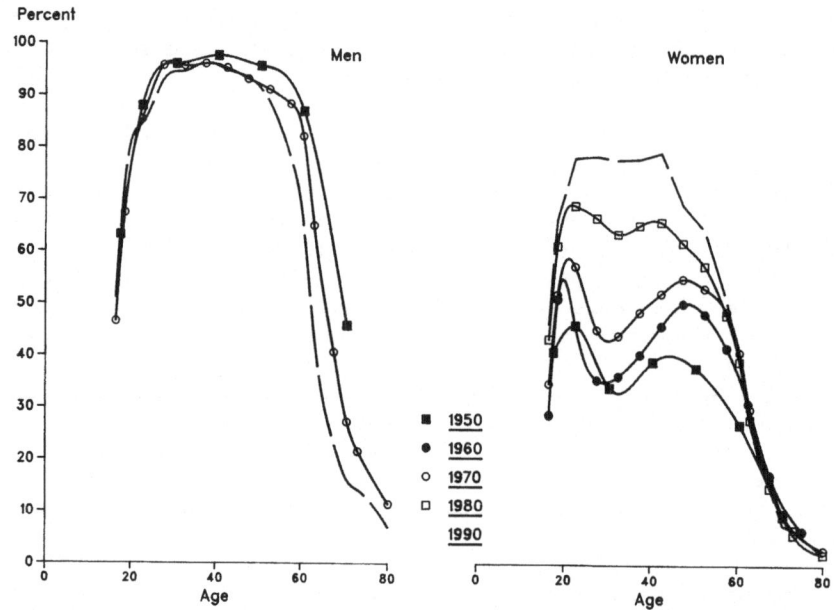

Figure 5. Labor force rates by age and sex, 1950-90.

in the late twenties while men's participation peaks in the early
thirties. One subtle change in women's participation is the slow
movement of the peak age of participation to an older age.

As Figure 5 and Table 3 indicate, participation varies by age;
as a large population group moves into an age with higher
participation, the labor force will grow more rapidly. When the
baby boom generation entered the prime working ages the overall
labor force grew rapidly. As this group moves into the ages where
participation is stable, the labor force will grow more slowly,
growth being dependent on increased participation as well as the
entries of the birth dearth. These changes lag the fluctuations in
births by fifteen to twenty years; youth's entry into the labor
force is spread over the ages from 16 to 24.

For the past thirty years, the participation of most groups of
women has been increasing among each of the subgroups of the
population just identified. Further, within each group, changes in
composition moved to those groups with higher participation, for
example, from married women to single women. During the fifties,
older women increased their participation, while during the late
sixties and seventies, younger women, particularly mothers,
increased their participation. The labor force rates for women ages
20 to 44 are projected to rise, but more slowly each year than the
previous year. The participation rates for males, ages 24 to 54,
are expected to continue declining, although not as fast as in the
seventies.

Table 3. Civilian Labor Force Participation Rate, by Sex and Age,
         1950-80 and Projected to 1995 (in percent)

Sex and age				Actual	Projected	
	1950	1960	1970	1980	1990	1995
Total, age 16 and over	59.2	59.4	60.4	63.8	66.9	67.5
Men	86.4	83.3	79.7	77.4	76.5	76.1
16 to 24	77.3	71.7	69.4	74.4	74.7	74.5
16 to 19	63.2	56.2	56.1	60.5	62.3	62.9
20 to 24	87.9	88.1	83.3	85.9	84.4	84.1
25 to 54	96.5	97.0	95.8	94.2	93.8	93.4
25 to 34	96.0	97.5	96.4	95.2	93.7	93.1
35 to 44	97.6	97.7	96.9	95.5	95.6	95.3
45 to 54	95.8	95.7	94.3	91.2	91.3	91.1
55 and over	68.6	60.9	55.7	45.6	37.4	35.3
55 to 64	86.9	86.8	83.0	72.1	65.5	64.5
65 and over	45.8	33.1	26.8	19.0	14.9	13.3
Women	33.9	37.7	43.3	51.5	58.3	60.3
16 to 24	43.9	42.8	51.3	61.9	69.1	71.6
16 to 19	40.9	39.3	44.0	52.9	56.8	58.2
20 to 24	46.0	46.1	57.7	68.9	78.1	82.0
25 to 54	36.8	42.9	50.1	64.0	75.6	78.7
25 to 34	34.0	36.0	45.0	65.5	78.1	81.7
35 to 44	39.1	43.4	51.1	65.5	78.6	82.8
45 to 54	37.9	49.8	54.4	59.9	67.1	69.5
55 and over	18.9	18.1	25.3	22.8	20.5	19.9
55 to 64	27.0	37.2	43.0	41.3	41.5	42.5
65 and over	9.7	10.8	9.7	8.1	7.4	7.0

Women are projected to account for about two-thirds of the
labor force growth during the 1980's and 1990's, about the same
proportion as the 1950's. Over the sixties and seventies, as more
and more of the men of the baby boom generation came into the labor
force, the proportion of growth attributed to women steadily dropped
despite the rapid increases in their rates. With the young men of
the baby boom now in the labor force, the share of labor force
growth attributed to women is projected to be greater over the next
decade.

Participation by Older Workers

Older workers leave the labor force rapidly: in 1980 white men
55 to 59 had a participation rate of 82.9 percent, by ages 65 to 69
this had dropped by 54.2 points, and those 75 and older had a rate

of 9.1 percent. White women's participation rates in 1980 at ages
55 to 59 were in fact less than the drop in male participation
between ages 55 and 65. Looking at the percent drop, the activity
rate for men 65 to 69 was 65 percent lower than for men ages 55 to
59; for women, the comparable figure was 69 percent. The percent
change in participation was even greater between ages 65 to 69 and
75 and older. This pattern is typical of other years. If we look
across time rather than age, we find participation is dropping for
men. Since 1950, participation by older men has dropped by 12
percentage points, while participation by women has increased by 4
points. Thus, labor force activity of older women has grown
slightly, though still significantly lower than men. The
participation rates for older men and women are expected to decline,
though more slowly than in the seventies.

Marital Status

Participation for women and men varies by marital status, as
well as by age. As marriage and divorce rates have changed, the
marital composition of the population has changed, affecting the
pattern of participation in the labor force. Participation of
married women has been increasing, and appears to be converging with
that of other women. Married women and men now make up a smaller
proportion of the total population. Married men have higher
participation rates than other men; this was considered to reflect
the need for higher participation--if the wife did not work, then he
needed to support several household members. Also, marriage was
considered a selective process for men, so married men had different
characteristics than other men, characteristics that led to higher
labor force participation. However, married men's participation is
dropping, toward that of other men, which is rising. (See Table 4.)

Table 4.   Civilian Labor Force Participation Rate, by Sex and
           Marital Status, 1960-1980 (In percent)

Year	Women			Men		
	Married women (as a percent of all women)	Labor force participation rate		Married men (as a percent of all men)	Labor force participation rate	
		married (percent)	other (percent)		married (percent)	other (percent)
1960	62.2	30.5	42.1	66.1	88.9	56.4
1965	59.9	34.7	39.7	65.0	87.7	51.6
1970	61.5	40.8	45.6	68.1	86.9	59.1
1975	59.8	44.4	48.2	66.0	83.1	66.6
1980	56.7	50.2	52.3	62.3	81.0	69.3

Although the 5 percent annual change in the proportion married
over a 20 year period appears small, applied to the 87 million women
in 1980 who were 16 or older, this is almost 5 million fewer married
women. The result of this shift in the proportion of married women
is a greater number of women in the labor force, even though, as the
table indicates, the difference in the participation rates of
married and of others has converged as the proportion of those
married has dropped. The decrease in the proportion of those
married reflects i) the "marriage squeeze," ii) the increased
divorce rate, and iii) the postponement of marriage. (The term
"marriage squeeze" refers to the joint effect of increasing births
with the phenomenon of women marrying men about two years older.
About twenty years after the period of increasing births, there
would be fewer men than women of marriageable age.) The three
factors which caused a drop in the proportion of married women also
lowered the birth rate and the proportion of women with young
children.

Parental Status

Aside from the different levels of participation, the most
striking difference between women and men's participation is the
drop in participation occasioned by childbearing responsibilities.
As would be expected from the earlier discussion on fertility
patterns, the bottom of the "valley" is getting higher. Recently,
the valley has started to fill in. One future would have women's
and men's participation similar, with no valley in women's rates.
While participation of mothers has been rising sharply, the
proportion of women with young children has been dropping, as the
lower fertility rates imply. (See Table 5.)

The postponement of marriage and the lower incident of marriage
has contributed to the lowering of the percent of women with young
children. In addition, the concentration of childbearing in a
shorter period allows women to reduce the period they are out of the
labor force for childbearing and the nurture of young children. We
have seen a dramatic increase in the participation rates of the

Table 5. Ever-married Women with Children under the Age of 6

Year	As a percent of all married women	Labor force participation rate
1960	27.4	20.2
1965	26.0	25.3
1970	23.6	32.2
1975	21.5	38.9
1980	19.0	45.1

mothers of young children. In a twenty year period, participation
has risen 25 percentage points. One effect of this is to increase
the need for childcare provide outside of the home. In 1970, much
childcare was provided either in the home or informally. By 1980,
this was provided by day care centers. One effect, then, of women's
entry into the market economy has been the provision by the market
economy of services which had been provided in the home.

This applies to other household services, including the care of
older family members. As the trend of the labor force rate of women
increases, then the need for these care services to be provided
through the market will increase. One can look at this either from
the point of providing the services to increase the ability of
household providers to enter the labor force (and possibly work in
an occupation where they are more skilled than that of a personal
services provider) or from the point of providing needed services
that used to be produced in the home.

Education

Americans steadily increased their years of formal schooling
between 1965 and 1980 (Table 6). Thus, the older population has
lower educational attainment than the younger population.

One noticeable difference in patterns of educational attainment
by sex is that a greater proportion of women complete high school;
but a greater proportion of men complete college. By 1980, less
than a quarter of all women in the labor force had not completed a
high school education. A fifth of the men had at least a college
education. It appears that these proportions will converge. As
education increases, the labor force participation of Americans also
increases (Table 7).

Table 6. Proportion of the Labor Force Completing High School or
          College[1]

Year	Women with at least		Men with at least	
	high school	college	high school	college
1959	55.9	8.1	46.8	10.5
1965	62.3	10.0	54.9	12.4
1970	69.4	10.7	62.7	14.1
1975	73.3	13.2	69.1	17.3
1980	78.9	15.8	72.2	20.0

[1] The percentages apply to those 18 and older for 1965 and to
those 16 and older for 1970, 1973, and 1976. All percentages are
for March of the specified year.

Table 7. Participation Rate by Level of School Completed[1]

Year	Women completed		Men completed	
	high school	college	high school	college
1965	45.0	53.9	91.3	92.6
1970	50.3	55.6	90.1	89.8
1973	51.3	57.8	88.5	90.3
1976	52.9	62.5	86.3	88.8
1980	57.7	65.4	85.2	89.5

[1]All percentages are for March of the specified year.

Participation by women of both educational levels has been
rising, with that of high school graduates rising more (Table 7).
In 1976, Jaffe and Ridley reported that the changes in white women's
participation could not be attributed to compositional changes in
educational attainment. The effect is to make women of different
marital and educational attainment status similar in labor force
participation.

Change in Age Composition

The low births of the thirties resulted in low labor force
growth in the fifties; the high births of the fifties translated
into high labor force growth in the seventies. (See Table 8.)
Because of the changing age structure, increases in the labor force
over the next 8 to 12 years will be influenced by i) the baby boom
generation, which will attain those ages at which both men and women
have their highest participation and by ii) the continued, but
slower, rise in participation among women ages 20 to 44. By
contrast, the increases in the labor force during the 1970's were
influenced by i) the initial entrance of the baby boom generation
into the labor force and by ii) the very rapid increases in labor
force activity of women, particularly married women, ages 20 to 44.
Thus, different age groups of the labor force will grow more rapidly
during some periods than others. The younger labor force scarcely
grew during the fifties, grew rapidly during the sixties and early
seventies, and will actually drop during the eighties and nineties.
The prime age labor force, ages 25 to 54, grew at the same rate as
the labor force during the fifties, as women entered the labor
force, but was the slowest growing component during the sixties;
during the seventies, it grew more rapidly than the labor force as a
whole. For the eighties and nineties, only this age group is
projected to grow.

As Table 9 indicates, growth of various age groups of the labor
force is not uniform; the younger labor force did not grow until the

Table 8.   Distribution of the Actual and Projected Labor Force,
           by Sex, and Age, 1950 to 1995 (In percent)

Sex and age	Percent distribution					
	1950	1960	1970	1980	1990	1995
Total, age 16 and over	100.0	100.0	100.0	100.0	100.0	100.0
Men	70.4	66.6	61.9	57.5	54.2	52.3
16 to 24	11.4	9.9	11.7	12.7	9.0	8.0
16 to 19	4.0	4.0	4.8	4.7	3.3	3.1
20 to 24	7.4	5.9	6.9	8.0	5.7	5.0
25 to 54	45.7	44.2	38.9	36.2	38.6	39.1
25 to 34	16.9	14.7	13.7	15.9	15.7	13.8
35 to 44	15.7	15.8	12.6	11.1	14.0	14.8
45 to 54	13.0	13.8	12.6	9.3	8.9	10.5
55 and over	13.3	12.5	11.2	8.5	6.6	6.1
55 to 64	9.3	9.2	8.6	6.8	5.1	4.8
65 and over	3.9	3.3	2.6	1.8	1.5	1.3
Women	29.6	33.4	38.1	42.5	45.8	46.7
16 to 24	7.1	6.7	9.8	10.9	8.7	8.0
16 to 19	2.8	2.9	3.9	4.1	3.0	2.9
20 to 24	4.3	3.7	5.9	6.8	5.6	5.2
25 to 54	18.6	21.1	22.0	26.1	32.4	34.1
25 to 34	6.6	5.9	6.9	11.5	13.4	12.4
35 to 44	6.7	7.6	7.2	8.1	12.0	13.3
45 to 54	5.3	7.6	7.9	6.5	7.0	8.5
55 and over	3.9	5.6	6.3	5.5	4.8	4.6
55 to 64	3.0	4.3	5.0	4.4	3.7	3.6
65 and over	0.9	1.3	1.3	1.1	1.1	1.0

Table 9.   Labor Force Growth

Period	1950-60	1960-70	1970-80	1980-1990	1990-2000
Total	1.3	1.7	2.6	1.6	1.0
Ages 16 to 24	.0	4.5	3.6	-1.3	-.8
Ages 25 to 54	1.3	1.0	2.8	2.9	1.6
Ages 55 and over	1.6	1.4	.4	-.6	-.2

late sixties and seventies; during the nineties, their numbers
should drop. On the other hand, the prime working age labor force
grew at the same rate as the labor force during the fifties; as most
were from the period of low births associated with the Great
Depression, their labor force grew slowly during the sixties. As
the younger members of the baby boom entered the prime working ages,
and as more women entered the labor force, the rate of growth of the
prime age work force has increased. Presumably, this should make
matters better for carrying social responsibilites, but these are
also capital-using ages, acquiring a home and educating children;
consequently, the abundance of such people may not be as helpful in
carrying support activities as has been speculated earlier
(Fullerton, 1980). The older labor force is projected to decrease
in size between now and the end of the century. On the other hand,
the older population is projected to increase in size and as a
proportion of the total. This raises questions about the ability of
the population to support the increasing number of older
dependents.

The Economic Dependency Ratio

Historically, there have been more dependents than labor force
participants. In the immediate future there is the possibility that
there will be more in the labor force than dependent. The economic
dependency ratio, which measures this relation between the structure
of the population and of the labor force, is the ratio of those
(including babies) not in the labor force to the labor force.
Around 1986, more of the total U.S. population should be in the
labor force than not in the labor force. This ratio declined
sharply through the 1970's as the baby-boom generation and women
entered the labor force in large numbers. (See Figure 6.) During
the 1980's and 1990's, the ratio should continue to decline, albeit
at a considerably more moderate pace, reflecting only the continued
increases in female labor force participation rates. The dependency
ratio can be decomposed to indicate the dependency of children
(below the age of 16), of working age (those 16 to 64) and of the
older population (Table 10). We can, for example, follow the career
of the baby boom generation by tracking their progress from
dependent to labor force status.

The steady drop (from 50 persons per hundred workers to 36
persons per hundred workers) in the ratio attributed to those 16 to
64 reflects the steady entry of women into the workforce. The ratio
for persons under 16 has declined steadily over the 1960 to 1980
period as the baby-boom generation and women entered the labor
market. During the next decade the ratio should be unchanged
despite the "echo" of the baby-boom, their own children. The ratio
for older people is expected to rise slightly over the next decade
and should continue to rise into the middle of the next century.
However, older people are the smallest of the groups of dependents.

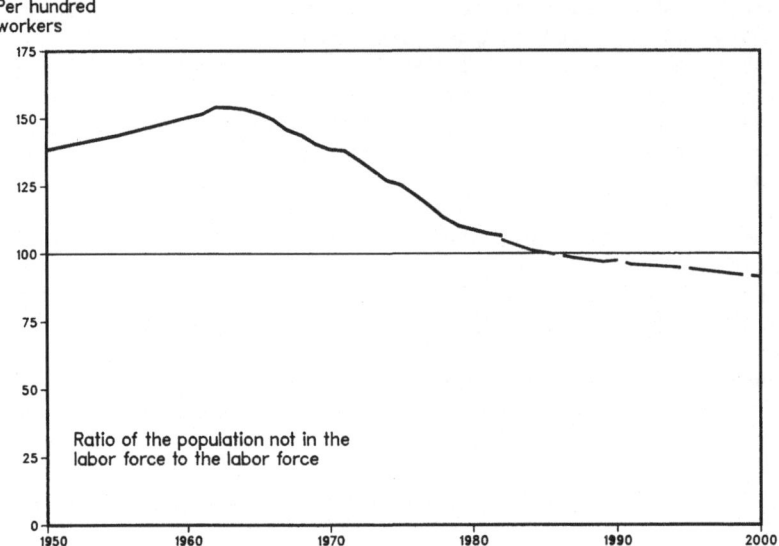

Figure 6. Economic dependency ratios, 1950 to 2000.

Participation Rate versus Composition Changes

At this stage, we have looked at the effects of composition
changes in the population on the labor force and at the effects of
rising participation rates. This leaves the question of the
relative importance of the two effects. We do not have all the
cells to investigate the effects of each composition change between
1950 and 1980, but we can apply the 1980 participation rates by age
and sex to the 1950 population to see if composition changes or
participation rate changes were more important (Table 11).

The 1950 population with the 1980 rates is much higher; the
effect of participation rate changes is more important than the
composition changes. By this exercise, the women's labor force is 50

Table 10. Economic Dependency Ratio

Year	1960	1970	1982	1990	1995
All persons	150.4	138.5	108.8	96.4	93.1
Under 16	81.5	72.1	50.7	45.2	43.1
16 to 64	50.2	46.8	37.3	28.4	26.9
Over 64	18.7	19.6	20.8	22.5	23.1

Table 11. Labor Force Participation Rates of 1980 Applied to
         1950 Population[1]

	Actual labor force	With 1980 participation rates	difference
Total	62,208	70,123	7,915
Men	43,819	40,492	-3,327
Women	18,389	29,630	-11,241

[1] Numbers in thousands

percent larger. Men's labor force is smaller. As we have seen,
between now and 2000, the composition changes favor high labor force
activity.

Thus, although there were significant changes in population
composition occasioned by fluctuating births, and by changes in
marriage patterns, both of which will continue into the future,
changes in labor force activity had a more important effect on the
level of the workforce. The compositional changes in the population
had the effect of increasing women's participation; all the changes
except the increase in educational attainment worked to decrease
men's participation. As changes in composition moderate, the
changes in labor force participation should tend to moderate.

REFERENCES

Campbell, Arthur A., 1975, "Beyond the Demographic Transition,"
     Demography, p.549-561.
Campbell, Arthur A., 1978, "Baby Boom to Birth Dearth and Beyond,"
     Annals, American Academy of Political and Social Sciences,
     p. 40-60.
Crimmins, E.M., 1981, "The changing pattern of American mortality
     decline, 1940-77, and its implications for the future,"
     Population and Development Review, 7:229-254.
Easterlin, Richard A., 1968, Population, Labor Force, and Long
     Swings in Economic Growth, General Series 86, National Bureau
     of Economic Research, New York.
Easterlin, Richard A., 1980, Birth and Fortune: The Impact of
     Numbers on Personal Welfare, Basic Books, New York.
Fries, James F., 1980, "Aging, natural death, and the compression of
     morbidity," New England Journal of Medicine, 303:130-135.
Fullerton, H.N., Jr., 1980, "The 1995 labor force: A first look,"
     Monthly Labor Review, p. 11-21.
Jaffe, A.J. and Jeanne Clare Ridley, 1976, "Educational Attainment
     and the Labor Force Participation of White Women in the United

States, 1930-1970," Proceedings of the Social Statistics
Section, 1976, American Statistics Association, Washington,
D.C., 1976, pp. 434-439.

U.S. Bureau of the Census, 1982, "Projections of the Population of
the United States: 1982 to 2040," Series P-25, No. 922.

Verbrugge, L.M., 1983, "Longer life but worsening health? Trends in
health and mortality of middle-aged and older persons," Center
for Social Research, Ann Arbor, Michigan.

OLDER WORKERS: FORCE OF THE FUTURE?

Cynthia Taeuber

Population Division, Bureau of the Census
Washington, D.C.

Thirty years ago, half of all elderly (65 years and over) men in the United States worked. Today, less than a fifth do. Among elderly women, there has been little change: 10 percent were in the labor force in 1950 and about 8 percent were in 1981. The labor force participation of older men 55 to 64 years dropped also but women between the ages of 55 and 64 increasingly joined the labor force, from just over a fourth in 1950 to 42 percent today.

What will happen to these trends thirty years from now when the baby-boom generation becomes elderly? Projections of the cost of retirement to the individual and to society have been so high that proposals to extend worklife are becoming commonplace. Will the baby-boom elderly be required to work longer than today's elderly? Will they need or want to work longer? Will they be able to work longer? This paper provides an overview of today's elderly worker and then discusses these questions in light of the possible characteristics of elderly workers 30 years from now, based on the current characteristics of workers 25 to 34 years old.

TODAY'S ELDERLY WORKERS

There are only about 3 million workers 65 years old or over (most of whom are 65 to 69 years old) and about 12 million 55 to 64 years old. Workers 55 years and over constitute about 14 percent of the nation's 110 million total labor force but workers 65 years and over represent less than 3 percent. The 65 and over group accounted for about 5 percent of the labor force in 1950. Of the 3 million elderly workers, 1.8 million were men, half of whom worked full-time, and 1.2 million women, most of whom worked part-time.

Today's elderly grew up in an environment which encouraged older persons to retire to open up jobs for young people. In 1935, the Social Security Act was passed with the express purpose of opening jobs for millions of unemployed by providing income security for the older population who in turn could vacate jobs. Since then, both public and private pension plans have encouraged early retirement during times of high unemployment, and retirement has become a commonly used mechanism for regulating the supply of labor over the business cycle. Retirement at least by the age of 65, if not sooner, has come to be viewed by many, including younger workers, as an earned right.

Labor Force Participation Rates

The labor force participation of elderly men has dropped rapidly over the last 30 years (Figure 1). In 1950, almost half of all elderly men were in the labor force; by 1960, only a third were working or looking for work; by 1970, only a fourth; and by 1982, less than a fifth (17.6 percent). The decreases are due in part to an increase in voluntary early retirement because of the availability of private/public pensions, desire for leisure, policies which push older persons out of the labor force to create jobs for younger persons, limited part-time employment opportunities, and a drop in self employment. The decrease in male labor force participation extends even to men in their fifties and early sixties. In 1960, over 88 percent of men aged 55 to 59 were in the labor force; by 1981, it had declined to about 80 percent. In 1960, 77 percent of men aged 60 to 64 worked, but by 1981, less than 60 percent did. At age 70 and over, in 1960, one out of four men worked, but by 1981, the proportion had dropped to one out of eight.

Labor force participation of elderly women, on the other hand, has varied little. In 1950, about 10 percent of elderly women worked, and by 1982, the percentage had dropped to just under 8 percent (1.2 million). For women over the age of 70, labor force participation dropped from 6 percent in 1950 to just under 5 percent today. But women between the ages of 55 and 64 have increasingly joined the labor force: in 1950, only 27 percent of these women worked but in 1982, 42 percent were in the labor force.

Historically, among older Black women, labor force participation has been distinguished by much higher rates than those for White women. Over the last 30 years, however, the rates have converged so rapidly that, by 1981, only a few percentage points separated the two groups. The extent of labor force participation for older Black males is somewhat lower today than the rate for older White men, and it has fallen more rapidly, probably because of differences in the occupational distribution. Elderly Blacks tend to work part-time more often than do Whites.

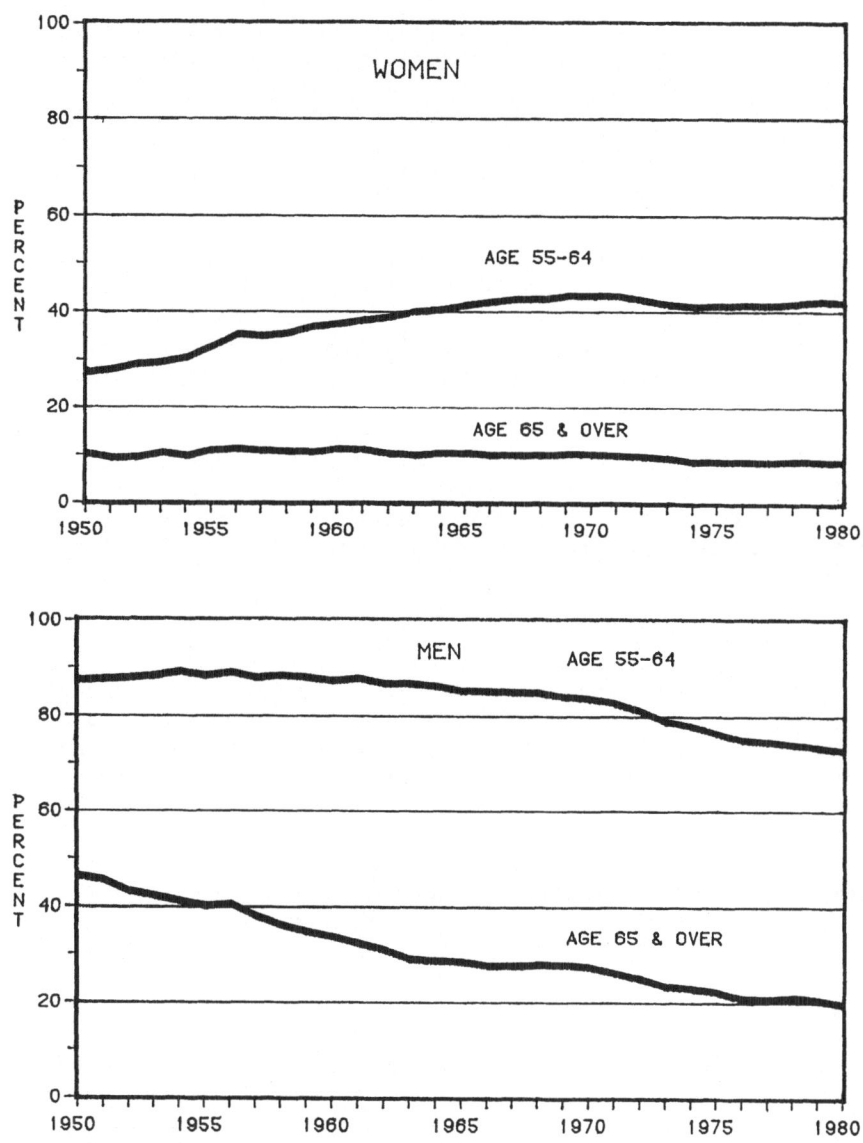

Figure 1. Labor force participation of persons aged 55-64 and
          65 & over: 1950-1980.

Industrial and Occupational Distribution

Among the 3 million elderly workers, half were in white-collar
occupations. The major occupations for elderly men were
professional and technical workers, managers and administrators,
farm workers, service, and craft workers. The majority of women

were service or clerical workers; about one in eight were in sales
and another one in eight were professional or technical workers.

Two-thirds of elderly male workers were in two industries:
trade (primarily retail) and services (primarily business and
repair, personal, and other personal), compared with 40 percent of
all other age groups in these two industries.

Sex and race are important determinants of the occupations of
the employed elderly. Three-fifths of elderly White women workers
were in white-collar professions and about two-thirds of Black women
workers were service workers, predominantly in private households.
About one-half of elderly White male workers were in white-collar
occupations and one-quarter in blue-collar work. Over a third of
elderly Black males were blue-collar workers with nearly a fourth in
white-collar jobs and another quarter in service jobs. Farm
occupations were more common among the oldest men; nearly a fifth of
Black and a sixth of White working males 70 and over were
farmworkers, compared with less than 4 percent for all males 25
years and over.

Part-Time Employment

Part-time employment is an increasingly important type of
employment for the elderly. In 1981, of the elderly who were at
work in nonagricultural industries, 48 percent of the men and 60
percent of the women were on part-time schedules as compared with 30
percent of the elderly men and 43 percent of the women in 1960.
Most who were on part-time schedules report that it is their choice
to work part-time, rather than being forced to work part-time for
economic reasons.[2] Over the last decade, elderly men have made
up 5 to 6 percent of all persons on voluntary part-time work
schedules and elderly women have made up about 4 percent, as
compared with women 18 to 64 years old who have made up about 50 to
60 percent of such workers.[3]

Surveys (such as a 1981 Harris Survey) indicate a preference
for some type of continuous employment after retirement that is
similar to the longest held job, but part-time. Most firms do not
have flexible employment policies and so most older persons must
retire permanently. Few elderly secure other employment after
retiring.[4]

Unemployment

The unemployment rate for the elderly in 1982 was 4.7 percent,
about half that of the population 16 years and over. The
unemployment rate for elderly White males was 2.4 percent but was
closer to 8 percent for elderly Black males; among women, the racial
differences were similar but not as pronounced. Unemployment among

older workers (55 years and over) at the close of 1982 was 6
percent, the highest since the government began measuring
joblessness after World War II. More than 770,000 Americans 55 and
over were out of work. This figure increases to 1.1 million if
discouraged workers who have stopped looking actively for work are
included.[5] Some early retirement is actually a result of
unemployment.

    Older workers, once they lose their jobs, stay unemployed
longer than younger workers, earn less in a subsequent job than
younger workers, and are more likely to give up looking for another
job following a layoff. Persons 55 and over are out of work on the
average nearly 20 weeks before being reemployed. That compares to
the 15.5 weeks on the average for all unemployed Americans.
Likewise, the older worker who successfully finds another job will,
on the average, earn $1,500 less than he or she got earlier.[6]
Finally, older workers are more than twice as likely as others to
give up searching for a new job. In 1982, there were about 334,000
discouraged workers 55 years and older who were no longer counted as
unemployed because they had stopped searching for work.[7]

ELDERLY WORKERS IN THE FUTURE

    At the beginning of this century, about 7.1 million persons,
less than 10 percent of the total population, were age 55 and over.
In 1982, over one-fifth of the American population was 55 or older,
an estimated 48.9 million persons. Through the year 2000, the
population age 55 and over is expected to remain at just over
one-fifth of the total population. By 2010, because of the
maturation of the baby boom group, the proportion of older to
younger will rise dramatically--one fourth of the total U.S.
population (74.1 million) is projected to be at least 55. In
addition, one out of seven Americans are expected to be 65 and over
(34.3 million). By the year 2030, it is likely that one out of five
Americans will be 65 or older (64.3 million) which will represent an
87-percent increase in a 20-year span.

    It is commonly assumed that the growth of the older population
is due to increased longevity; the prime cause, however, is the
number of annual births. The post-World War II baby boom
(1946-1964) accounts for the projected rapid rise in the number of
elderly from 2010 until 2030. After that, the growth will slow
again because of low birth rates during the "baby bust" period from
1965 to 1973.

    The papers are filled today with news about the likelihood of
deficits in public and private pension plans. The substantial
increases in the number of elderly yet to come and the substantial
decreases in the size of the working-age population who have

traditionally supported the nonworking older population with their
contributions to pension plans is bringing the problems of aging
more and more to the attention of policymakers. In 1900 there were
about 7 elderly persons for every 100 persons 18 to 64 years of age;
by 1982, the ratio was 19 per 100 but by 2010, the ratio will jump
to 22 and then increase rapidly to 37 by the year 2030. Much of the
increase in the support ratio will be for persons in the oldest age
groups (Figure 2). It is these changes in the age structure which
will be the single most significant factor in the escalation of the
cost of pension plans under current policies.[8]

    Projections such as these have forced a reexamination of
pension policy and have caused policymakers to consider various
options for extending worklife including the controversial idea of
increasing the age of retirement to help meet the mounting costs of
retirement benefits. It is the aging of the baby-boom cohort which
causes the most concern, just by virtue of its relative size. Will
this group want, need, and be able to work when they are elderly?
Certain factors relevant to these questions are known and can be
reasonably forecast 30 years into the future; e.g., their likely
size and educational level. Many other factors are very difficult
to predict because they are subject to the whims of public policy or
the changes that can occur from major breakthroughs, disasters, the
economy, health status, industrial and occupational shifts,

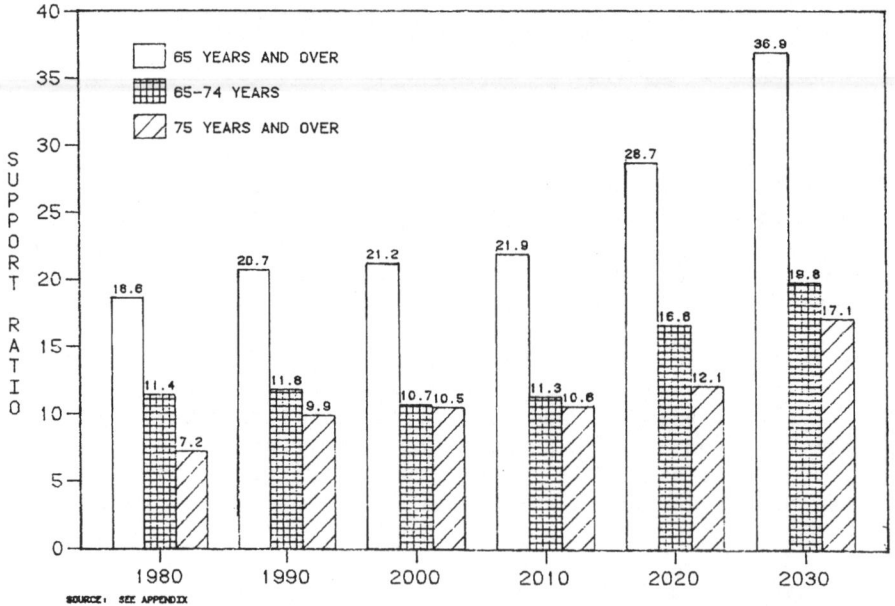

Figure 2. Support ratio of persons 65 years and over by age:
           1980-2030 (Number of persons per 100 aged 18-64 years)

etc. The following discussion deals with a number of possible
scenarios about whether the baby-boom group will want, need or be
able to extend their worklife but such forecasts, especially when
dealing with long-term complex interrelationships, are, in the end,
only guesses and by no means exhaust all the possibilities. The
discussion will focus on the aging of the cohort which is now 25-34
years old.

Will the Elderly Want to Work?

Today, most elderly persons do not work. Some don't want to
work and consider retirement a reward for years of work. For many,
it doesn't pay to work. For others, poor health prevents working.
But there is also evidence that many would like to work (usually
part-time), especially those with low-expected pension incomes, but
are discouraged. Even though the elderly have relatively low levels
of job search and relatively high levels of withdrawal from the
labor force, this could be a result of a labor market which is[9]
unresponsive to the needs and preferences of the older worker.
Older workers are less likely to find a job than are younger persons
and when they do find a job, the wage level is usually considerably
lower than with previous employment. Fringe benefits are
practically nonexistent.[10] Older workers must compete with
younger people and women of all ages for part-time work. The rates
of worker discouragement among those 65 years and over are almost as
large as the unemployment rates. Among women, discouragement is more
common than for men regardless of age but is especially high for
elderly women.[11]

What will happen with the baby-boom group? One important clue
is the occupational distribution of today's working elderly compared
with that of 25 to 34 year olds. About 57 percent of 25 to 34 year
olds are in white-collar occupations and another 11 percent are in
service jobs. The Bureau of Labor Statistics projects that 70
percent of the overall increase in employment through 1990 is
expected to occur in professional, technical, clerical and service
occupations and given current knowledge about technological trends,
it is reasonable to expect those to continue as the major areas of
job expansion well into the next century. Among today's 3 million
elderly workers, half are in white-collar occupations and another
fifth are in service occupations. Since the baby-boom group hold a
large share of all white-collar jobs, and thus will theoretically
have a background in the skills required for the work available in
the years 2000 through 2030, and since flexible working arrangements
and part-time work are more available in white-collar occupations
than in blue-collar work,[12] it is not unreasonable to assume
that a larger proportion of elderly may want to work in the next
century. Even if the proportion remained the same as that of today,
the number of elderly workers would be much larger because the base
is so large.

82   C. TAEUBER

Will the Elderly Need to Work?

Several factors lead to the conclusion that the elderly of the
next century will need to work. The sheer size of the baby-boom and
the relatively small size of the working age group behind them
already make it clear that problems are on the horizon for pension
plans. Since the 1960's, there has been an increased reliance on
Social Security and the related medical benefits. Today, Social
Security benefits are the single largest source of money income for
the elderly and the single source on which the largest proportion is
most dependent. For over half of those now receiving Social
Security, the benefits comprise at least half of their income. For
some, Social Security is vital: a fifth of the total elderly
population received virtually all of their income from Social
Security. But Social Security is in trouble and it seems unlikely
that the baby-boom group can count on it as much as the elderly of
today do. It is questionable whether Social Security income alone
will be sufficient. Some are saving for retirement through IRA and
Keogh accounts and other types of investments, but for most, it is
likely that earnings as an elderly worker will make the greatest
difference in their economic position.

If the age for receipt of Social Security is raised, it is
likely that the age for receipt of medical benefits will also be
raised. Older workers may need to continue working just to maintain
the medical insurance available in the workplace.

It is also reasonable to expect that tomorrow's elderly will
have more economic responsibilities than today's elderly. There
will be a dramatic surge in the size of the older segments of the
elderly population. In 1980, about 2.2 million persons were 85
years and over but by 2010, it is projected that 6.8 million will be
that age and by 2030, there will be 8.8 million persons 85 years and
over. Increasingly, we will become a four-generation society in
which oldsters will face the concern and expense of caring for their
very old parents. Because of the longer life expectancy of women,
it may be mostly elderly daughters caring for their aged mothers.
The women of the baby-boom cohort are more likely than their mothers
to have been in the workforce continuously, and thus this factor
should be helpful for those women who need to continue in the labor
force.

Will the Elderly be Able to Work?

While the indications are strong that more of the elderly
population will want and need to work 30 years from now than is true
today, and public policy appears headed in a direction that will
require them to work longer, whether they will be able to work
depends on two major factors: their health and the work options
actually available.

Will the health of the elderly, especially men, allow them to work longer? Definitive long-term data on health status are not available and thus this is a controversial area. In the March 1981 report of the National Commission on Social Security normal retirement age was linked to the age at which people are no longer "sufficiently healthy to function in their jobs."[13] The majority position was that "the Commission anticipates that increased longevity will be accompanied by a corresponding increase in active life. . ."[14] A minority of the Commission held that "the evidence does not support any claims that longer life is equivalent to longer years of good health. . .The evidence certainly does not support speculation that the incidence of good health is increasing."[15] These two opposite viewpoints can be seen throughout the literature on health status.

There are indicators, however, that the improvements in death rates may not have signified an improvement in work disability rates. Death rates for women have been declining rapidly and steadily since 1935 and for men for the last 15 years.[16] There is a popular notion that mortality and morbidity improve together. Definitive long-term trend data on health status are not available but in a statement before the National Commission on Social Security Reform, Jacob Feldman[17] showed that in the last decade, there has been an increase in the proportion of men aged 50 to 69 who reported themselves unable to work because of illness, the same period when the death rates were declining. Part of this difference could be attributed to changes in the occupational structure which give a sick person fewer work options (e.g., less self-employment and fewer jobs with reduced physical demands, such as a former laborer converting to a doorman) as well as the availability of disability payments for those under age 65. Nevertheless, there has been some increase in men 65 to 74 who report that they cannot work because of illness even though they are no longer eligible for disability payments but can convert to Social Security. In earlier times, most of those disabled at age 55 would have died before reaching 65 but since the late 1960's, this has changed. The death rates for heart disease, for example, have declined rapidly for men over 50 and at the same time, work disability due to heart disease has increased, especially for men 60 to 69 years.[18] Feldman also noted that "a great deal of disability is caused by conditions that are not lethal. Musculoskeletal conditions are the cause of a large proportion of work disability. Arthritis, for instance, does not appear to shorten one's life to any great extent."[19]

There is a wide differential in work disability rates according to educational attainment. Among men 65 to 67 years old, 31 percent of the men with less than a high school education were unable to work because of health problems, but for those with at least some college education, only 12 percent were unable to work because of a health problem, probably because of differences in the type of work

done (this finding bodes well for the baby-boom group who have
received more years of formal education than the current generation
of oldsters). Overall, over a third of the men in this group were
limited in the kind or amount of work they could do (13 percent) or
were prevented from working at all (24 percent). Thus, nearly
two-thirds of men 65-67 reported that health did not prevent them
from working. At each age over 65, women are reported as having a
higher prevalence of work disability than are men.[20] Survey data
suggest that one-third to one-half of early retirees accept reduced
benefits because of health and/or unemployment.

There is much conjecture about the "compression of
infirmity"[21] which suggests there will be widespread and
permanent change in health habits and technological advances that
will combine to reduce the prevalence of work disability. Advances
in the prevention, treatment, and rehabilitation of musculoskeletal
conditions could make a great difference. But there is considerable
uncertainty and too many intangibles about the future course of
disability prevalence rates to make firm recommendations.[22]

It seems that the most important point is that a majority of
persons in their mid-sixties will not be prevented from working
because of health. Others could continue working if there was more
flexibility about how long they worked or if they could find less
physically demanding work. But there is also a "sizable minority
with a rather wide discrepancy between the demands of their regular
jobs and their remaining functional capacities . . ."[23] Even if
the majority are able to compress the time of illness in the future
(and there is no evidence that this will occur rapidly), this
"sizable minority" will not go away. Across-the-board plans to
increase retirement age and medical benefits have not taken this
group into account.

If it is true that many of the elderly of the next century will
want and need to work and will not be prevented from working because
of their health, the most crucial question is whether jobs will be
available. The indications are that we are moving from a
goods-producing society to an information-producing society.
Traditionally, the elderly have been unable to compete effectively
for jobs and many of the jobs have been physically demanding. There
have been problems of providing meaningful jobs, competition with
younger persons and women for jobs, and fast technological change
which requires a high level of skills and frequent retraining.

Some contend that because of the low birth rates there will be
a reduced labor force and, consequently, a demand for older
workers.[24] The contention, however, is unlikely given that
because of fast technological change, many jobs require a
significant skill level and frequent retraining so that the worker
does not become outdated. It also appears that a larger proportion

of the reduced younger workforce and middle-aged women will participate in the labor force and many will have had to remain in entry-level jobs longer. It is likely that the older workforce can be easily overwhelmed by this type of competition.[25]

Policymakers have examined a number of ways of extending worklife.[26] They include:

(1) expansion of job opportunities through job redesign, phased retirement, job sharing, and part-time or part-year work;
(2) legislation to outlaw age discrimination in employment;
(3) training, retraining and "second careers" programs;
(4) subsidized employment; and
(5) pension reforms (i.e., raising the age for pension eligibility and eliminating the Social Security earnings test).

Pension reforms are the most likely way to force work but there is now such a scarcity of flexible work options for the older population that many lessen their job search or leave the labor force entirely. In fact, it appears that the low unemployment rates among the elderly are to some extent caused by the difficulties and limited opportunities the older worker finds in the market place.[27] The ultimate result of delaying pension benefits could be increased use of the unemployment insurance, disability insurance, and welfare systems. As things are now, the older population cannot compete effectively in the labor market even though they may need and want to do so.

The baby-boom cohort does have a big advantage that the current generation of oldsters does not have: they are well-educated and have a relatively high skill level. Nearly 87 percent have at least graduated from high school and a fourth have 4 or more years of college.

CONCLUSIONS

In 30 years, a large aging society will be upon us whether we have prepared for it or not. The elderly will be the baby-boom cohort, a group that has restructured the traditions of American society at every point in their maturation. There are many indications that restructuring will occur again, including labor force participation. Inflexible work options and a sluggish economy are likely to be the biggest barriers to work for an aging baby-boom cohort. There is time now to begin experimenting with the different options. If the retirement age is increased, protection for the substantial minority who cannot work so long will have to be devised.

The magnitude of change required is a challenge to our capacity to adapt public policy far enough in advance to be successful and sets the overall context for decisions made today regarding the aged and aging in America.

NOTE

The opinions expressed are those of the author and do not necessarily reflect the views or policy of the Bureau of the Census.)

ACKNOWLEDGEMENTS: The author would like to express her appreciation for their helpful comments to Paula Schneider, Paul Ryscavage, and Tom Palumbo of the Bureau of the Census.

SOURCES FOR FIGURES

Fig. 1. Bureau of the census and Bureau of Labor Statistics.
Fig. 2. U.S. Department of Commerce, Bureau of the Census, 1980 Decennial Census and "Projections of the Population of the United States: 1982 to 2050 (Advance Report), Series P-25, No. 922, October 1982. Projections are the middle series.

REFERENCES

1. Dean Morse, The Utilization of Older Workers, Special Report No. 33, National Commission for Manpower Policy, Washington, D. C., 1979.
2. U. S. Department of Labor, Bureau of Labor Statistics, Employment and Earnings for January 1961, 1971, and 1982.
3. Employment and Training Report to the President, 1981, Table A-25, p. 158.
4. Malcolm H. Morrison, Department of Labor, National Studies of Mandatory Retirement, "The Aging of the U. S. Population: Human Resources Implications," Monthly Labor Review, May 1983, pp. 13-19.
5. U. S. Department of Labor, Bureau of Labor Statistics, unpublished data, November 1982.
6. J. Mincer and H. Ofek, Interrupted Work Careers: Depreciation and Restoration of Human Capital, Journal of Human Resources, vol. 17, Winter 1982, pp. 1-24.
7. U. S. Department of Labor, Bureau of Labor Statistics, unpublished data, November 1982.
8. James R. Storey and Gary Hendricks, Retirement Income Issues in an Aging Society, The Urban Institute, Washington, D.C., 1979.

9. Philip L. Rones, "The Labor Market Problems of Older Workers," Monthly Labor Review, May 1983, pp. 3-12.

10. Herbert S. Parnes, Mary G. Gagen, and Randall H. King, "Job Loss Among Long Service Workers," in Herbert S., Parnes, ed., Work and Retirement: A Longitudinal Survey of Men (Cambridge, Mass, MIT Press), 1981, pp. 65-92; and Dean W. Morse, Anna B. Dutka, Susan H. Gray, "Retirement Experience of Non-Supervisory Personnel: A Study of Three Large Corporations," draft final report, (New York, Columbia University, Conservation of Human Resources), 1981.

11. Rones, op. cit., p.8.

12. In 1980, one-fifth of all employment in professional, technical, and clerical occupations was part-time.

13. National Commission on Social Security, Social Security in America's Future, Washington, D. C., 1981.

14. Ibid., p.126.

15. Ibid., p.331.

16. Jacob Feldman, "Work Ability of the Aged Under Conditions of Improving Mortality," Statement before the National Commission on Social Security Reform, June 21, 1982, p.2.

17. Feldman is the Associate Director for Analysis and Epidemiology, National Center for Health Statistics, Department of Health and Human Services.

18. Feldman, op. cit., pp.2-4, 8.

19. Ibid., p.8.

20. Ibid., pp.4-7.

21. J.F. Fries, "Aging, Natural Death, and the Compression of Morbidity," New England Journal of Medicine, July 17, 1980, 303, pp.130-135. Also Lois M. Verbrugge, "Longer Life But Worsening Health? Trends in Health and Mortality of Middle-Aged and Older Persons," paper presented at the Population Association of America, Pittsbugh, Pa., April 1983.

22. Feldman, op. cit., pp.9-11.

23. Ibid., p.7.

24. Lawrence Olson and others, The Elderly and the Future Economy, (Lexington, Mass., Lexington Books), 1981.

25. Morrison, op. cit., pp.16-19; and Kahl, op. cit., p.13.

26. James H. Schultz, The Economics of Aging, 2d ed., Wadsworth Publishing Company, Inc., Belmont, California, 1980.

27. Rones, op. cit., p.10.

CHANGES IN LABOR FORCE PARTICIPATION OF PERSONS 55 AND OVER SINCE

WORLD WAR II: THEIR NATURE AND CAUSES

Richard A. Easterlin, Eileen M. Crimmins, and
Lee Ohanian

University of Southern California
Los Angeles, California

NATURE OF TRENDS

In the United States trends in labor force participation since 1940 have varied widely among age-sex groups in the population 55 and over (Figure 1). Setting aside the temporary World War II upsurge, at one extreme are men 65 and over whose rate of participation trended sharply downward throughout the entire period. The participation rate of men aged 55-64 was fairly constant through the late 1950s, but then started downward and by the 1970s was falling at a rate similar to that of men 65 and over. In contrast to the declining rates for men, rates for older women were either constant or rising. The rate for women 65 and over was virtually unchanged throughout the entire period, remaining at the low level of 10 percent or less. Among women 55-64 labor force participation rose at an unprecedented rate until around 1970, after which it leveled off. At present among persons 55-64, 70 percent of men are in the labor force and about 40 percent of women; among those 65 and over, less than 20 percent of men are in the labor force, and under 10 percent of women.

Considering the population aged 55 and over as a whole, one finds that these disparate movements in age-specific labor force rates resulted, on balance, in a mild decline in the percentage in the labor force--from about 41 percent in 1940 to 33 percent at present. More startling, perhaps, is the shift in sex composition of older workers--between 1940 and the present, the share of women among older workers rose from a little over 15 percent to almost 40.

89

For the most part, trends in labor force rates since 1940 continue those in the preceding half century, although the decline for men 65 and over and the rise for women 55-64 have both accelerated. The one new development is the start of the downtrend for men 55-64 in the late 1950s.

THEORETICAL VIEWPOINT

In view of the diversity in movements among these age-sex groups, it is unlikely that any single factor, such as growth of real GNP per capita, technological change, the establishment and expansion of Social Security benefits, or health trends among the older population is responsible for the patterns observed. Unfortunately, there are surprisingly few attempts to construct

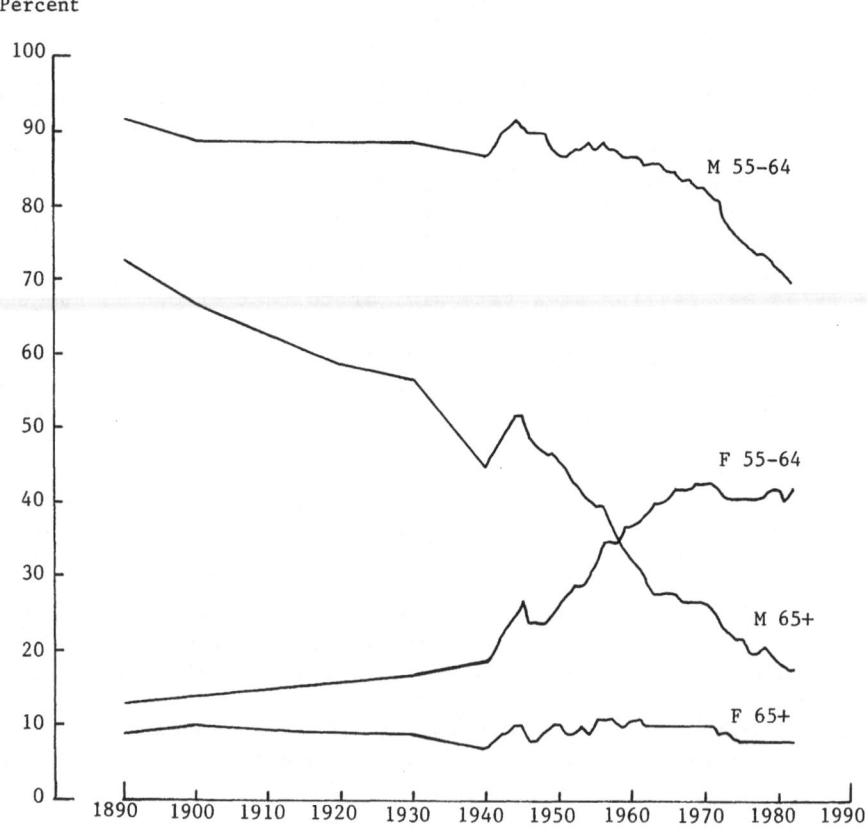

Figure 1.   Labor force participation rate, persons aged 55 and over, by sex and age, 1890-1982.

multivariate models of these time series movements. There is a
larger number of cross-sectional analyses but these are plagued by
the inevitable problem of the uncertain applicability of
cross-sectional analyses. Health frequently plays an important part
in withdrawal from the labor force, but it is doubtful that this has
been a dominant factor in the changes in labor force rates over
time. Moreover, both time series and cross-sectional studies have
focussed almost wholly on males, and are frequently limited to a
single age group, usually men 65 and older. Curiously, there is
almost no discussion, let alone analysis, of the striking rise in
labor force participation among females 55-64.

The present discussion of causes stresses longer term trends
rather than annual changes. Although it is necessarily speculative,
it aims for consistency with observed experience since the beginning
of this century and across age-sex groups.

Let us start with a brief sketch of the theoretical viewpoint.
In general, one would expect that among the older population of a
given age and sex, the probability of being in the labor force would
vary directly with their current earnings possibilities and living
level aspirations, and inversely with their potential sources of
support outside the labor market. Thus, other things equal, the
higher the group's earnings possibilities or living level
aspirations, the higher would be the group's expected labor force
participation rate; the higher the group's potential sources of
support outside the labor market, the lower would be the expected
labor force participation rate.

The trend in a group's rate would be a function of the trend in
these three conditions. The trend in earnings possibilities would
depend on the trend in rates of pay for given jobs and the prospect
of employment at these jobs. The latter would, in turn, depend on
trends in the general state of the economy (i.e., prosperity or
depression), age-sex specific demand conditions, reflecting, say,
technological changes, and age-sex specific supply conditions, such
as health trends among the older population. The trend in living
level aspirations would be affected, among other things, by
education, media exposure, and, of special importance for the
present analysis, the trend in the real GNP per capita. If the
economic socialization experience of a group plays an important part
in shaping its material aspirations, then one would expect that a
group raised in a richer economic environment (i.e., higher real GNP
per capita) would have correspondingly higher living level
aspirations. Finally, the trend in sources of support outside the
labor market would be importantly affected, among other things, by
the establishment of governmental and privately funded retirement or
disability programs, and variations in provisions regarding
eligibility and benefits under such programs, once established.

CAUSES OF TRENDS

Viewing the striking downtrend in labor force participation of males 65 and over since World War II, a hasty observer might jump to the conclusion that Social Security is the dominant force at work (see, for example, the curve showing the marked rise in eligibility for Social Security in Figure 2--here plotted inverted so that a downward movement implies a downward movement in the labor force rate). Such a conclusion, however, runs up against the fact

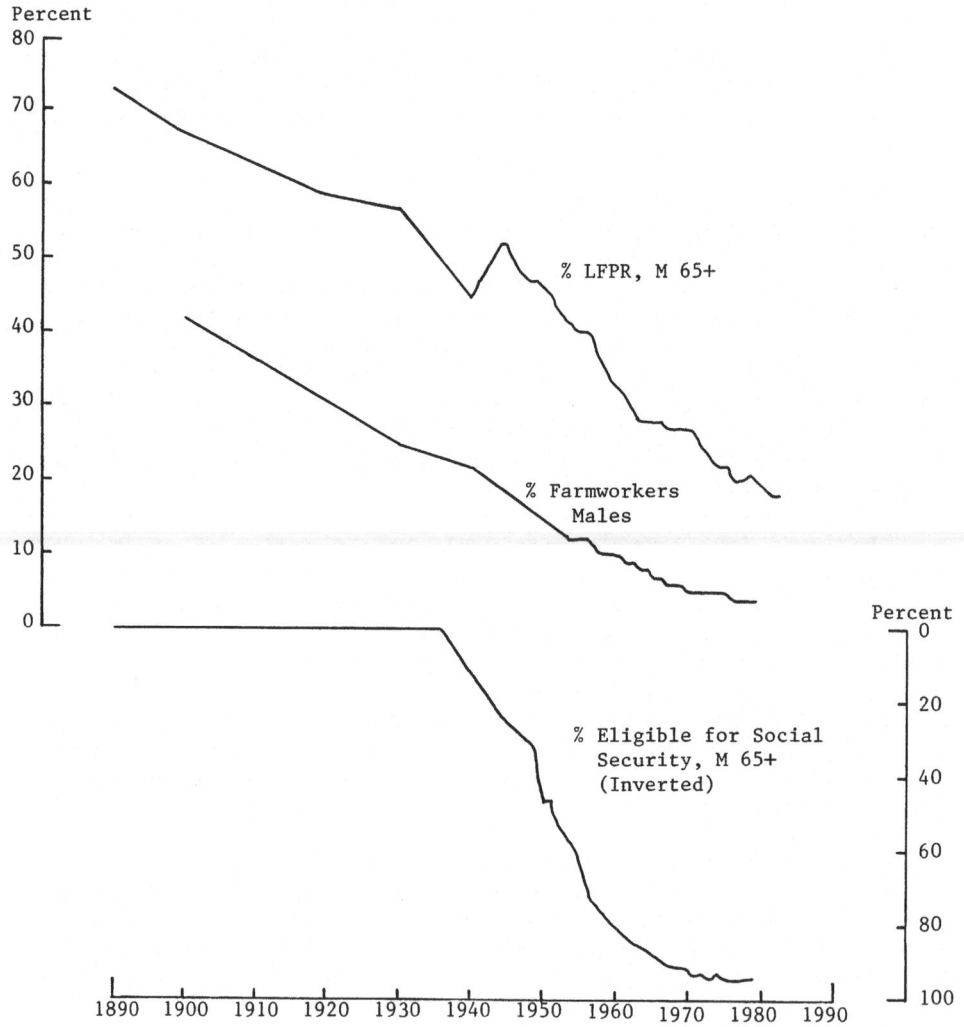

Figure 2.   Labor force participation rate, males 65+, percent of male labor force in farm occupations, and percent of males 65+ eligible for Social Security, 1890-1980.

that labor force participation of these men trends almost as sharply downward before the advent of Social Security. To be sure, the rate of decline accelerates somewhat after World War II, and it seems reasonable to suppose that this acceleration does reflect chiefly the impact of Social Security, but clearly some other force has been at work over the long run.

In this case, that force seems to be declining earnings possibilities, associated especially with the decline in self-employment opportunities. As real income per capita has trended upward, the demand for nonagricultural goods has benefited at the expense of agricultural. Also, technological change has brought a shift from small to large scale establishments in manufacturing, retail trade, and other sectors. Both of these developments have led to a shift from self-employed to employee status in the work force, and with it, diminished opportunity for continued labor force participation with the onset of old age. As an index of this, Figure 2 shows the long term trend in the percentage of the male labor force that are farm workers. This curve parallels closely that in the participation rate of men 65 and over, though a discerning eye will note a more rapid decline in the participation rate curve as eligibility for Social Security becomes widespread. In general, then, an historical perspective leads to emphasis on the changing structure of employment opportunities in the decline of labor force participation among men 65 and over, with perhaps a mild assist from Social Security.

For men 55-64, unlike those 65 and over, there is nothing in the long term trend of labor force participation to suggest an adverse trend in demand--note that the participation rate for this group in the mid-1950s was the same as at the beginning of the century (Figure 3). For these men, the onset of a downtrend in the late 1950s seems to be clearly associated with new governmental programs--in 1957 the initiation of disability coverage and in 1961 the lowering of the retirement age under Social Security to 62. These developments have been especially important to that segment of the group with poorer health, who now have an income option other than work not previously available.

To turn to females 55-64, their story must be seen in the context of the trends in labor force participation among women more generally. As Figure 4 shows, for almost a century now the entire age range of women between 20 and 64 has benefited from a growing demand for female labor, associated particularly with the rise of clerical and service jobs due to industrialization and urbaniza-tion. There are, however, some notable contrasts between the younger and older age groups. In the period 1890-1940 younger women were the primary beneficiaries; from after World War II to 1960, older women were; since 1960 there has been a switch back to younger women. These shifts can be attributed chiefly to the high degree of

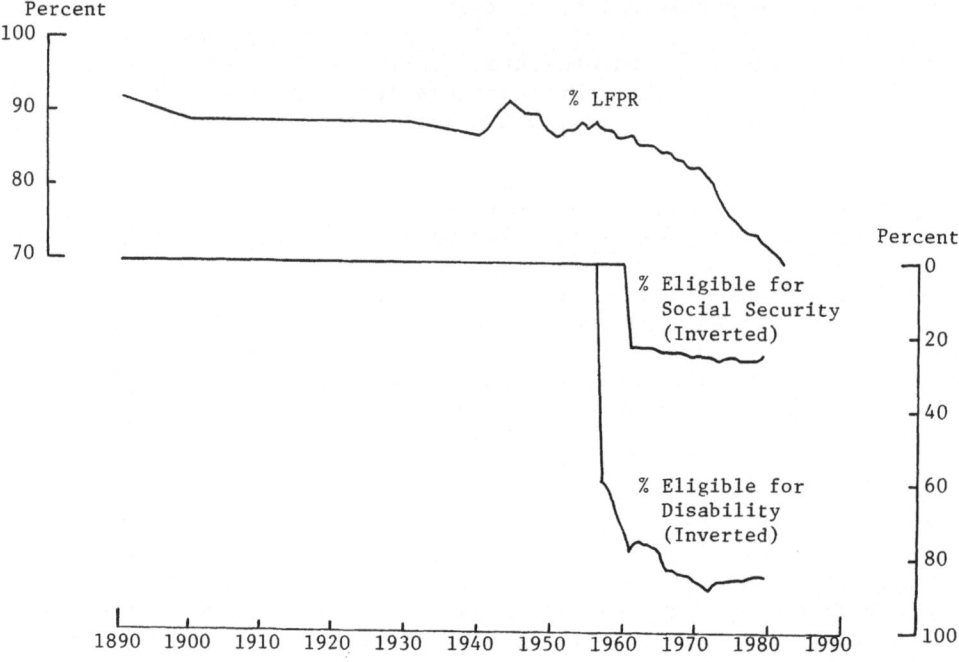

Figure 3.   Labor force participation rate and percent eligible for
            retirement and for disability, males 55-64, 1890-1980.

substitution between older and younger females in the labor force,
in contrast to the situation for men, who more typically follow a
career ladder, and to the tendency of employers traditionally to
favor younger over older females because of their better education
and other factors. Thus, the typical pattern is that found before
1940 and after 1960 when both younger and older women enter the
labor force at increasing rates, but younger women, whom employers
favor, enter more rapidly than older. A break in this pattern
occurred between the end of World War II and 1960, when there were
unusual opportunities for marrying and starting families among
younger adults. This development interrupted the long term rise in
labor force participation of younger women, and the jobs thereby
left unfilled created unusual opportunities for older women. After
1960 as the family prospects of young adults took a turn for the
worse, and younger women resumed their rise in labor force
participation, there was a corresponding slowing in the growth of
labor force participation among older women. Thus, the demand for
older women has been a direct function of the demand for female
labor generally and also (inversely) of the responsiveness of
younger women to this demand.

     As Figure 4 shows, recent years have seen a divergence in trend
between the two older groups. Here, it seems reasonable to suppose

one is witnessing the same factors at work as among men 55-64, namely, the negative impact of disability and retirement programs on labor force participation of females 55-64 (see the two bottom curves in figure 4). Indeed, the recent upturn in labor force participation among females 45-54 may reflecet a response to opportunities created by the withdrawal of those 55-64.

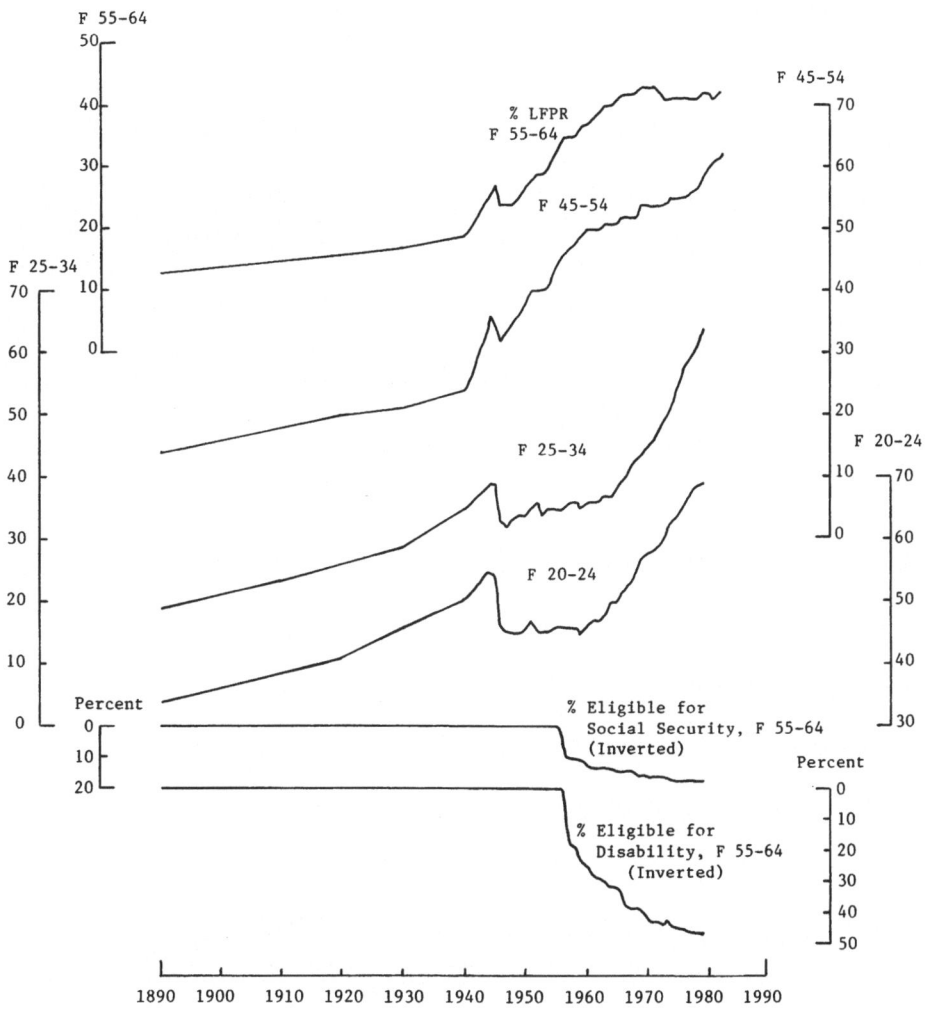

Figure 4. Labor force participation rate, females 20-64, by age and percent eligible for retirement and disability, females 55-64, 1890-1980.

Finally, for females 65 and over, the stability in their labor
force participation before 1940 may be attributed to the fact that
as the oldest age group of females, they benefited very little from
the pre-1940 expansion of demand for female labor (compare the small
increase for the next youngest age group, females 55-64 in Figure
5). And, by the time of World War II, when older women's
opportunities expanded rapidly, the countervailing influence of
Social Security--especially in the form of survivors' benefits--came
into play. (Actually, a close reader of the pattern since 1940 will
discern a faint echo of that for females 55-64--a slight upswing
from 1940 through 1960, then, a leveling off and downturn preceding
that for females 55-64, in keeping with the earlier availability of

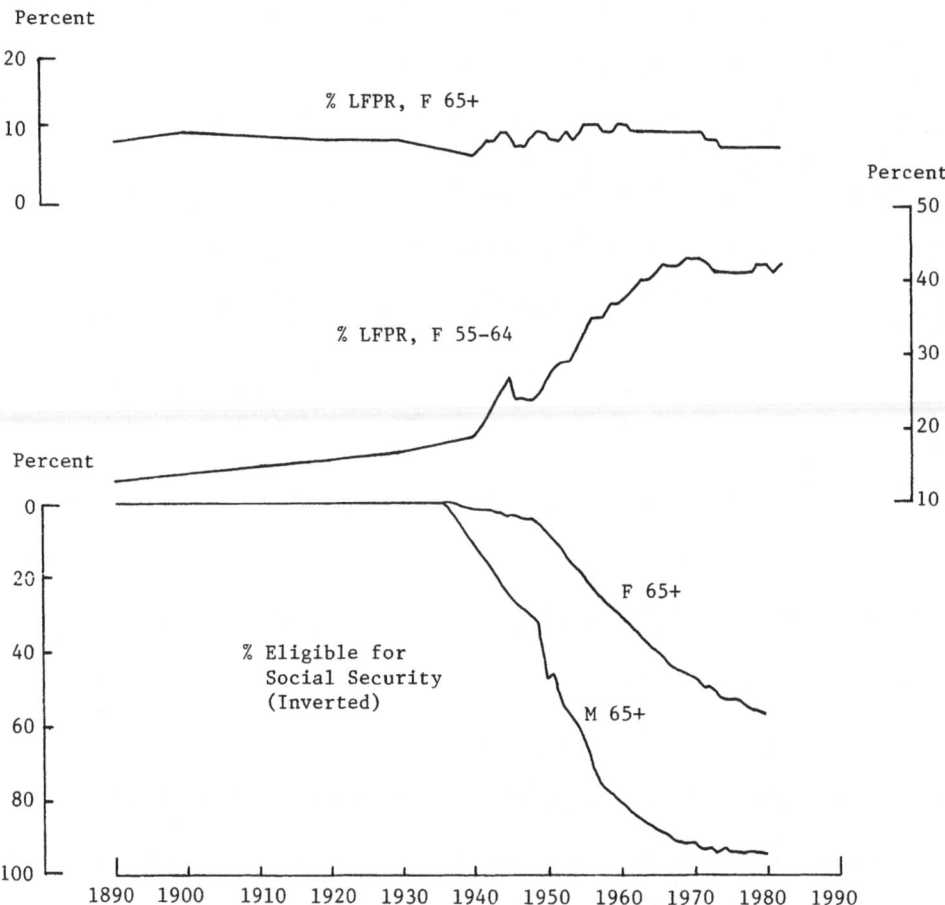

Figure 5.   Labor force participation rate, females 55-64 and 65+,
            and percent eligible for Social Security, males 65+
            and females 65+, 1890-1980.

Social Security benefits for those 65 and over.) As a result, the pattern for this group of women has been one essentially of stability in labor force participation over the long term.

    In general, then, the disparate patterns among the four age-sex groups reflect two dominant forces. One is the trend in job opportunities associated with long term economic growth--generally favorable for women (but more so for younger than older) and unfavorable for men 65 and over. The other is the impact of disability and retirement programs under Social Security which has had the common effect of lowering labor force participation, though in differing degrees.

TECHNOLOGICAL CHANGE AND THE LABOR MARKET SITUATION

OF OLDER WORKERS

Paul J. Andrisani
Temple University
Philadelphia, Pennsylvania

Steven H. Sandell
National Commission for Employment Policy
Washington, D.C.

Introduction and Overview

Research on labor markets has shown that much about age patterns of earnings can be understood in terms of the accumulation of job skills over a lifetime, particularly through investments in schooling and on-the-job training (Becker, 1975; Mincer, 1970, 1974). In this framework, individuals enhance their productivity and earnings before and during their work careers by accumulating job skills through investments in schooling and on-the-job training. Jobs are seen as offering two types of payoffs that are relevant for future earnings--an earnings component and a training component. The training component, which provides no immediate financial reward but increases a worker's future productivity and earnings capacity, can be a significant part of the worker's total compensation package. For example, Lazear (1976) estimated that training constituted about one-third of the total compensation package for young men.

Research and common sense both imply that younger workers will make greater investments in training than older workers. Not only will they have a longer time period until expected retirement over which to amortize the costs and reap the financial benefits of their investments, but they also generally bear lower costs in terms of earnings lost during training (e.g., Becker, 1975; Ben-Porath, 1967; Rosen, 1972). The same is also true of employers' incentives to invest in the training of their workers. Thus, it follows that earnings will rise rather quickly early in careers as productivity increases due to investments in schooling and training, and more

slowly or not at all later when workers and their employers make
little or no investment in training (and hence productivity growth),
due to rising opportunity costs and shorter payback periods.
Earnings may even decline later in the life cycle because of the
obsolescence of schooling and training.

A worker's productive capacity and earnings also can change
over the life cycle for reasons other than schooling, on-the-job
training, and skills obsolescence. Some of these may be labeled as
the effects of aging per se (Lazear, 1976). In the early stages of
the life cycle, these reflect the effects of maturation and learning
in non-labor market activities that also increase productivity and
earnings. In later stages, they reflect physiological changes
associated with the aging process--e.g., the onset of chronic health
problems that often result in a decline in physical capacity.
Furthermore, preferences and attitudes regarding work also may
change over the life cycle. These too may cause changes in
productivity and earnings. As workers age, for instance, so do
their children, thus diminishing the burdens of dependency. Perhaps
more importantly, many older workers have greater availability of
non-earned income through social security, personal savings, and
private pensions than younger workers, thus increasing their
preferences for leisure.

Finally, earnings may also decline with age due to employment
discrimination. Drawing an analogy to race and sex discrimination,
ageism or age discrimination exists when an employer offers
different wages or conditions of employment to older workers because
of their age, having no economic or productivity-related basis.
Studies have shown that at least some managers ascribe to
traditional age stereotypes that lead to discriminatory personnel
decisions. In a survey of subscribers to the Harvard Business
Review, for example, respondents tended to make different personnel
decisions about fictitious older and younger employees described as
having identical characteristics aside from age (Rosen and Jerdee,
1977). Older workers were more likely than younger workers to be
transferred than counselled when their performance was judged to be
low, were less likely to be provided retraining and development
opportunities, and were less likely to be promoted to positions
involving innovative thinking or stress. In addition, results from
attitude surveys suggest that, in general, both younger and older
workers as well as managers are convinced that most employers
discriminate against older people (Harris, 1981).

Purpose of Paper

The purpose of this paper is to summarize six recently
completed studies conducted under the auspices of the National
Commission for Employment Policy (NCEP) as part of its recent

Project on National Employment Policy and Older Americans.* Two of the six studies deal with innovative practices designed by firms to ameliorate the employment problems of older workers, including those which are the direct result of technological advances. A third study deals with part-time employment as a potential solution to at least some of the employment problems of older workers who would prefer this alternative to full-time work, retirement, or unemployment. The fourth and fifth deal directly with the employment problems of older workers who were permanently displaced from their jobs. The sixth and final study deals directly with the effects of technological change and skills obsolescence on the productivity and earnings of older workers in "High Tech" industries. This particular study simulates the labor market situation of older workers who are in a growth industry as well as the forefront of technological change.

Although seemingly obvious, the simple fact that technological change may produce changed circumstances for older persons' employment situations warrants reiteration. Furthermore, much can be inferred from careful examination of similarly situated current older and younger workers. First, for example, some older workers will lose their jobs. Two of the studies from the Project on National Employment Policy and Older Americans summarized here deal directly with this. Second, policies might be needed to deal with an expected reduction in the demand for labor which might accompany technological advances. These may include increased options for part-time work and innovative programs and policies by employers to meet the special needs of potential older workers. One study from the Project on National Employment Policy and Older Americans critically examines the market for part-time work while two others take an analytical look at innovative employment practices in the private sector. Third, technological change implies a need for retraining. Yet another study from the Project takes a case study approach in examining earnings and human capital development and obsolescence in high-tech industries.

Taken together, these six studies can enlighten the policy decisions that will need to be made to respond to continued technological progress. While the magnitude and the time dimensions of these problems are uncertain, it is doubtless true that displacement of older workers, a reduced demand for some types of labor, and retraining will be key aspects of the effects of continued technological advances in the world economy. Thus, public and corporate policymakers can become better prepared for such impending labor market developments in the future by careful consideration of the studies that are summarized here. Unfortunately, the development of policy prescriptions is beyond the scope of the present study.

Alternative Work Options for Older Americans

Several types of employment options tailored for middle-aged
and older workers are used by employers. These include job sharing,
phased retirement, labor pools for part-time work, active job
recruitment targeted toward older workers, job redesign, job
transfers, and retraining programs. While structured programs
exist, often the options have become available through informal
personnel practices. Not many older workers use them, however, due
in part to low levels of communication by management about the
options and in part to the reduction of salaries and benefits
associated with part-time work, even when these are combined with
the availability of other sources of retirement income.

The first study, "A Human Resource Management Perspective on
Work Alternatives for Older Americans," by Carolyn E. Paul of the
Andrus Gerontology Center of The University of Southern California,
examines the issue of alternative work options for older workers in
considerably greater detail than had heretofore been accomplished
(Paul, 1983). The existing literature on which this study builds
contains mainly descriptive information, often simply highlighting
only the positive aspects of the various work alterntive programs
studied. The Paul paper, however, analyzes the implementation and
outcomes of various work alternatives. Furthermore, it also
examines the factors that influenced management to adopt such work
alternatives. In the process, the study focuses on the
characteristics of employers that were associated with the
development and administration of the employment options and on the
public policy variables that influenced the availability and use of
the options by older persons. The information used in the study was
obtained from telephone discussions with managers responsible for
the administration of the work programs in their organizations.

The Paul study shows that employment options are offered by
employers when they are seen to be profitable. Businesses gave
several reasons for increasing their employment of older workers:
attracting older consumers, stabilizing the younger work force with
older role models, gaining experience in working with a generally
older work force in anticipation of demographic changes, hiring
workers who would accept lower wages, and responding to government
policies.

Firms that use these programs reported that they were making a
conscious effort through them to project a positive image with older
persons, especially older consumers. They also noted that such
options were implemented in order to meet productivity and labor
supply needs, and that the costs were perceived along the same lines
as other employee benefits that hopefully contribute to higher
morale and improved worker performance--e.g., health insurance.

However, and not surprisingly, business conditions affect the implementation of alternative work programs, with economically hard-hit employers offering fewer and less attractive options. Nonetheless, in general, these programs were seen to represent an inexpensive tool for managing older workers.

Paul's study also found that government policies, and federal legislation in particular, have affected the offering of these options by employers in three ways: (1) stimulating their creation, (2) influencing their design, and (3) affecting their use by older workers and retirees. One third of the employers interviewed stated that they were motivated to offer employment options by the Age Discrimination in Employment Act or by state policies eliminating mandatory retirement. Because of the "1,000 hour rule," the provision of the Employee Retirement Income Security Act of 1974 (ERISA) mandating pension vesting for employees working 1,000 hours or more annually, part-time work hours are usually restricted by management to fewer than 1,000 hours. Additionally, the Social Security earnings test, which currently provides for a reduction of benefits by 50 percent for earnings exceeding $6,000, and the ERISA provision that allows for a firm's pension benefits to be suspended for retirees who are reemployed by the firm for 40 hours or more in a calendar month, each operate to limit the number of hours retirees wish to work. Thus, pension income together with these rules provide a strong reason for older Americans to desire part-time work.

The second study, "Innovative Employment Practices for Older Americans," by Lawrence S. Root and Laura H. Zarrugh of the Institute of Gerontology at the University of Michigan, also examined alternative work options, but with the purpose of determining the extent to which they address different employment problems and meet the needs of the companies which offer them (Root and Zarrugh, 1983). The University of Michigan's National Older Workers Information System (NOWIS), a computerized information system containing descriptions of company programs and practices, provided the data for the study.

In their study, Root and Zarrugh found that private sector programs can be expected to expand with the growth in the proportion of older persons in the labor force. These efforts may be limited, of course, to situations in which a program or practice works to the mutual advantage of the employer and its older workers. That is, programs and practices to increase employment options for older Americans are successful when they are "symbiotic"--benefiting both the worker and the company. Most programs presently involve mainly white collar workers, with blue collar production workers inadequately served. Those programs currently in existence for blue collar workers primarily involve employees in service occupations.

Root and Zarrugh provide examples of programs and policies by corporations designed to meet and mitigate the well known barriers to employment historically faced by older workers. These barriers include: (1) negative stereotypes that limit employment and advancement on the job, (2) physical disabilities that interfere with regular work routines, (3) limited training opportunities, and (4) limited part-time employment opportunities. The NOWIS data show that present programs are not distributed in a way to deal adequately with these problems. For example, part-time employment accounts for more than one-half the programs reported, while programs that seek older workers for regular full-time positions are only 12 percent of the examples listed. Job redesign, including both physical changes and changes in job responsibility, make up less than 9 percent of the cases and flexible scheduling accounts for only 10 percent of the programs. The remaining 13 percent of the NOWIS programs involve training that has particular application to older workers.

Comparing the occupational distribution of older workers with the distribution of the programs reported, provides information on the extent to which private sector programs serve certain types of older workers. For instance, 28 percent of employed persons aged 45-64 in 1979 were in professional and managerial occupations and 32 percent of the NOWIS programs were directed toward this group. However, clerical and other white collar workers were "overserved" to a greater extent, since they represented less than 30 percent of older workers but more than 40 percent of the programs. Semi-skilled or unskilled blue collar workers may be underrepresented, being served by 20 percent of the programs, while this group (including non-household service workers) comprised 28 percent of the work force.

Root and Zarrugh reached these conclusions on the role of the private sector's programs and practices: (1) There are relatively few employment programs and practices for older workers; (2) these programs often affect only a limited portion of a company's work force; (3) though varied, these approaches address only particular segments of the national work force and certain problems of older workers; (4) the existence of many programs appears to be tenuous, with some companies suspending operation in the face of adverse economic conditions.

Thus, based on both of these studies, it seems fair to conclude that innovative employment programs appear to work best when they make good business sense and meet the needs of older workers. As Paul noted: "The actual need for the options in the workplace was the trigger mechanism that motivated management to create them." Both studies also report, however, that few older workers actually take advantage of these programs, even when they were promoted by employers. Paul noted that low participation is generally caused by "the low level of management communication of the availability...and thé reduction in salary and employee benefits often attached to the part-time work schedules and job transfers." In other situations, management "prefers not to openly discuss the availability of job redesign and job transfer options but rather to use them informally when individual worker situations call for their implementation."

Part-Time Employment for Older Workers

Older Americans often express the desire for part-time work, but the normal pattern is to move from full-time work to complete retirement. While more than half of those over 65 who are employed work part-time, the vast majority of persons over 65 do not work at all. In 1981, for example, 88 percent of men and women aged 65 and over did not work, 6.3 percent worked part-time, and 5.7 percent worked full-time. There are two key policy questions associated with the explanation of the rarity of part-time employment. First, do federal policies impose employment costs that make part-time workers more costly than full-time workers? Second, should the government do anything to encourage the part-time employment of older Americans? These were the research questions addressed by the study of James Jondrow, Frank Brechling, and Alan Marcus of the Center for Naval Analyses.

The Jondrow et al. (1983) study, "Older Workers in the Market for Part-Time Employment," attempted to explain the discrepancy between older Americans' often expressed desire for part-time work and the reality of the usual pattern of moving from full-time work to complete retirement. The key reason for the usual pattern is the disproportionately low pay for part-time work. This, in turn, is explained at least in part by employment costs that are recouped by employers paying part-time workers less than full-time workers.

Current federal policies seem to have little effect on either these costs or businesses providing older workers with part-time employment opportunities. The major federally mandated expenditures vary, for the most part, with workers' pay and thus are approximately the same proportion of total employment costs for part- and full-time workers--about 8 percent of total employment costs. The federal government's share of total employment costs in 1981 was 6.1 percent for FICA (social security taxes), 0.9 percent for UI (unemployment insurance), and 0.8 percent for workman's compensation.

The Jondrow et al. study concludes that "even though the government probably could change the part-time market for older workers, the theory...and the evidence that supports it do not suggest that this would necessarily be beneficial. That so few older workers hold part-time jobs is not due to an unavailability of part-time work; it simply reflects the reality that part-time work does not pay very well. There is no obvious reason to believe that older workers are kept out of this market or that they would be better off if they were eased into it by Federal policy. Instead, older workers rationally choose sudden retirement over part-time work." While individual older Americans may not have the opportunity to keep working part-time at career jobs, it may not be useful to create legislation that does more than encourage employers to provide the opportunities when it is in their own interest to do so. Doing more would impose additional costs on employers or taxpayers or both.

Older Displaced Workers

Displaced older workers, defined as persons who left their previous job involuntarily and have no expectation of returning to that job, may face special labor market problems. Thus, two research studies concentrated on this group. Terry Johnson, Katherine Dickinson, and Richard West, of the Stanford Research Institute (SRI), compared the characteristics of displaced older workers to younger displaced workers and other unemployed older workers, examined the short-term reemployment experiences of older workers, and examined the role of the Employment Service in assisting them in obtaining jobs. David Shapiro and Steven Sandell examined several aspects of one key question: How do older men who involuntarily lose their jobs fare in the labor market relative to younger job losers?

Some of the findings of the Johnson et al. (1983) study, entitled "Older Workers' Responses to Job Displacement and the Assistance Provided by the Employment Service," run counter to conventional wisdom. They find "no evidence that our sample of displaced (45 years and older) workers is at all disadvantaged relative to other unemployed workers." Displaced men had more education (11.2 compared to 10.6 years), were less likely to have health problems (16 compared to 22 percent), were more likely to be unemployment insurance applicants (66 compared to 52 percent), and had more favorable previous employment histories (i.e., earnings the previous year of $13,400 compared to $11,700) than other older unemployed men. Displaced women also were less likely to have a health problem (8 compared to 13 percent), were more likely to be UI applicants (69 compared to 54 percent), and had better previous employment histories (previous year earnings of $7,500 compared to $5,800) than other unemployed older women.

After application to the Employment Service, however, the median duration of nonemployment (the number of weeks unemployed plus those spent out of the labor force) for displaced older male workers was 20 weeks compared to 13 weeks for other older male job-seekers. This compared to median lengths of non-employment of 13 weeks for displaced and 9 weeks for other nonemployed younger

men. Similar differences are reported in other measures of labor
market success, including earnings during a 6 month interval and
whether the individual is employed at the end of 6 months.

Women 45 and older had reasonably similar frustrations in
finding new jobs whether they were displaced or not (21 compared to
19 weeks from date of application to the Employment Service). Their
experience also was poorer than unemployed younger women who
remained without jobs for 16 weeks if displaced or 11 weeks if
unemployed for other reasons.

Among older workers in the sample, age itself affected the
labor market success of women but not men after other factors (i.e.,
education, previous work experience) were controlled. The number of
weeks transpiring from the survey date until a subsequent job was
obtained was 15 for 45-54 year old women, 17 for 55-61 year-olds,
and 24 for 62 to 64 year olds. Men's duration was approximately 16
weeks for all men 45 and over. Earnings over the 6 month period
were about $2,400 for men age 45-61, but only $1,500 for those aged
62-64. Women's earnings were $1,500 for those aged 45-54, $1,000
for those aged 55-61, and less than $300 for women aged 62-64.

Older men and women received fewer job referrals from the
Employment Service than their younger counterparts. Receipt of an
ES referral did not result in any significant impact on any measure
of men's labor market performance. However, receipt of a referral
for women led to about $500 more in earnings in the subsequent 6
month period and a reduction of about 7 weeks in the length of time
between jobs, after controlling for other characteristics that
affect the reemployment experience.

The second study of older displaced workers analyzed the
earnings of older job losers after they found new jobs. The
starting wages of these workers are more likely to represent
unencumbered market forces than the wages of all older workers,
since the wages of "stayers" with many years of seniority can
reflect workplace features such as annual cost-of-living adjustments
and the effects of seniority provisions and firm-specific on-the-job
training of little relevance to new employers.

"Age Discrimination and Labor Market Problems of Displaced
Older Male Workers," by David Shapiro of Pennsylvania State
University and Steven Sandell of the NCEP staff, also was designed
to increase our knowledge of age discrimination in the labor market
(Shapiro and Sandell, 1983). If the wages of older workers who do
not lose their jobs are protected from potential age discrimination
by factors such as seniority provisions and across-the-board annual
wage increases, it is appropriate to examine the relationship
between age and wages, and possible age discrimination, using a
sample of older workers who are forced to look for new jobs.

The study examined the age/wage relationship among male workers 45 years and older who were displaced and subsequently found new jobs between 1966 and 1978. The Shapiro-Sandell study established that the reduction of earnings of displaced older workers can be largely attributed to the loss of firm-specific training useful only on the previous job and, in some cases, to searching for a new job when labor market conditions are not propitious. By the same token, job search during periods of low unemployment provides a means of recouping possible wage losses resulting from the loss of firm-specific training.

The loss of firm-specific skills and knowledge associated with seniority on the pre-displacement job accounts for a 3 and 1/2 percent drop in the average hourly earnings of men, representing nearly 90 percent of the average value of earnings loss for the sample. Furthermore, since older workers have more seniority than younger workers, older men who become displaced lose more firm-specific experience than do younger workers who become displaced. Consequently, the loss in average hourly earnings of displaced older workers is correspondingly greater. For example, workers in our sample who were over age 60 when displaced averaged more than 11 years of job tenure, compared with 6 years of seniority for displaced workers age 45-49. Since this firm-specific component of experience is not useful on the new job, the results suggest an average wage loss of nearly 6 percent for displaced men over age 60, compared to a 3 percent wage loss of displaced men age 45-49. Thus, there is an age-related drop in earnings, but since it stems from the loss of firm-specific human capital, it should not be considered to be age discrimination.

Those workers who lost their jobs between 1966 and 1969 when the national unemployment rate was relatively low did not experience wage loss, on average, while those who were displaced during the subsequent period of higher unemployment had an average loss of 6 percent of their pre-displacement average hourly earnings.

Finally, age discrimination in wages was not evident among these displaced older workers, except, perhaps, for workers over age 65. Some of the age-related loss in earnings can be attributed to changes in occupations and hours of work of displaced workers. Older workers who return to work are more likely to change occupations and to work part-time than younger workers in the sample, perhaps due to pension restrictions which limit their ability to work for their previous employers and the high marginal tax brackets they may find themselves in due to the Social Security earnings test. It should be noted that the age dimension of earnings losses during the often lengthy period of unemployment was not measured in this study.

Age and Earnings: Discrimination and Other Explanations

Two of the sponsored research studies examined in depth the relationship between age and earnings. The first, by David Shapiro and Steven Sandell, which analyzed the earnings of displaced older workers after they found new jobs, was summarized in the preceding section. The second, "Age, Productivity, and Earnings," by Paul Andrisani and Thomas Daymont of the Center for Labor and Human Resource Studies and Institute on Aging at Temple University, examined the extent to which age differences in earnings are due to age differences in productivity (Andrisani and Daymont, 1983). More specifically, they examined earnings differences between older and younger professional and managerial workers in the high tech industry and estimated how much was due to relative differences in schooling, different types of work experience, skills obsolescence, and a number of additional characteristics of the workers themselves rather than ageism, physiological aging, or changing attitudes toward work, which of course are much more difficult to measure.

The Andrisani and Daymont study also examined changes in earnings over the 1974-1980 period to estimate age differences in earnings growth, protection from inflation, and changes in productivity, and the extent to which skills obsolescence influences the relative productivity of older workers employed specifically as engineers and scientists. The data used in their study were from an anonymous company's Human Resource Information System (HRIS) and were longitudinal in nature. The data included complete work histories on all professional and managerial employees of the firm. The majority of the workers in the study are technical and scientific personnel with the remainder in administrative positions.

The findings of the study can be briefly summarized. There is a strong association between age and earnings, and presumably age and productivity, as they were measured. At age 30, and at each age thereafter, increasing age is associated with less schooling and/or schooling in the less well-rewarded fields of study. Older workers' possession of relatively more time with the employer, however, or firm-specific training, is a source of economic advantage, although age differences in "time in grade" are apparently of no consequence.

Older workers' greater quantity of pre-hire work experience is economically helpful, at least up to age 50. However, the greater likelihood of skills obsolescence (because of greater elapsed time since schooling) presumably makes this disadvantageous after age 50 due to depreciation of human capital. The results concerning engineers and scientists, who might be expected to be most susceptible to skills obsolescence, are more dramatic. Age differences in skills obsolescence beyond age 50 thus appear to be

an important source of age differences in earnings over those years,
at least for professional/managerial and especially scientific
workers in the high tech field.

Even at age 30, and up to and past age 60, older workers also
are disadvantaged relative to younger workers in terms of their
leaves of absence, gaps in work experience, health problems,
military history, and/or geographic mobility. Older workers do
possess very substantial productivity and earnings advantages in
their more valuable types of pre-hire work experience brought with
them at time of hire (or 1965). The quality of pre-hire skills and
work experience has a continuing effect on the earnings and economic
position of workers as they age over the course of the entire life
cycle.

External labor market forces in wage-setting appear to be
important, as is likely, due to the extremely competitive nature of
the product and labor markets in which high tech industries as a
whole operate. Education and major field of study continue to
strongly influence earnings, and presumably productivity, beyond
their effects on initial assignments and starting salaries.
Scientific and managerial skills and advanced degrees are apparently
of greatest relevance, much to the detriment of older workers, since
they are less likely to possess these factors than younger workers.

The most important factor associated with earnings and
productivity changes over the life cycle that were observed,
however, is the incentive to invest in further training. When the
opportunity costs or foregone earnings associated with
employer/employee incentives to invest in further training are
considered in the analysis, virtually all of the association between
age and earnings changes over the life cycle disappears. This
suggests that physiological aging, changing work attitudes, and age
discrimination are of little importance here, relative to the
effects of skills accumulation and obsolescence and
employer/employee incentives to invest further in such training.

NOTE

* The National Commission for Employment Policy is an independent
  Federal agency, established by the Comprehensive Employment and
  Training Act and continued under the Job Training Partnership
  Act.

REFERENCES

Andrisani, P. and T. Daymont. 1983. Age, Productivity, and
    Earnings. Project on National Employment Policy and Older
    Americans. Washington: National Commission for Employment
    Policy.
Becker, G. 1975. Human Capital. New York: National Bureau for
    Economic Research.
Ben-Porath, Y. 1967. The production of human capital and the life
    cycle of earnings. Journal of Political Economy 75: 352-365.
Harris, L. 1981. Remarks at Press Conference on Aging, November 13,
    1981. Washington, D.C.
Johnson, T., K. Dickinson, and R. West. 1983. Older Workers'
    Responses to Job Displacement and the Assistance Provided by
    the Employment Service. Project on National Employment Policy
    and Older Americans. Washington: National Commission for
    Employment Policy.
Jondrow, J., F. Brechling, and A. Marcus. 1983. Older Workers in
    the Market for Part-Time Employment. Project on National
    Employment Policy and Older Americans. Washington: National
    Commission for Employment Policy.
Lazear, E. 1976. Age, experience, and wage growth. American
    Economic Review 66: 548-558.
Mincer, J. 1970. The distribution of labor earnings: A survey with
    special reference to the human capital approach. Journal of
    Economic Literature 8 (March): 1-26.
Mincer, J. 1974. Schooling, Experience, and Earnings. New York:
    Columbia University Press.
Paul, C. 1983. A Human Resource Management Perspective on Work
    Alternatives for Older Americans. Project on National
    Employment Policy and Older Americans. Washington: National
    Commission for Employment Policy.
Root, L. and L. Zarrugh. 1983. Innovative Employment Practices for
    Older Americans. Project on National Employment Policy and
    Older Americans. Washington: National Commission for
    Employment Policy.
Rosen, S. 1972. Learning and experience in the labor market.
    Journal of Human Resources 7 (Summer): 326-342.
Rosen, B. and T.H. Jerdee. 1977. Too old or not too old? Harvard
    Business Review 55 (November/December): 97-106.
Shapiro, D. and S. Sandell. 1983. Age Discrimination and Labor
    Market Problems of Displaced Older Male Workers. Project on
    National Employment Policy and Older Americans. Washington:
    National Commission for Employment Policy.

IMPACT OF TECHNOLOGICAL CHANGE ON MIDDLE-AGED AND OLDER WORKERS:

PARALLELS DRAWN FROM A STRUCTURAL PERSPECTIVE

Jon Hendricks

Department of Sociology
University of Kentucky
Lexington, Kentucky

In the last three decades labor force participation rates for middle-aged and older workers have shown a consistent linear decline. While it is true that female employment patterns have remained fairly stable, even rising somewhat among those women over the age of forty-five who are active in the labor market, male participation has not kept pace. To explain what has been happening to older workers, labor economists, sociologists, and social scientists of all stripes must focus not only on life cycle factors or human capital variables characteristic of individuals, but on structural determinants of employment conditions as well. To adequately understand the broad ranging changes which spell displacement and dislocation for many workers, it is essential that a sociology of labor markets examine business cycles, product market characteristics, firm or sectoral placement and so on. Permeating each of these should be a concern with the differential impact of innovations in technology, automation and modes of production in general. According to the Commissioner of Labor Statistics for the New York federal region, there is already sufficient evidence accumulating to suggest technological displacement is being disproportionately settled on workers, especially males, over the age of forty-five (Ehrenhalt, 1983).

In the discussion to follow it will be contended that technological changes already on the horizon will be a major force in the ways our lives take shape. The impact of technology will reach well beyond the way productive processes are organized in the work-a-day world and give new meaning to the post-industrial age. As the transformation in modes of production moves along the continuum from labor to capital to technology intensive processes, both long-run gains and short-term hardships might well be

113

anticipated. One thing is fairly certain: the exponential growth of high technology will not automatically lead to a vast expansion in employment opportunities. Some jobs will be gained, many will be lost, productivity will ultimately increase but the quality of life, especially in the workplace, will inexorably be altered. Among the immediate challenges are the social and psychological costs facing those who must accommodate themselves to the impact of technology. For the middle aged worker these costs will pose a major specter, as they are faced with the task of either up grading their technological skills to fit with sophisticated automation or accepting lateral or downward transfers to remain employed. The kinds of things people do on the job will be different and this in turn will color nearly every other aspect of their lives. As Marx and Veblen pointed out so long ago, the way work is organized to adapt to machines will determine how people think, act and even dream (Ginzberg, 1982). The computer-driven wave of technology currently making itself felt in all economic pursuits will bring changes in job security, autonomy, pacing, decision-making and the entire range of social parameters growing out of the way work is done. In short, it will alter the way we conceive of our lives. Accordingly, technological innovations, largely revolving around computer-mediated work, portend changes in the way social scientists of all stripes must conceive of the complex of roles and statuses which give meaning to life. By this I mean to say that neo-classical economic theory and such basic analytic templates as social class will be hard pressed to be of much utility in explaining the life situation of the new labor force. A structural interpretation of the impact of technology on the work role is advanced as a viable perspective.

TECHNOLOGY AND MID-LIFE EMPLOYMENT

According to Glenn Watts, President of the Communications Workers of America, serious disruptions and dislocations will eddy through the American workforce in the next twenty years (Watts, 1983). As has been pointed out repeatedly, few if any labor economists doubt there will be a continuing decline in the labor force participation rates among older workers. Even for those who remain employed there is every likelihood that a greater tendency toward occupational displacement will become all too commonplace. Two interrelated issues face middle-aged and older workers: first is the prospect of unemployment due to questionable labor demands for individuals in their age category, second is the quality of the work experience for those who are employed. Each of these will be addressed in turn.

As American industry embraces new forms of technology to insure its survival in an increasingly competitive global economy, cost effective strategies will tend to emphasize capital mobility at the expense of worker mobility, thus accelerating the process of skills obsolescence for older members of the labor force (Morrison, 1982). According to recent government statistics, unemployment rates for those over the age of fifty-five are climbing faster than for any other age group. In the first three-quarters of 1982, for example, there was an increase of 24 percent in the unemployed age 55-64; this compares to a 16 percent increase for the rest of the working population. Furthermore, once unemployed older workers are more than three times as likely as younger workers to simply give up the search for a replacement job (House Select Committee on Aging, 1982). Thus a kind of involuntary early retirement, well in advance of official retirement age, occurs as their niche in the labor market disappears.

Already certain patterns are becoming obvious. To date the greatest technological displacement has occurred in traditional manufacturing and smokestack industries. Nearly every indicator suggests many basic industries will continue to decline as the size of the automobile, rubber, and steel industries continues to shrink. Just to provide one illustration: in 1979, 90,000 people were employed in the steel industry in the greater Pittsburgh area; by 1983, fewer than half that number were still on the job. Much of the decline is due to the economic instability of the industry but automation is occuring simultaneously and lessening the probability of rehiring even if an upswing does take place. If plans in the appliance assembly lines at General Electric are carried through, half of the 37,000 line workers could be replaced by robotic equipment; those laid off by steel are not likely to be reemployed in parallel manufacturing firms (Joint Economic Committee, 1983). It may not be too farfetched to claim that blue collar workers may go the way of farm workers in the next few decades (House Committee on Education and Labor, 1982). In the next decade manufacturing employment may perhaps fall to half its current levels as the push toward virtually unmanned factories accelerates. Some estimates range as high as one million jobs disappearing before the year 2000 due to the burgeoning impact of technology and to lessened demand for goods stemming from shifts in the population due to declining birth rates.

Heretofore much of the effect has been dampened due to the off-setting growth in the service sector of the economy. In the period from 1969-1976, 90 percent of all new jobs created were in the service sector so that today it employs approximately 70 percent of all workers. According to current projections, of the potential 17.5 million jobs which may be created by economic expansion in the late 1980s and 1990s most will fall in the lower paying reaches of the service sector. While openings for 150,000 computer personnel may well develop, these will be far outstripped in the next decade by the fast-food industry where the need for workers and kitchen help will approach 800,000. Overall, high technology occupations

may account for no more than seven percent of the new jobs. More
significant opportunities will open in clerical and service
occupations--in those jobs that require only modest training beyond
the high school level (Rumberger and Levin, 1983).

     The question is not whether innovations will be adopted, but
how rapidly and how quickly they will begin to have a major impact
on management and related skilled career lines. White-collar jobs
will surely be affected dramatically in the next decade; already
something on the order of seven million workers spend their days in
front of a videoscreen, and this is scarcely more than the tip the
iceberg. By 1992 over half of the office labor force in the United
States will be affected by computers. Anything routine or repetitive
is likely to be automated and thus affect all other jobs as well.
As James A Baker, a vice-president at General Electric succinctly
put it, American manufacturers have little or no choice; if they are
to remain viable in the world economy they must "automate, emigrate
or evaporate" (N.Y. Times, June 2, 1983). While a consensus has yet
to emerge on the extent of long range displacement, since new jobs
will be created along the way, the question remains as to whether
middle aged and older worker can be reabsorbed (House Committee on
Education and Labor, 1982). To return to smokestack industries, for
an example, more than 86 percent of workers over the age of 45 have
not found reemployment since losing their jobs in the current
recession. Similarly, while one in six workers falls in the middle
aged or older category, only one in twenty-three jobs created in the
first half of 1983 was filled by someone from that age category.
Furthermore, among those who are working the majority may be
employed at lower levels than their skills would justify (USA Today
1983).

STRUCTURAL FACTORS AND WORK EXPERIENCE

     Beyond the immediate and pressing question of job displacement
and the short term consequences of lay-offs is the whole issue of
the nature of work, the work experience, and the quality of life.
On one hand, tasks that are laborious, tedious, and drearisome will
be taken over by machines, thus relieving workers of much that is
monotonous. Electronic technology also has the potential of
permitting both decentralization--workers may attend to their jobs
in their own homes or elsewhere--and flexible work schedules. At
the same time, it has been suggested that the price and quality of
consumer goods may also be made more favorable once capital
expenditures begin to pay off. Legislation, such as the 1981
Economic Recovery Tax Act, designed to foster increased productivity
through tax incentives based on investments in automated equipment,
will underwrite many of the changes which have the potential to
improve the quality of life for workers and consumers alike.

At the same time, however, technology has the potential to nurture developments which in many respects will put the average worker at a disadvantage. True, there will be some expansion of the job market for a relatively small category of elite and highly trained specialists, but for the bulk of American workers, the exigencies of "high-tech" economies may well alter their working conditions and the degree of control they are able to exercise in their own working lives. Some hint of concerns over just these issues can be gleaned form the so-called "New Technology Bill of Rights" (House Committee on Education and Labor, 1982), new collective bargaining arrangements, including life-time employment, retraining, and severance agreements modelled after European programs, and the upsurge in conferences focusing on job security in the face of structural unemployment stemming from the application of new technology.

TECHNOLOGY AS STRUCTURE

In the following discussion I propose one model for making sense of the actual work situation of future workers--including those in their middle and later years. It is not an alternative created ex nihilo, but ex cathedra--outside an orthodox perspective of the status attainment model. While there is no denying the importance of human capital in determining occupational patterns, it is my contention that such a model must be augmented if we are to make sense of what may be termed the new structuralism of technology in the world of labor.

In adopting a structural perspective in reference to future work conditions, the implication is that there are certain broad ranging factors which are inherent in technological applications in the workplace over which individual attributes will have relatively modest effect. That is to say, as technologically intensive productivity becomes predominant, the returns on human capital for the vast bulk of the labor force will be even further beyond their control than they are presently. In short, the exigencies of the labor market will be such that individual action will not be a palliative of great magnitude. For the average worker, the infrastructure of technology will alter his relative position vis-a-vis the productive process, increasing his dependency while simultaneously depreciating the human capital variables he brings to the workplace. Continuing employment at levels commensurate with one's track record or securing reemployment will be less dependent on experience than it is on one's ability to use new technology.

For the older workers especially, such changes will spell dependency via dislocation well in advance of recognized retirement. This is not because they cannot learn new techniques; they certainly can. Rather, the reasons reside in the same

retraining biases as now exist, in addition to the fact that the
basic assumption of retraining is that once skills are upgraded,
there will be higher status positions waiting. There may well be
jobs, but they will cluster in those occupations which are not
contingent on being highly skilled. None of this is to claim there
will not be a small core of elite workers who design and manipulate
the technology; they will exist, and they will definitely be
regarded as of superior status, but they will constitute less than
ten percent of the work force.

As productivity shifts to ever more technology intensive
processes, both the content and the style of work roles is likely to
undergo a radical change. This will be true not only for
technologically advanced occupations, but for jobs in the tertiary
or service sector as well. Without any particularly astute crystal
ball gazing, it is possible to predict a "de-skilling" of many
white- and blue-collar jobs, as well as a greater propensity for job
instability, underemployment, and loss of autonomy in many
occupations. Indeed, past research has already pointed to the
importance of structural components of the work world as they exist
apart from individual worker attributes (Gordon, 1972).

As technological applications move beyond relatively basic
conversion and transfer operations to the so-called "smart"
computerized processing which mimics human reasoning, complete with
tactile sensitivity, it may be hypothesized that a new social class
structure will be manifested, based on structural locale in terms of
technologically intensive pursuits or service industry employment
(including the public sector). Furthermore, it might be claimed
that structural constraints based on sectoral placement will
arguably carry over into all aspects of life. The importance of
sectoral differences is that they will shape social relations and
will be relevant contextual factors coloring, to a large extent,
what individual workers will experience despite their personal
attributes.

The development of technology or its application to production
operations is itself a social process. It is neither inevitable nor
non-partisan but reflects a managerial choice (Noble, 1978) made in
light of what is perceived to be the most advantageous cost-benefit
ratio. Regardless of the reasons underlying the use of the most
recent generation of intelligent machines, one consequence will be
that operators will further relinquish control over the immediate
conditions of work and their labor process. That is, with the
separation of conceptualization from execution there will be a
homogenization of work environments among those responsible for the
latter. This in turn will likely lead to a lessening of
occupational differentiation. To illustrate, the distinction
between engineer and draftsman will become less salient in the face
of the new technology. In fact, the decline in the numbers of new

professions and professionals is already evident, along with a loss of autonomy which is marked by an attenuation of their involvement in management decisions (Braverman, 1974; Browning and Singelman, 1978; Wright, 1979).

Challenges to neo-classical economic explanations of status attainment and reward have been on the rise for many years (for example, see Stolzenberg, 1975; Osterman, 1975; Bibb and Form, 1977; Beck, Horan and Tolbert, 1978; Kallenberg, Wallace and Althauser, 1981). With the advent of technology intensive productivity, it is likely that the human capital approach will come under even greater scrutiny. Neither structural unemployment nor the growth of the service and public sectors fit easily within traditional economic theory (Ginzberg, 1982). While the various structural perspectives differ among themselves, generally they all lay stress on the fact that a segmentalized industrial order exerts suzerainty over the characteristics of individual workers. So long as research strategies continue to look only at personal attributes, explaining differential rewards and opportunity structures in a technological society will remain problematic.

SOCIAL CONSEQUENCES OF SECTORAL PLACEMENT

Accordingly, it is here hypothesized that an economic model which includes as a primary focus a concern with the structure of technology as it defines the world of work is essential in explaining the quality of the work experience. It is maintained that in those industries where capital investments in technology and automation are greater than investments in labor capital, the work experience for all but a small elite cádre will parallel the work experience of workers presently employed in what is termed the peripheral--as opposed to the core--sector (Baran and Sweezy, 1966; Averitt, 1968). As noted by Bluestone, Murphy and Stevenson; core and periphery industries may be characterized thus:

> The core economy includes those industries that comprise the muscle of American economic and political power....Entrenched in durable manufacturing, the construction trades and to a lesser extent, the extraction industries, the firms in the core economy are noted for high productivity, high profits, intensive utilization of capital, and a high degree of unionization. What follows normally from such characteristics are high wages. The automobile, steel, rubber, aluminium, aerospace, and petroleum industries are ranking members of this part of the economy. Workers who are able to secure employment in these industries are, in most cases, assured of relatively high wages and better than average working conditions and fringe benefits.

Beyond the fringes of the core economy lies a set of industries
that lack almost all of the advantages normally found in center
firms.   Concentrated in agriculture, nondurable manufacturing,
retail trade, and sub-professional services, the peripheral
industries are noted for their small firm size, labor
intensity, low profits, low productivity, intensive product
market competition, lack of unionization, and low wages.
Unlike core sector industries, the periphery lacks the assets,
size, and political power to take advantage of economies of
scale or to spend large sums on research and development
(Bluestone, Murphy and Stephenson, 1973: 28-29).

Obviously such worker experiences as underemployment, job
instability, low wages, access to a wide range of opportunities both
during and after working life--including retirement
adjustment--could easily be affected by sectoral differences in
technological investments (see also, Dowd, 1980). The core will
provide those jobs in which technological facility is high, where
employment is stable; turnover, low; and wages high. For the most
part, positions within the core can be seen as part of an internal
mobility ladder wherein upward movement is possible and individual
attributes are often used to mediate entry. Positions in the
periphery can be characterized by relative instability as there is
less need to maintain continuity in the labor force. They will be
marked by sporadic unemployment, high turnover, low wages; they are
jobs which will not provide as ready access to pathways for upward
mobility. By its very nature, the core sector will provide the
higher returns, pension planning, insurance programs and fringe
benefits which promote optimum adjustment in later years. In
contrast, placement in the periphery, or low technology intensive
industries, will effect a "dampening" of these and a number of
related areas. Even so fundamental a consequence as lifetime
earnings will not just be based on eduction, occupational mobility,
aptitude, sex, race or any of the other individual attributes
traditionally pointed to as causal factors. Rather earnings will
reflect employment in industrial sectors as defined by varying
degrees of technological application.

If one assumes the bifurcation of industrial sectors, based on
vertical integration of supply and product markets as well as
control over technological resources, is indeed a valid depiction of
the economic realm of tomorrow, what else can be anticipated for the
average worker? Perhaps the best guide can be drawn from the
existing literature. If the political economy of today's labor
market does indeed make a difference in the lifeworld of
individuals, there is every reason to expect similar patterns will
continue. Beck, Horan and Tolbert, (1978) in an early effort to
identify effects of sectoral employment, found core workers to be
better paid, and reveal less internal discrepancy in income than
their peripheral counterparts. They were also more likely to work

full time, put in more hours per week, be white, male and belong to
a union. They concluded that significant differences exist between
core and peripheral workers which cannot be explained away by
reference to the quality of the labor force of any of the standard
human capital factors. In a subsequent attempt to look more closely
at the work situations and socioeconomic experiences of workers
these same investigators utilized a different data base but arrived
at the same conclusion (Tolbert, Horan and Beck, 1980).

In a subsequent investigation Tolbert (1982) examined mobility
patterns within and between sectors. His findings, derived from
longitudinal patterns during early, middle and late stages of
careers are relevant here. Mobility between sectors was infrequent,
fewer than 26 percent moved across sector boundaries, and even
within sectors the vast majority of the mobility took place early in
the career. By age 50-59 occupational prestige had ceased to
improve and even some slight downward drift was apparent. Not too
surprisingly, real dollar wages of peripheral worker workers eroded
in the later stages while core workers continued to improve their
purchasing power.

In still another analysis of tiered placement within sectors,
significant differences in the work experience and rewards accruing
to workers in the two realms were also identified. Analyzing
respondents in the five percent Neighborhood Characteristics sample,
generated from the 1970 Public Use Sample of the United States
Bureau of the Census, it was found that proportionately more
females, blacks and part time workers were employed in the periphery
at lower wages. Interestingly, within-sector placement of females
also revealed a distinctive pattern. In the core, women work
primarily in lower ranking occupations; in the periphery no such
patterns could be identified. These trends are not apparently due
to educational distribution, either between the two sectors or
between male and female participation in the labor force. Congruent
with the results reported by Tolbert, Horan and Beck, we found mean
income differences between peripheral workers of higher and lower
status were not as pronounced as those in the core. In explaining
the greater homogenity in income within the periphery it can only be
assumed that the status hierarchy does not necessarily carry
financial advantage (Hendricks and McAllister, 1983).

CONCLUSION

To date, the dominant paradigm in labor market research has
been one which casts individuals and their personal attributes as
the appropriate unit of analysis in nearly all investigations. In
recent years there have been increasing calls for an alternative
model which concentrates instead on the normative imperative of
structural arrangements. As the mode of production moves even

further along the continuum of labor intensive to capital intensive
to technology intensive processes the impact on the worker is not
likely to lessen. Similarly, the effects of inflation, recession
and taxation are likely to show sectoral differences. The
examination of the generalized social import of a dual economic
configuration which has been presented here may offer a concrete
alternative for analyzing financial well-being and its many
ramifications among older workers.

As the diffusion of what is now the fifth generation of
automated equipment quickens a wide ranging realignment in the
relations of production looms on the horizon. Among the many
consequences which might be anticipated is a further fragmentation
of the labor process. Historically, all available evidence suggests
that such changes carry with them a lessening of workers' abilities
to exercise control over their own work activities, a lowering of
the skill levels required to perform any particular task in the
production process and a disruption in the career trajectories open
to average workers. A corresponding change may well be levelling of
the occupational hierarchy thus leaving the bulk of workers in a
position analogous to that characteristic of those who are currently
employed in the peripheral economic sphere. While automation will
undoubtedly carry with it the creation of new positions, the best
evidence suggests that these will tend to cluster among the less
skilled categories. For those workers who are displaced by the new
technology, the alternative may well be employment in pursuits below
that of their previous position. Without meaning to sound overly
pessimistic, the impact of technology may portend a degradation of
the work experience for large segments of the labor force.

The data presented are themselves enough to buttress the claim
that something other than just human capital variables is affecting
the lifeworld of individuals. Our examination of sectoral
differences points to structural inequalities that arguably carry
over into old age and will help to locate old people within the
context of basic social and economic structures. It stands to
reason that in any earnings based system of social security, the
bulk of those workers retiring from the core are likely to receive
larger pensions/social security payments by virtue of their greater
annual income in the years immediately preceeding retirement and
more stable work histories. At the same time it may be hypothesized
that for those retiring from the periphery, the financial realities
of latter life will be harsher. As Estes, Swan and Gerard (1982)
make clear, it is not sufficient to point our finger at economic
short-falls or benefit loopholes as a major problem. We must
examine whether statutory provisions, benefit levels and most
importantly, social psychological well-being also reflect economic
structures. A dual economic model integrated into labor and
gerontological research may allow researchers to enhance their
predictive accuracy--which until now has not been particularly

admirable--and to develop a better sociology of later life by
pinpointing some of the macrolevel processes which serve to impose a
dependent status on many elderly almost irrespective of their
personal skill, attributes or ability to cope with the onset of old
age. Such an effort will move us beyond a rather enervated form of
class analysis derived from the tautology of occupational
hierarchies toward a dynamic structural perspective on the social
relations of later life.

REFERENCES

Averitt, R. T., 1968, The Dual Economy. New York: W. W. Norton and
    Company, Inc.
Baran, P. A., and P. M. Sweezy, 1966, Monopoly Capital. New York:
    Monthly Review Press.
Beck, E. M., P. M. Horan, and C. M. Tolbert II, 1978,
    "Stratification in a dual economy: A sectoral model of earning
    determination." American Sociological Review, 43, 704-720.
Bibb, R., and W. H. Form, 1977, "The effects of industrial
    occupational and sex stratification on wages in blue-collar
    markets." Social Forces, 55, 947-996.
Bluestone, B., W. M. Murphy, and M. Stevenson, 1973, Low Wages and
    The Working Poor. Ann Arbor: The Institute of Labor and
    Industrial Relations, University of Michigan.
Braverman, H., 1974, Labor and Monopoly Capital: The Degradation of
    Work in the Twentieth Century. New York: Monthly Review
    Press.
Browning, H., and J. Singlemann, 1978, "The transformation of the
    U.S. labor force: The interaction of industry and occupation."
    Politics and Society, 8, 41-509.
Dowd, J.; 1980, Stratification Among the Aged. Monterey, Ca.:
    Brooks/Cole Publishing Company.
Ehrenhalt, Samuel M., 1983, cited in M. McNamee, "Older Men Left in
    Job Squeeze." USA Today, November 28, 1983, p. 1.
Estes, C. L., J. H. Swan, and L. E. Gerard, 1982, "Dominant and
    competing paradigms in gerontology: Toward a political economy
    of ageing." Ageing and Society, 2, 151-163.
Ginzberg, Eli, 1982, "The mechanization of work." Scientific
    American, 247, 3 (September): 67-75.
Gordon, D. M., 1972, "From stearn whistles to coffee breaks,"
    Dissent, (Winter):197-210.
Hendricks, J., and C. M. McAllister, 1983, "An alternative
    perspective on retirement: A dual economic approach." Ageing
    and Society, 3.
House Committee on Education and Labor, 1982, New Technology in the
    American Workplace. Subcommittee on Labor Standards,
    Washington, D. C.: U. S. Government Printing Office.
House Select Committee on Aging, 1982, Unemployment Crisis Facing

Older Americans (97-367). Washington, D. C.: U. S. Government Printing Office.

Joint Economic Committee U.S. Senate, 1983, Robotics and the Economy. Washington, D. C.: U. S. Government Printing Office.

Kallenberg, A. L., M. Wallace, and R. P. Althauser, 1981, "Economic segmentation, worker power, and income inequality." American Journal of Sociology, 87, 651-683.

Morrison, Malcolm, 1982, Economics of Aging: Future of Retirement. New York: Van Nostrand Reinhold.

New York Times, 1983, "Technology." Steven J. Marcus, D2, June 2.

Nobel, D. F., 1978, "Social choice in machine design: The case of automatically controlled machine tools and the challenge for labor." Politics and Society, 8, 313-348.

Osterman, P., 1975, "An empirical study of labor market segmentation." Industrial Relations Review, 28, 508-523.

Rumberger, R., and H. Levin, 1983, "Skills for the Workplace in an Era of High Technology: Myths Realities and Implications." presented to Southern Regional Education Board, Asheville, North Carolina.

Stolzenberg, R. M., 1975, "Occupations, labor markets and the process of wage attainment." American Sociological Review, 40, 645-665.

Tolbert III, P. M., 1982, "Industrial Segmentation and Men's Career Mobility." American Sociological Review, 47, 457-477.

Tolbert, C., P. M. Horan, and E. M. Beck, 1980, "The structure of economic segmentation: The dual economic approach." American Journal of Sociology, 85, 1095-1116.

USA Today, 1983, "Older Men Left in Job Squeeze." November 28, 1983, 1.

Watts, G., 1983, "Training and retraining workers will be an important challenge to unions in the 21st century." Personnel Administrator, 28, 82-84.

Wright, E. O., 1979, Class Structure and Income Determination. New York: Academic Press.

RECENT TRENDS IN RETIREMENT POLICY AND PRACTICE IN EUROPE
AND THE USA: AN OVERVIEW OF PROGRAMMES DIRECTED TO THE
EXCLUSION OF OLDER WORKERS AND A SUGGESTION FOR AN
ALTERNATIVE STRATEGY[1]

Bernard Casey

International Institute of Management/Labour Market Policy
Science Centre Berlin
Platz der Luftbruecke 2, D-1000 Berlin 42

INTRODUCTION

In all of the industrialised market economies, advancing age
tends to bring with it a greater vulnerability on the labour
market. A precise and universally applicable age threshold from
which this increase of vulnerability can be determined to start does
not exist, but most commentators do agree that between ages 40 and
45 labour market problems begin to appear, and between ages 55 and
65 they worsen considerably. The particular vulnerability of older
workers can, in principle, be ascribed to the difference between
their diminished capacities and employers' requirements - usually
based on the capacities of younger workers. In addition, older
workers are generally less able to adapt to new requirements imposed
by technological change; whilst, in terms of human capital
investment, it is more "profitable" for firms to hire younger
workers, because the time available for amortisation of the cost of
recruitment and on-the-job training is longer.

The degree of vulnerability of older workers has varied very
markedly in accordance with the general state of the labour market.
Already in the 1960s, discussion in most OECD countries had
identified them as a problem group. Then, at a time of full
employment, when labour market policy was principally concerned with
the mobilisation of labour reserves, the most urgent issue in many
countries was how older workers' lower productivity could be
countered or compensated for, and what adjustments to working
conditions might be necessary to prevent a premature cessation of
employment. It was no coincidence that, at this time, the OECD
released a series of studies concerned with how the greatest
possible number of older workers could be maintained in work for the

125

longest possible time, and covering such issues as job re-design and
retraining for those in employment and possible policies for the
reintegration of the older unemployed.

By the mid-1970s the situation of labour shortage had largely
been replaced by one of labour surplus, a product both of
demographic trends and the increasing pressure for rationalisation
induced by the slowdown of growth in the years following the 1974/75
oil crisis. Under these circumstances, whilst the policies of
integration developed in the previous decade were not abandoned, and
whilst indeed new protective measures were developed, both of these
efforts paled into insignificance alongside new provisions whose
purpose was to facilitate the exclusion of other workers from the
labour market. It is these provisions, which have led to a
progressive and often substantial lowering of the effective age of
retirement, which are the subject of this paper.

EARLY RETIREMENT IN EUROPE

At the formal level the actual instruments of exclusion are
highly varied. Some have been introduced as explicit elements of
labour market policy. Others were conceived as essentially social
policy measures, but have subsequently been increasingly utilized to
serve labour market policy purposes.

In a number of countries the link to labour market policy is
made very clear by the fact that early retirement schemes are
financed directly by the unemployment compensation system. France
provides a particularly good example of this. As early as 1972 the
parity-controlled. unemployment insurance fund established a special
higher level of benefit available to all persons over 60 becoming
involuntarily unemployed. Recipients of this benefit were excluded
from the need to register with the employment office as job seekers
and henceforth were counted as "inactive." In 1977 the same "income
guarantee" payment was made available to all over 60 year olds
resigning voluntarily although, since to a considerable extent this
new provision also increased enterprises' ability to skirt around
legal procedures regulating redundancies, the term "voluntarily"
should be carefully interpreted. In the same year adjustments to
the regular unemployment compensation scheme were made which enabled
persons losing their jobs from as young as 56 and 2 months, after
having received benefits on an extended basis, to pass into the
"income guarantee" system on reaching 60. This contributed to a
further lowering of the de facto, if not de jure, age of early
retirement. At the end of the decade the authorities went one step
further, introducing yet another new benefit for employees as young
as 55 who were dismissed for economic reasons from "problem"
industries and freeing them from the obligation to register as job
seekers. By 1981 the number of de jure "early retirees" exceeded

300,000, to be compared to 1.7m registered unemployed. In addition
there existed a further 170,000 55-59 year old unemployed who,
although having to wait until they reached the aged 60 before being
officially reclassified as early retirees, to all intents and
purposes already "enjoyed" this status.

Belgium also provides an example of a country where, following
a national collective agreement of 1975, the unemployment insurance
system provides benefits on a higher and extended basis to older
persons dismissed for economic reasons. Whilst within the national
system entitlement is open only to those within five years of the
normal retirement age, the relevant provisions can be supplemented
by branch or enterprise collective agreements to cover employees as
young as 55 (males) or 50 (females). Once again the requirement to
register as a job seeker is dropped and the persons concerned are
considered as "early retirees". A special provision introduced in
1976 in the Netherlands allows unemployment benefits to be payed
until the age of 65 is reached to those losing their jobs and aged
at least 57 and 6 months. Although such persons still appear in the
total registered unemployed, active consideration is now being given
to freeing them from this requirement. This is related to the major
increase in utilization of these benefits in the past year and a
half. Until the start of 1982 dismissal protection legislation
required that candidates for redundancy had, in terms of age, to be
representative of the total labour force of the firm. Subsequently,
under pressure of a rapid deterioration in the labour market since
1980, new regulations were drawn up explicitly permitting age
selective redundancies and allowing the burden of redundancies to be
borne by older workers.

Another means of linking early withdrawal directly to
unemployment is to be found in Austria, Sweden and Germany. In all
three countries the public pension system makes provisions for an
unreduced early old age pension at the age of 60 to persons who have
been unemployed for at least one year (1 year and 9 months in
Sweden). Because of the simultaneous existence of other early
retirement possibilities (described below), the quantitative
significance of these measures is relatively limited. However, for
particular industry branches experiencing substantial economic
difficulties and/or engaged in major restructuring operations (steel
and coal or iron-ore mining in all three countries, motor
manufacturing and shipbuilding in Germany and Sweden) they form a
key component in "social plans" regulating labour force reduction.
In such cases enterprises engage in age selective dismissals and
agree to top up unemployment benefits/pensions until the normal age
of retirement is reached. Some collective agreements, drawn up at
the level of the enterprise rather than the branch, provide for the
early retirement of all workers aged over 55 and not just those over
59.

Compared to the countries so far described, public compensation provisions for the older unemployed in Great Britian are very underdeveloped. This deficit is, however, at least in part compensated for by private provisions, normally regulated at the level of the individual firm. Enterprises obliged to make redundancies will often make arrangements whereby, with a combination of lump sum payments and the granting of unreduced company pensions, those members of the labour force in the last few years of their working life can be "bought out." Examples of such arrangements, affecting employees from the age of 55 onwards, are to be found. Furthermore, whilst in order to maintain their rights to various social insurance benefits or (in the case of those older unemployed in receipt of no or only inadequate company pensions) to receive income support payments, many of these de facto "early retirees" continued to register as unemployed, a recent series of administrative reforms have freed most over 60 years old from this obligation. The present total of some 3m unemployed has now been reduced by about 160,000 as a consequence.

All of the strictly market policy oriented early retirement programmes discussed so far, although affecting in the first instance older workers, are of course also intended to benefit younger persons whose own employment might otherwise have been threatened. The link with providing employment opportunities for otherwise unemployed (young) persons is made even more explicit in those schemes which seek to link voluntary early retirement with the compensating recruitment by the enterprise of a registered job seeker. Such measures have been operative in Belgium and Great Britain since 1977, in the first country covering persons in the last five years before the normal retirement age and offering a relatively high replacement income, in the second only persons in the last two or three years and with a much less attractive level of "bridging payment." Reflecting the relatively more favourable conditions accorded to the early retiree, take-up under the Belgian system is nearly four times as high (compared to the size of the total labour force) as under its British counterpart. The most extensive application of the "compensating recruitment" principle, however, has occured in France. There, the new socialist government, as well as announcing a reduction of the normal retirement age to 60 from spring 1983, legislated for the signing of "solidarity contracts" permitting publicly financed voluntary early retirement between 55 and 59 in firms which committed themselves to replace (directly or indirectly) the early retirees by (young) unemployed persons. In the course of 1982 nearly 20,000 such contracts were signed, affecting 310,000 persons, approximately one third of the private sector labour force in the 55-59 age range.

Some investigations have been made of the workings of such "compensatory recruitment" schemes. In Great Britain, it is estimated that the actual replacement rate has been in the order of

80%. In France, case studies have suggested that for firms undergoing major restructuring (geographical, product or technological), the opportunities provided by "solidarity contracts" can be highly advantageous. Older, essentially superfluous employees can be eased out, whilst a considerable share of the supposedly compensating recruitments might well have been made in any case. Overall, the "real" rate of replacement is thought to be only some 50%.

Finally, within the context of early retirement programmes specifically instituted for "work sharing" objectives, we should mention those developed in the Netherlands since 1977 by branch level collective agreements. Such agreements, now covering more than 80% of the dependent labour force, provide for voluntary withdrawals, first from the age of 63 or 64, now from the age of 60 or 61, in receipt of a "bridging payment" funded by a part of the negotiated wage increase. Take up rates lie between 40 and 60%. In a few cases, when negotiating a lowering of the age of such voluntary early retirement, the parties have included an obligation for compensating recruitments from amongst younger persons. On average, however, the replacement rate of early retirees is only of the order of 25%, emphasizing that the scheme's principal function in practice is that of facilitating labour force procedures.

Somewhat more indirect in their working, but quantitatively at least as significant, are those mechanisms which ostensibly relate early retirement to "disability." Particularly notable are the schemes operative in Sweden, Holland, and Germany, insofar as all of these countries included not just medical factors but also "labour market chances" in the definition of "disability."

In Sweden, at the end of the 1960s, concern began to be raised about the effects at the workplace of the processed rationalization and technical and organizational change. It was pointed out that this had contributed to a steady disappearance of those "retraite" jobs into which older workers had been traditionally transferred towards the end of their working lives and where work-demands could be better adapted to their changing capacities. A subsequent revision of the disability pension law determined that, in making a decision about the pension eligibility of workers over 60 years old, whilst medical factors are still to be taken into account, the authorities "are mainly to concern themselves with the applicant's ability and possibility of obtaining a continued income through such work as he has previously done or through other suitable or available work." In the course of the 1970s, and particularly in years immediately after this reform, the incidence of disability pension of older workers increased significantly. By 1979 over a quarter of the male population aged 60-64 were in receipt of such benefits, whilst in 1981 one in twenty of active males in this age group left the labour force for this reason.

Exactly how far this phenomenon is to be explained by the deterioration of the labour market is still a source of considerable debate in Sweden. However, one indication that a considerable share can be ascribed to such developments can be seen by reference to the situation in the Netherlands. Here, a person with only a minimal medical disability can be classified as totally disabled if it is considered that he would be unable to find "suitable" work within the local labour market. In addition, a decision of the relevant parity regulatory body dating from 1973 provides for the automatic award of a full pension to any involuntarily unemployed person with minimum medical disability. Frequency of entry into a disability pension is strongly correlated to age, and by 1979 over half of the insured 60-64 and a third of insured 55-59 year old age groups were in receipt of such benefits. Sector by sector analysis has shown a clear relationship between a decline in job opportunities and the rate of entry into the disability pension, with this being the strongest for those branches where contraction has been greatest.

In the case of Germany, the link between labour market chances and disability is a slightly more complex one, although essentially similar principles apply. Here a person can be declared partially disabled (which in practice means not being able to work more than part-time), and he is then entitled to a partial disability pension which is intended to top up earning from a reduced (i.e. part-time) employment. Until 1969 it was immaterial if suitable part-time employment could be found, if not the person could claim partial unemployment benefit. However, in that year the Federal Social Court introduced the notion of a "closed labour market" for suitable part-time work by declaring that if such employment was not available, the person was not partially but fully disabled (unable to work at all) and entitled to full disability pension. This move from an "abstract" to a "concrete" principle of judging work incapacity opened up the possibility for a full pension not only for those who would previously have received only partial benefits, but also for those whose degree of disability would not even have entitled them to a partial pension. This is because although such persons were adjudged capable of working at least part-time (although less than full time), not even suitable part-time jobs were available to them. One estimate suggests that, as a consequence of the Social Court's decision, by 1977 over a quarter of a million persons were no longer in the labour force. Again disability pensioning is strongly related to age, with over 60% of new awards of male blue collar workers in 1980 going to over 55 year olds.

Finally reference should be made to the opportunities presented by the availability of an early old age pension in situations other than unemployment or disability. At the beginning of 1973, almost purely for social policy reasons, the German government introduced the possibility to draw a full pension at age 63, rather than 65, for persons satisfying the requirement of having made 35 years of contributions. The timing of this reform of pension legislation proved extremely fortuitous, and a major increase in the incidence of early retirement provided (alongside the reduction of the foreign labour force) one more possibility for the country to adjust, at least with respect to the level of "open" unemployment, to the sharp downturn in the number of jobs occuring in the years immediately subsequent to 1974. In Austria a similar, although much more longstanding, early pension opportunity allowing retirement at 60, played a similar role at the moment of downturn.

THE SITUATION IN AMERICA

Having surveyed European development for ever earlier retirement, I want to turn briefly to consider the situation in the USA, since there exists in many quarters the conception that somehow "things are different there." It is certainly the case that the pursuit of an early retirement strategy as a response to current labour market problems has not been an element of public policy in the USA and indeed has been firmly rejected. Furthermore, many commentators have pointed to the amendment of the Age Discrimination in Employment Act in 1978, lifting the limit at which enterprises can impose retirement on the ground of age from 65 to 70, as a step[3] towards strengthening the employment position of older workers. More recently, and more or less simultaneously with the taking effect of the new lower legal retirement age in France, Congress actually voted an increase of the age at which the full public pension could be drawn (albeit to be operative a considerable way in the future).[4]

As the following table showing participation rates of older males indicates, however, retirement practices in America are generally very similar to those in European countries, with the same tendency for increasingly early withdrawals from working life being apparent.

The mechanisms by which early retirements are facilitated tend to be less generous, in terms of the replacement incomes they provide, than their European counterparts. Thus only at the price of an actuarial reduction in its value is an early pension available at age 62 under the public old age insurance system. Nevertheless, by 1981 some 38% of eligible males in the 62-64 year age group were in receipt of such an early pension. An explanation could be that the American unemployment compensation system is also relatively ungenerous, the duration of payment of benefits being very much shorter than in most European countries. Thus many older unemployed have no other option, once these have expired, but to claim an early pension. The increase in retirement on grounds of disability that

Table 1.   Labour Force Participation Rates of Older Males

	55-59		60-64	
	1970	1981	1970	1981
Austria	88.4	85.7	47.7	29.8
Belgium	82.3	70.5	63.8	34.1
France	82.5	79.8	65.2	42.4
Germany	89.2	81.9	74.7	44.5
Great Britain	95.3	91.1	86.6	72.0
Netherlands	86.9	71.1	73.9	48.4
Sweden	90.8	87.8	79.5	68.2
U.S.A.	89.5	81.3	75.0	58.7

Austria 55-59 = 50-59, Great Britain and Netherlands 1970=1971
Source: National Data (Labour Force Sample Surveys)

has been observed in recent years, such that by 1981 some 15% of eligible 62-64 year old males were in receipt of a disability pension, also needs to be viewed in the context of a deteriorating labour market. Although, according to formal regulations, inability to work is defined in "abstract" rather than "concrete" terms (i.e., without chances of actually being able to find suitable work being taken into account), with a decline in employment opportunities many disabled persons who were previously able to compete on the labour market are now no longer able to do so, and are forced instead to make a claim for a disability pension.

On top of these public provisions, and indeed often arising to compensate for their relatively ungenerous nature, are a myriad of private early retirement programmes, grounded either in collective agreements (e.g., the "30 and out" of the automobile industry) or within the "vesting" provisions of company pension schemes. As in Great Britain, the practice of granting an unreduced company pension when early retirement is at the request of the enterprise, is used to "smooth" labour force reductions and to enable labour force rejuvenations to be effected.

CONSEQUENCES OF EARLY RETIREMENT

To conclude this description of the various mechanisms for
exclusion, I want now to consider a few of the issues raised by the
pursuit of an early retirement strategy.

From the point of view of the enterprise, such a strategy
provides one of the most satisfactory means of effecting labour
force reductions, and one that provides a way around often
relatively strict seniority/age related dismissal protection
provisions. To impose redundancies according to the criteria these
latter set means that the enterprise tends to left with its least
skilled/least adaptable older workers. At the same time, it will
find it more difficult to hire the most desirable labour in the
future if it has a reputation for singling out its younger employees
for dismissal in a situation of downturn. Equally, to rely upon
"natural wastage" will often mean that the enterprise in fact looses
its most skilled/most adaptable younger workers, since it is these
who can most easily find employment elsewhere. For those enterprises
faced by changing markets or technology and seeking to restructure
their labour forces, early retirement (even when facilitated by
those public programmes where the replacement of every retiree is
obligatory) allows younger/more productive/ lower-wage workers to be
substituted for older/ less productive/ higher-wage workers.

Among trade unions, who are likely to be opposed to redundancy
in general, there is likely to be a greater readiness to accept
particular instances of it when those who loose their jobs are
generously compensated. Thus, there is a willingness to see
dismissal protection provisions circumvented. Furthermore the early
retirement option is most acceptable to the core group of the trade
unions' membership--prime age workers--whose employment situation is
unaffected as a consequence of its practice.

Governments too, faced with the politically fraught problem of
large scale youth unemployment, are prepared to tolerate and often
largely to finance systems which redistribute unemployment away from
this particularly viable group and on to others for whom the
socially acceptable "alternative role" of "early retiree" is
available. In addition, the possibility to reclassify such persons
as "out of the labour force" has the additional political attraction
of serving, sometimes in a not insubstantial fashion, to reduce the
level of "open" unemployment.

On the other hand, the very existence of generous compensation
provisions and a socially acceptable "alternative role" in practice
considerably weakens the employment security of older workers.
Whilst early retirement opportunities were first introduced to
provide financial security to a category of persons who had already
lost their jobs, and whose chances of reintegration into work were

acknowledged to be particularly low, the extension of these
opportunities now means that older workers are the first to be
singled out for dismissal. Furthermore, the opportunity for
enterprises to largely externalize the human costs involved might
well mean that they are able to engage in rationalizations that
would not otherwise be contemplated. In the same way, trade unions,
rather than pressing for greater efforts to humanize working
conditions, or to provide retraining or alternative sources of
employment for displaced workers, might well be prepared to take a
more passive stance if the financial situation of the persons
affected by such developments is not thereby unduly prejudiced.

       The pursuit of an early retirement strategy can also be
considered as giving rise to potentially serious social problems.
The concept of "rationing jobs," that is in a time of labour market
slack of giving priority of employment to particular groups, has
largely been rejected by policy makers if it is seen to involve
denying married women the chance to work. However, if such sexual
discrimination on the labour market is deemed unacceptable, age
based discrimination is obviously not so. The notion that older
workers have "had their turn", that they must "stand aside and make
way for younger persons" is obviously a highly attractive one for
all parties exercising a regulatory function on the labour market.
Nevertheless, although for certain persons whose working lives have
been particularly hard or who have genuine and severe health
problems, early retirement does provide a sought after relief, it is
to be questioned if a sudden and sometimes highly premature
withdrawal from economic activity is beneficial to the majority of
those affected. In a society where one's employment provides an
important source of self identity, such persons might well be left
"stranded." As has been pointed out by one observer, the consequence
of a policy of exclusion might well be the creation of a new
"problems group" lacking both social and occupational status. Early
retirees are neither workers, nor unemployed, nor retirees; "they
are too young to be considered as old, but too old to be treated or
accepted as young."

       Finally, a policy of exclusion is also a rather costly one.
Whether or not in terms of its impact upon the level of open
unemployment such a "compensatory" strategy is more expensive than
"integrative" or "reintegrative" policies is rather difficult to
say, since the general absence of such efforts makes any
quantification difficult. Nevertheless, it should be noted that
budgetary consequences of an ever expanding population of early
retirees have obliged a number of governments, (e.g. in Belgium,
France, Germany, and Holland) to at least reconsider their stances
and to take some steps to reduce the financial attractiveness of
early retirement for the retiree, to increase the financial participation of enterprises making age specific redundancies in the costs
involved, or to reduce the access to a disability pension.[5]

AN ALTERNATIVE TO EXCLUSION

     Given that the labour market problems and the need to adapt to
the rapid technological changes that have given rise to the
phenomenon I have been describing in this paper are likely to
persist well into the medium term, it is worth asking whether there
exists some alternative to a policy of simply excluding older
workers. On the basis of a general overview of labour market and
social policy experiences relating to older workers, one path in
particular suggests itself as fruitful to explore--that offered by a
system of "partial retirement". Experiments in this direction have
been made in a number of European countries, and in America too,
although in almost all cases they have been limited to initiatives
taken by individual firms and in quantitative terms have been of
very little significance. Sweden, however, provides an exception in
that, alongside the early retirement programmes already described,
there also has existed since 1976 a national partial pension
system. This provides the opportunity for all 60-65 year olds to
reduce their hours of work and to draw state benefits to compensate
for a substantial part of foregone income. At present some 30% of
the eligible work force is in receipt of such a partial pension.

     Studies of partial retirement systems (particularly of that in
Sweden) show that they can serve a wide variety of functions. From
the social policy point of view, they enable a smoother transition
from employment to inactivity and thus help avoid possible problems
of "pension shock". At the same time, given the reduced
capabilities of many older workers, they provide for a meaningful
term of "work adjustment", and unlike many others one that is not
rendered rapidly unusable by changing technology. Equally, they can
also be used for labour market purposes, facilitating labour force
reductions or, where compensatory recruitment takes place, allowing
the objective of the provision of jobs for the otherwise unemployed
to be fulfilled. In cost terms, too, the institution of a system of
partial retirement is likely to be advantageous. Partial retirees,
as opposed to early retirees, still contribute toward their upkeep,
and indeed to the extent part-time employment can substitute for
"retraite" jobs or other employment opportunities suitable for
workers not able to meet the demands of regular work, it might also
serve to maintain in an active state those who would otherwise be
obliged to retire. There is some evidence that in Sweden the rate
of growth of disability pensioning declined after the introduction
of the partial pension possibility.

     A study of Swedish partial retirees has shown that such
persons, although enjoying a state of health that was inferior to
persons continuing in full-time work at the moment of their partial
withdrawal, subsequently experienced a significant slow down in the
role of deterioration of their physical and mental well being.
Those persons taking full early retirement, however, enjoyed no such

relative improvement. At the same time the social lives of partial
pensioners appeared superior to those of persons who left employment
altogether.

Finally, the experience of Sweden also indicates that the
introduction of partial retirement opportunities does not present
enterprises with any major organizational problems. Most partial
pensioners remain in their original jobs, and only a very small
number of applications by employees for reduced working schedules
have ever been refused.

CONCLUSIONS

The tendency for a progressive lowering of the de facto age of
retirement is common to all of the Western European economies and is
to be found in America too. This phenomenon is, we have argued,
intimately linked to the deterioration of the labour market and the
rapid pressure for rationalization of work experienced in the course
of the last decade. Most labour market actors collude in the
process of excluding older workers from the labour market, often
without being fully aware of the social and financial costs
involved. An alternative strategy, which would meet many of the
same objectives, would be free of many of the disadvantages, and
which would indeed have additional advantages, is conceivable--that
of partial retirement. Close consideration should be given to a
reform of national retirement systems that would permit such gradual
withdrawals to be a realistic, large scale possiblity.

REFERENCES

[1] The author would like to thank Andreas Hoff and Harald
Russig for helpful comments during the preparation of this
paper. Because of its highly condensed nature and limitations
of space, no references are given in the text. A considerable
amount of the material presented here is to be found in B.
Casey and G. Bruche, Work or Retirement? Labour Market and
Social Policy for Older Workers in France, Great Britain, the
Netherlands, Sweden and the USA (Gower Press, Aldershot, UK,
1983), which also has an extensive bibliography. Much of the
more up to date information has been reported in the
Internationale Chronik zur Arbeitsmarktpolitik (published
quarterly by the Policy Information Group of the IIM/Labour
Market Policy). The author is very willing to supply further
references on request.

[2] At one extreme it has been suggested that the major
determinant of regional variations in the incidence of
disability pensioning are variations in regional unemployment

rates. At the other, it has been argued that most of the
growth of disability pensioning in Sweden in the 1970s can be
attributed to a change in the administrative practices of the
pension authorities who, for bugetary reasons, sought to
transfer persons from receipt of sickness benefits (with a
higher income replacement rate) to a disability pension.

3
Whether in fact ADEA has had this outcome is less certain.
In so far as studies have shown that, in order to defend
themselves against age discrimination suits, employers have
developed extensive performance appraisal systems, there are
grounds for suggesting the opposite. Such a shift toward more
performance oriented personnel policies might well increase the
employment insecurity of older/less productive members of the
labour force, since ADEA, whilst prohibiting discrimination
based on age, could be said to legitimize discrimination based
on performance.

4
Interestingly the French move, whilst ostensibly introduced
at least in part for labour market reasons, was also motivated
by an attempt to solve the budgetary problems of the social
security system. As a consequence of the various early
retirement systems operative, the average age of retirement was
already 60 or below. However, whilst the early retirement
systems paid benefits worth 70% of last gross income, the old
age pensions system pays benefits worth a maximum of only
62-64%.

5
Whether such efforts will actually lead to a reduction in
the incidence of early retirement is a matter of some
speculation. Previous experience, both within and across
countries, suggests a relatively high substitutability among
various types of public early pension, between early pensions
and unemployment compensation, and between public and private
(or collectively bargained) provisions. If labour market slack
persists, the consequence may well be only that early
retirement manifests itself in different external forms.

THE AUSTRALIAN LABOUR MARKET FOR OLDER WORKERS

Andre J. Kaspura and Lynne S. Williams

Bureau of Labour Market Research
Canberra, A.C.T. Australia

INTRODUCTION

The publication of the Borrie Report (Population and Australia) in 1975 marked a water-shed in Australian social research. Till then everyone was convinced that ours was a young society and that our problems were demonstrably different from those being experienced throughout the western world. This proved not to be the case. Instead, the difference was simply a matter of degree. The Australian population is ageing, but quite slowly, and compared to other countries, the Australian population is still quite young. Nevertheless, the realization that things were not as everyone thought they were provoked a great deal of interest in both the research and policy communities.

In the context of the labour market, ageing of the population requires us to deliberate upon two inter-related policy issues: what are the implications of an older population for the labour market, and how does the labour market impinge on the increasing numbers in the older cohorts? Below, we will show that the consequences of the former are comparatively minor in Australia, but will assume growing significance, approaching the dimensions of the ageing problem in other western countries around the turn of the century.

Over the past decade or so, the Australian economy has been forced through a period of traumatic adjustment. The initiating impetus came from a 25 per cent across the board tariff reduction in July 1973. The competitive pressures from the subsequent rise in imports caused many Australian industries to look carefully at the technologies they employed. The pressures continued in the form of the oil crisis, and the recession from which the country is still

suffering. To all this must be added the continued relative growth
of service industries and the structural pressures from a
short-lived mining boom which occurred in the late 1970s and
extended into the first eighteen months or so of the 1980s.

Parallel changes occurred in the structure of the labour
market. However, of equal significance were several strong changes
which emerged for other reasons. There were strong rises in married
female labour force participation. On the other hand, the labour
force participation rates for males trended downwards, a pattern
also displayed by unmarried females. The impact of the recession
accelerated these down trends, dampened the married female rises,
and caused the rates in the oldest age groups to fall a little.
What we currently have on our hands in Australia is a complex
picture of long term trends upon which has been superimposed the
shorter term consequences of the recession.

As a result of the recession the employment position of older
workers has deteriorated. While recorded unemployment rates are low
in comparison to other age groups, the average length of an
unemployment spell increased much more for workers ages 45 and over
than for younger workers. The recession has also seen an upsurge in
"early retirement" offers from firms in trouble. In many cases the
distinction between these offers and redundancy has been blurred.
This has led to the view that "disguised unemployment" could exist
at levels well above recorded unemployment rates and raises the
question--to what extent are the falls in labour force participation
the outcome of voluntary decisions by older workers and, to what
extent are they involuntary? Quite different policy prescriptions
apply to the two situations and therefore we need to know clearly
what the problem is.

In this paper we review research undertaken by the Bureau of
Labour Market Research to reflect on these issues. The second
section establishes the effects of changing age structure on the
Australian population. Trends in labour force participation rates,
employment patterns and the unemployment position of older workers
are also reviewed. Whenever appropriate, reference is provided to
the important issues and the literature which reflects on these in
greater detail. The final section briefly summarizes the main
points made in the paper.

THE OLDER POPULATION AND LABOUR FORCE PARTICIPATION

Ageing and the Australian Population

Perceptions concerning the definition of an older worker derive
from social conventions concerning retirement age. In Australia

eligibility for the age pension occurs at age 60 for women and 65 for men and these ages have become common benchmarks. However, the importance of planning for retirement, the preference for earlier retirement, and indeed the harsh reality of the recession, suggest that analysis should commence from a much younger boundary. Inspection of Australian data led us to adopt age 55 for most purposes, but to investigate, in a more limited way, the 45 to 54 age group in order to reflect on some of the particular problems experienced by this group which will feed into the older age groups over the next twenty years. Generally, labour market experience and behavior varies substantially between groups; our analysis segmented the target group by five-year age groups, sex, and for females, marital status and degree of labour force participation.

Between 1961 and 1981 the Australian population increased by around four million people, or 39 per cent. In comparison the population aged 45 and over increased by one and a quarter million or 42 per cent. This picture is somewhat distorted by the presence of the low fertility cohort from the Great Depression in the 45 to 54 age group in 1981. Table 1 illustrates this point and shows that much larger increases occurred in older age groups.

While numerically the changes are quite large (for Australia), the corresponding changes in population shares were small. For example, the share for the 65 and over group increased from 8.5 per cent in 1961 to 10.0 per cent in 1981. Typically, we observe that changes in age structure were of some significance for the groups aged 55 and above, but in contrast, population change in the 45 to 54 age group was a quarter million less than might have been expected had the age structure of the population remained constant.

The significance of this point is that the progression of this cohort into the older age groups over the next twenty years will distort the underlying trend towards an ageing population. In essence, it will restrict the problem for Australia to a numbers problem and buy us some time to plan ahead for the real structural problems which will inevitably follow.

Labour Force Participation

The size of the Australian male labour force increased from 3.16 million in 1961 to 4.17 million in 1981, an increase of 32 per cent, or a little less than population growth. On the other hand, the female labour force increased from 1.05 million to 2.52 million. In other words, it more than doubled. The proportion of the labour force aged 45 and over fell for both sexes. However, in the case of males this took the form of a steady reduction over time, while the female share rose until around 1976 and the reduction has occurred since then.

Tabel 1. The Size and Age structure of the Older Australian
Population, Actual 1961-81 and Projected to 2001

('000, per cent of population in brackets)

Year (June 30)	Age group (years)				Total all ages
	45-54	55-59	60-64	65 & over	
1961	1 225.9 (11.7)	463.4 (4.4)	400.9 (3.8)	894.3 (8.5)	10 508.2 (100.0)
1971	1 444.4 (11.3)	605.4 (4.8)	501.5 (3.9)	1 065.0 (8.4)	12 775.6 (100.0)
1981	1 474.6 (10.0)	723.7 (5.0)	598.2 (4.1)	1 461.6 (10.0)	14 574.5 (100.0)
1991	1 821.1 (10.7)	707.4 (4.2)	718.8 (4.2)	1 938.4 (11.4)	17 008.6 (100.0)
2001	2 527.7 (13.4)	957.9 (5.1)	768.2 (4.1)	2 255.1 (11.9)	18 917.7 (100.0)
Growth (%) 1961-81	20.3	56.2	49.2	63.4	38.7
Projected growth (%) 1981-2001	71.4	32.4	28.4	54.3	29.8

Sources: ABS 1961 Census, Volumne viii, Detailed Tables; 1971
Census Bulletin 3.9; 1981 Census of Population and Housing
Preliminary Counts - Australia, States and Territories
(2209.0); Projections of the Population of Australia 1981
to 2021 (3204.0).

A much better impression of the situation can be gleaned from
an inspection of Figures 1 through 3. Figures 1 and 2 show that
there is a great deal of similarity in the labour force
participation trends for males and unmarried females. The most
important difference is that participation by males was typically
higher for all ages. Apart from this, the level of participation

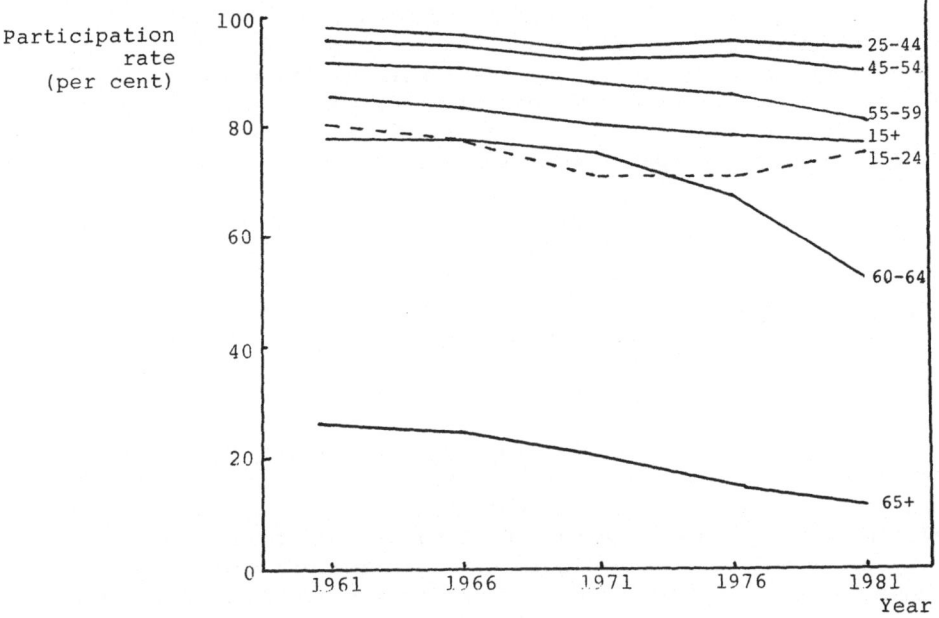

Figure 1. Labour force participation rates by selected age groups,
males, census 1961-1981.

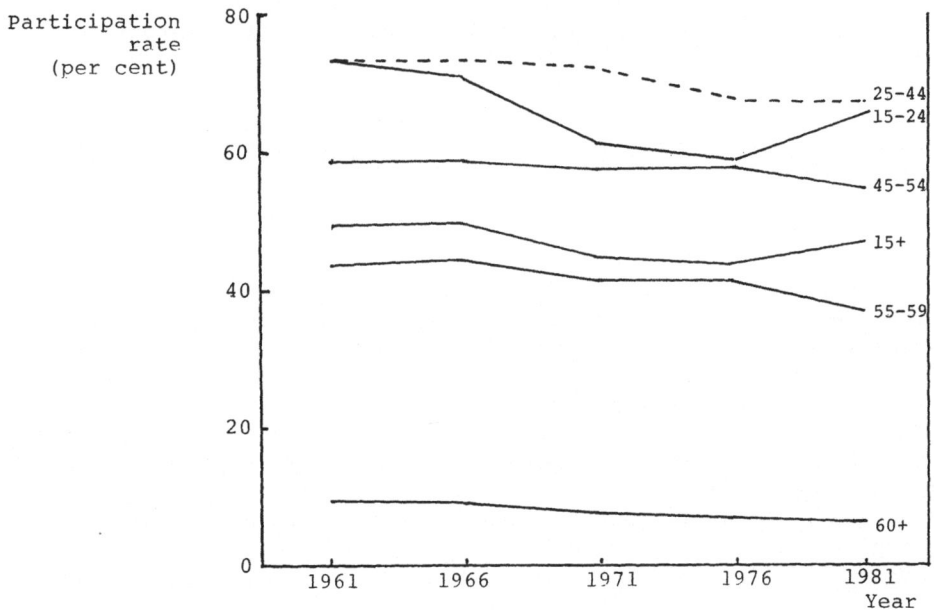

Figure 2. Labour force participation rates by selected age groups,
unmarried females, census 1961-1981.

falls with age over 25, the differential between labour force
participation for successive age groups increases with age, and the
general trend over time is towards lower participation, with a
perceptible acceleration in the trend for most age groups around
1976. This change in trend was particularly acute for the older age
groups, and the sharpest reduction was for men aged 60 to 64.

The results of econometric investigations (Dunlop and Williams
1982, 1983) show that for men, real wages, steadily increasing
superannuation cover, and income from other private assets were
important explanatory factors for each of the three age groups over
55. Economic conditions were important for the groups aged over
60. As might be expected, government pensions proved important for
those aged 65 and over, but of much greater interest was the strong
influence of war veterans on the behavior of the 60 to 64 group.
group. Since eligibility for the war service pension occurs five
years earlier than eligibility for the old age pension, this result
can be taken as indicative of what might happen if the age pension
eligibility age were lowered. Real wages, superannuation, and
private income were also significant explanatory variables for
unmarried women aged 55 and over, while government pensions were
important for the group aged 60 and over.

The behavior of married females was characterised by rising
labour force participation (Figure 3). Married female labour force
participation was typically lower than for unmarried females, with
the level of participation being lower and the differential being
greater as age increased. The impact of the recession was to
moderate participation growth for groups up to age 54, and to cause
small reductions in participation for older age groups. Econometric
analysis for this group has not been nearly as conclusive as for
males (Dunlop and Williams 1982, 1983), but suggests that real wages
are an important influence on full-time labour force participation,
and that spouses' behavior could not be overlooked.

The econometric work referred to above has been used to reflect
upon the distinction between voluntary and involuntary labour force
separation. The results show that most of the falls in labour force
participation rates have been due to voluntary factors. More
precisely, 75 per cent of the fall for men aged 55 to 59 can be
attributed to voluntary factors. The corresponding figures for the
60 to 64, and 65 and over groups were 90 per cent and 85 per cent
respectively. Nevertheless, the estimated size of the discouraged
worker effect was still sufficient to more than double the recorded
unemployment rates reported below. Comparison with other Australian
studies (Stricker and Sheehan 1981, Merrilees 1982, BLMR 1983)
suggests that these estimates are best treated as lower bound
estimates rather than precise estimates of the numbers of older
"discouraged workers."

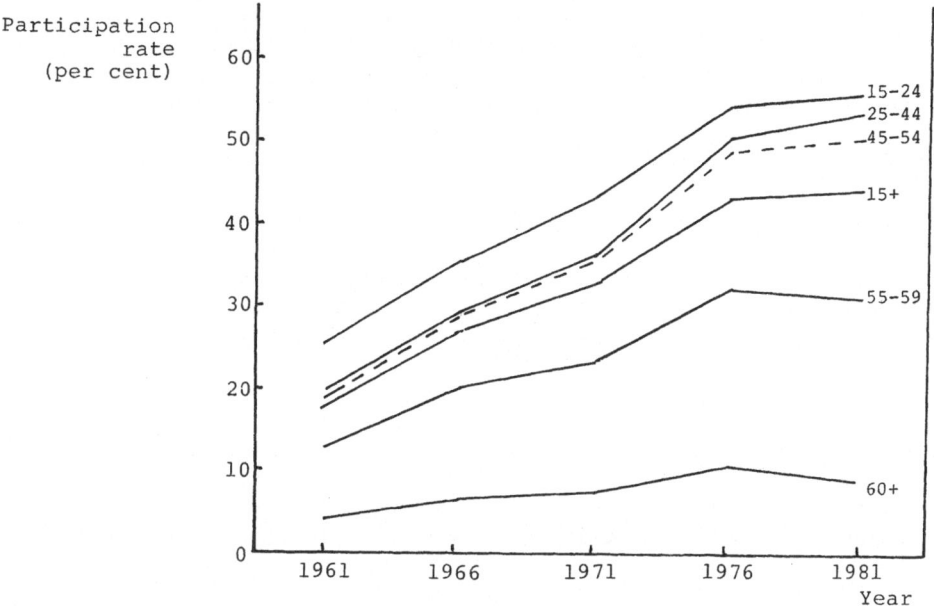

Figure 3.   Labour force participation rates by selected age groups,
            married females, census 1961-1981.

Recorded Unemployment

        Unemployment in Australia has steadily worsened since the mid
1970s and is currently running at around 10 per cent of the labour
force. However, the incidence of recorded unemployment has been
biased towards younger age groups. While unemployment rates for
older workers have increased over time, the share of workers 45 and
over in the unemployment pool has fallen (Table 2). Given the
relative deterioration in the employment position of older workers,
this evidence suggests support for the "discouraged worker" effect
identified above.

        The impact of the recession on older workers depends on how the
industries in which they work fare and on the method used by firms
to identify workers to be laid off. On the first point it seems
that older workers are over-represented in industries which have
experienced sharp workforce reductions over the decade 1971 to
1981.

        In particular, older workers are over-represented in
agriculture, which is continuing its historical trend towards the
use of labour-shedding technology, and in manufacturing industries
such as textiles, clothing, footwear, chemicals and transport
equipment, which have done very badly as a result of re-structuring
forced by import competition (BLMR 1983).

Further evidence of the relatively disadvantaged position of
unemployed older workers is provided by gross flows data. For
1979-80, these data show that only about 11 per cent of persons aged
55 years or older who were unemployed in one month would be likely
to be employed in the next: This compares with 21 per cent for the
workforce as a whole.

Employment Patterns

Since 1970 annual growth in full-time employment in Australia
has averaged only 0.8 per cent. In contrast, part-time employment
has increased on average by 7.5 per cent each year. Against this
background we note that since 1970 the changes in full-time
employment for men, unmarried women and married women aged 45 years
and over have been a fall of 92,900, a fall of 12,200 and a rise of
15,100 respectively, amounting to a net reduction of 90,000 in
full-time employment. The corresponding changes in part-time
employment were an increase of 13,300, a fall of 4,100 and an
increase of 60,500 respectively. Overall older worker employment
fell by 20,300.

Table 2.   Recorded Unemployment Among Older Australian
            Workers, 1971-82

August		1971	1976	1981	1982
Number	45-54	10.9	29.1	31.0	38.8
Unemployed	55+	7.3	14.3	20.7	20.5
('000)	All Ages	92.7	292.7	375.6	458.5
Unemployment	45-54	1.1	2.6	2.9	3.6
Rate	55+	1.0	2.0	3.0	3.1
(%)	All Ages	1.7	4.7	5.7	6.7
Proportion	45-54	11.7	10.0	8.2	8.5
of Total	55+	7.9	4.9	5.8	4.5
Unemployed (%)	All Ages	100.0	100.0	100.0	100.0

Source: Kalisch and Williams (1982a).

The highest older male employment losses occurred in the manufacturing (23,300), construction (23,900) and wholesale and retail trade (20,300) industries. For older females the highest employment losses were in the manufacturing (9,200) and wholesale and retail trade (12,800) industries. These have been the industries subjected to the greatest adjustment pressures, and industries in which, historically, technological development has been slower than that experienced by countries producing Australia's imports.

These changes have resulted in a significant downwards trend in the proportion of older workers in full-time employment. The problem associated with this was explored by Stretton (1983). He found that the taxation system and social security provisions governing income which could be earned before pension entitlement fell effectively increased the marginal tax rate for pensioners or private-income recipients working part-time, to twice the regular level, constituting a serious impediment to people who wanted to retire progressively.

In line with observations elsewhere, there is a high degree of self-employment among older workers in Australia. Among those aged 45 and over, job tenure increases significantly, labour turnover is much lower, and labour mobility decreases. An important adjunct to the latter phenomenon is Stretton's (1983) finding that superannuation coverage, highly favoured by the tax laws in Australia, is not uniform. Coverage in some sectors of the labour force lags well behind the average, (45 per cent of men and 20 per cent of women in 1979) and is typically not portable.

Income Support

Everyone requires some means of supporting themselves. For older workers the alternatives are wages and salaries from employment, income derived from private assets including superannuation policies, and the government transfer system. The main components of the latter of relevance are the aged, invalid, war service and wive's and widow's pensions. Essentially, most pensions are paid at the same rate, but there are important differences in fringe benefits and age eligibility requirements.

Coinciding with reductions in the proportion of older workers employed, particularly in full-time jobs, pension take-up rates have increased substantially. The main features of the shifts here are that full-time employment has always been the main form of income support, but its significance has diminished over time, particularly among men aged between 55 and 64. Take-up rates for invalid pensions increased in all older age groups and there was a dramatic increase in the take-up rate for the war service pension by men aged 60 to 64. Finally, the importance of older couples is reflected in

the higher proportion of women than men neither working nor
receiving a government transfer.

SUMMARY

     The problems associated with the ageing of the Australian
population are not likely to assume the same level of importance as
in other western countries, and countries such as Japan, until the
turn of the century. The main problem for Australia is not so much
one of age structure, but one of coping with the increasing numbers
of older people relative to the current social infra-structure. An
important element is the passage of the low fertility cohort of the
Depression through the older age groups over the next twenty years.
Once this group passes through, Australia will begin to feel the
true effects of ageing.

     The concern in the Bureau of Labour Market Research has been
with the labour market circumstances of older workers. The
background here is declining labour force participation on the part
of older men and older unmarried women, and rising participation on
the part of older married women. The recession has superimposed
itself on this situation to accelerate the falling trends and to
halt the rising ones. The evidence suggests that most of the falls
in participation of older males and unmarried females are due to
voluntary factors as part of a longer term trend. However, the
impact of the recession in causing "hidden unemployment" among older
workers, and the impact of the structural adjustment on job
opportunities, constitute important problems of immediate concern.

     Recorded unemployment rates for older workers are relatively
low and their share of the unemployment pool has fallen over the
course of the recession. Those who remain unemployed have found
that the duration of a spell of unemployment has progressively
lengthened and that it is now over a year. At the same time pension
take-up rates have increased rapidly, particularly among those
eligible for war service pensions, that is, the group five years
younger than the conventional retirement age.

     In Australia, the important influences on older age labour
force participation appear to be real wages, real pension values,
eligibility for the war service pension, the recession (for men) and
spouses' income (for women). Quite important differences were
observed between men and women; men in different age groups; and
women in different age groups. Similarly, whether women were in the
part-time or full-time labour market mattered. These results
highlight the importance of segmenting analysis rather than treating
older workers as a homogeneous group.

REFERENCES

Bureau of Labour Market Research (1983), Retired, Unemployed or at Risk: Changes in the Australian Labour Market for Older Workers, Research Report No.4, Australian Government Publishing Service, Canberra, Australia.

Dunlop, Y. and Williams, L. S. (1982), A Time Series Model for the Projection of Labour Force Participation Rates at Older Ages, Bureau of Labour Market Research Working Paper No. 18, Canberra, Australia.

Dunlop, Y. and Williams, L. S., (1983), Declining Labour Force Participation Rates at Older Ages, Bureau of Labour Market Research Conference Paper No. 24, Canberra, Australia.

Merrilees, W. J., (1982), "The mass exodus of older males from the labour force: an exploratory analysis", Australian Bulletin of Labour 8(2), 81-94.

Population and Australia; A Demographic Analysis and Projection; First Report of the National Population Inquiry, (1975), W.D. Borrie (Chairman), Australian Government Publishing Service.

Stretton, A., (1983), A Review of Policies Affecting the Labour Market Experiences of Older Workers, Bureau of Labour Market Research Working Paper No. 21, Canberra, Australia.

Stricker, P. and Sheehan, P., (1981), Hidden Unemployment: The Australian Experience, Institute of Applied Economic and Social Research, University of Melbourne, Australia.

THE SOCIAL CONSEQUENCES OF RAPID TECHNOLOGICAL CHANGE

Margaret E. Kuhn

Gray Panthers
Philadelphia, Pennsylvania

The symposium on Aging and Technological Advances symbolizes the possibility of continuing international collaboration and cooperation--not in defense and military expansion, but in sharing and applying modern technologies to world wide social and economic issues and the eradication of poverty, oppression, and disease in the human family.

As a next step, I can see countries coming together to pool their technologies and convert their vast military budgets and resources to peace-making--socially useful, life-enhancing ways to stop the arms race. The super technology that produces nuclear weaponry has the terrifying potential for global destruction. Limited nuclear wars and "first strike" strategies are genocidal and unwinnable, our scientists agree.

The word "technology" prompts very different kinds of response in people. Technological change has the potential for great benefits in the health and well being and creativity of human beings. It also has the potential for dislocation and social conflict, as well as ultimate destruction. Technological change linked with demographic change introduces new elements of social change which profoundly influence older populations in the United States and throughout the world, changing their roles, status, widening social class divisions between the rich and the poor, and further separating the old from the young.

There are different definitions of technology just as there are divergent attitudes and opinions about it. This presentation defines technology as "the practical application of organized bodies of knowledge."

But the scope and the speed of current technological change
have created world wide human problems for which we have no "quick
fix" solutions or even perception of the issue. Increased
automation, the robotizing of work, plant closings, and the
exportation of industrial operations to Europe and the developing
world have caused the layoff and displacement of millions of skilled
American workers and may indeed waste a whole generation of workers.
Possibly more than a million unemployed workers may never return to
their old jobs in steel plants, automobile manufacturing and other
shrunken industries.

Efforts to retrain and relocate displaced workers are in my
view too little and too late. The typical displaced worker in
Michigan or Pennsylvania, for example, is middle-aged with a family,
a mortgaged home, neighborhood ties and loyalties. What forms of
retraining and relocation assistance can make a dent in the
problem?

Furthermore, super technology in offices and service
organizations has computerized jobs and displaced tens of thousands
of office workers. The new computers expose the employed to new
hazards. They are just beginning to organize to safeguard their
health and their jobs--against strong opposition from top managers.

As we know, multinational corporate investment and
industrialization are spreading the high technology of the Western
nations to all parts of the developing world. Great sweeps of
change have followed. Peasants have left the land, young people
have left older family members for jobs in the new industries.
Sanitation and living have been improved, but many societies have
been destabilized and are in social conflict as the traditional
social fabric is torn apart.

We do not have adequate and effective structures or methods for
defining what technologies are "appropriate" for the enhancement of
the lives of older people, nor the public will to put the decisions
in place in our U.S. economic system and in the economic and social
systems which are spreading to the developing world.

To summarize, shock waves of technological change are changing
lives and changing society. The turbulence in the wake of these
changes requires: (1) Enlarged public awareness of the scope of
technological change. (2) Thorough assessment of the effects of
technological change in the lives of people over forty by
governmental and non-governmental groups. (3) Analysis of the roots
of technological change in the United States and elsewhere in our
profit centered economic system. (4) Concern for the long-range
social consequences of the technological revolution in the U.S.A.
and outside the U.S.A. (5) Some agreement on how new technologies

can be made "appropriate" and socially responsible for the common
good and human and peaceful ends.

What I have reported is a somewhat negative view of technology.
Out of my frustration and outrage I have been reflecting on what
might be socially responsible, appropriate technology, and propose
the following six criteria/guidelines for consideration. Socially
responsible technology: (1) enhances and safeguards the health,
safety and well-being of people; (2) maintains continuity and
quality of life; (3) provides for freedom of choice, options for
persons and groups affected by change; (4) is concerned for social
and economic justice as well as efficiency and profitability; (5)
protects the environment from exploitation, pollution, ravage, and
waste; (6) maintains balance between the public interest and special
private interests which stand to gain enormous financial advantages
and wealth.

It is urgent to experiment with new structures to achieve the
above criteria. One such structure has been established within the
United States: the Office of Technological Assessment (OTA). This
is a non-partisan, analytical, support agency which serves the
United States Congress by providing objective analysis of major
public policy issues relating to scientific and technological
changes. OTA conducts most of its assessments at the request of one
or more committees of Congress. Thus a continuing dialogue between
OTA and congressional committees is essential for OTA to keep
abreast of congressional needs and the upcoming legislative agenda,
and for the committees to understand OTA's capabilities and
expertise.

I have been privileged to serve on a panel for assessing the
positive and adverse effects of technology on the lives and well
being of older Americans. The OTA model could be improved if it
were studied and evaluated by old people, their employers, and their
families, as well as by scholars and policy analysts.

The linkage between technological change and demographic change
has prompted Gray Panthers' concern about the tragic waste of
experience and historical perspective of old people in the U.S. We
have opposed U.S. policies that segregate and isolate the young from
the old and violate the essential wholeness and continuity of human
life; we have demonstrated in our organization how old and young
people can work together. When our society loses the continuity and
interplay between the generations we lose the ability to develop
broader historical perspectives and essential critique of rapid
social change. In our organizational goals we have defined and
tested new roles and responsibilities for old people. These roles
involve cooperative effort, critical analysis and willingness to
take risks demanded for peace and justice. We suggest, as specific
ways to encourage technology that is appropriate and socially
responsible, the following roles:

Educators of ourselves and our peers for new roles and responses to change. Education is essential to understanding the complexities of technology, enlarging our world view. We are also mentors of the young--through new arrangements of intergenerational living and monitoring to provide continuity and historical perspective, continuity and cohesion in turbulent change. We ave survival skills to share with the young who fear the future. We can give warning signals and advise.

Social historians and critics, as we remember the great depression, the Dust Bowl, two world wars, windmills whirling on farms, and clean lakes and rivers. We perceive the contradiction between labor saving technology and making jobs for the workers displaced by automation. Learning from the past can avoid past mistakes in the future. The Past is Prologue.

Organizers of neighborhood groups to work on safe, renewable forms of energy with appropriate techniques--solar greenhouses and solar heating, intergenerational learning centers, using new computerized data and information systems. Residents' councils in retirement communities could be enlisted in exploring and evaluating technological changes.

Monitors of public bodies, for example, city councils, township commissions, public utility commissions, planning and transportation authorities which should be exploring and utilizing technologies to contain energy costs, providing assistance for the handicapped. We can monitor environmental standards and clean air regulations, guard the public interest, and sound the alarm when public interest is threatened.

Advocates of patients' rights in nursing homes and hospitals, especially pertaining to life extending technologies, and their ethical dimensions; speaking up for consumer protection and safety; going to bat for environmental protection with tighter control of hazardous waste; advocates of change and safety in the workplace.

Ethical counselors in places of corporate power and privilege; working with insiders in bureaucracies to humanize systems that are impervious to change.

Simone de Beauvoir in Coming of Age (1973) reminds us that "The issues of age challenge the whole society and put the whole society to the test."

REFERENCE

de Beauvoir, S. 1973. The Coming of Age. New York: Warner Books.

AGING AND TECHNOLOGICAL ADVANCES:

LABOR FORCE PARTICIPATION

Sally Coberly
Andrus Gerontology Center
University of Southern California

Malcolm Morrison
The Wharton School
University of Pennsylvania

INTRODUCTION

The purpose of this chapter is to summarize the discussions among the participants who attended the sessions on labor force participation. The summary draws most heavily on the discussion which took place around research reports dealing with the topics of demography and labor force participation, retirement age choices, adaptation to new technology, and the use of technology in performance appraisal. The chapter is organized into two parts. The first presents a summary of the major topics and issues; points of disagreement or contention are highlighted. Future research needs are identified in the second portion of the chapter.

SUMMARY OF MAJOR TOPICS AND ISSUES

Most of the discussion centered on identifying and explaining labor force participation trends, particularly the early retirement trend among middle aged and older men, and differing retirement trends for women. Additional topics included the response of older persons to technological change in the workplace, the use of technology in performance appraisal, and the effects of technology on employment opportunities for older workers. Information on population and labor force trends and their consequences for future dependency ratios provided the background for the discussions. Major topics, and the issues which arose during the sessions, are presented below.

Labor Force Participation Trends

Declining labor force participation among men aged 65 and older is a consistent pattern across developed countries. Numerous factors have contributed to the early retirement trend among older men. These include real economic growth (due in part to technological advances) which has allowed individuals to trade-off work for leisure; the availability of public and private pension and disability income; diminishing opportunities for self-employment; and adverse cyclical economic conditions such as recessions. While poor health is often a factor in very early retirement, it is less likely that it has been the most important factor influencing the general early retirement trend.

There may be danger in using recent labor force participation trend data to draw conclusions about the relative contribution of each of the above factors to the early retirement trend. For example, recent data from the United States suggest that the availability of social security and disability income has played a major role in encouraging early labor force withdrawal among men age 65 and older. Historical data, however, reveal that the decline in labor force participation among men in this age group began before the turn of the century, well prior to the introduction of social security. An alternative explanation for the drop in market work of these older men is that technological change has brought a shift from small to large scale establishments, resulting in relatively rigid retirement age policies and diminishing the opportunities for earnings from self-employment. In contrast, historical data suggest that the drop in labor force participation of men aged 55 to 64 is more closely linked to the availability of income supports such as early social security retirement and disability benefits.

Although the labor force participation rates of older women in most developed countries have risen, relatively little attention has been focused on this segment of the labor force. In the United States, the labor force participation of women aged 20 to 64 has risen significantly. Given the high degree of substitution between younger and older women and the presumed preference of some employers for younger women, the demand for older women may depend on the responsiveness of younger women to the demand for female labor in general. That is, if younger women increase their rates of participation in the labor force, employment opportunities for middle-aged and older women may diminish. However, the more recent cohorts of women who are committed to career employment may lead to higher labor force participation during late middle age.

By and large, early retirement has been voluntary rather than involuntary. While very early total withdrawal from the labor force is not necessarily fully welcomed by employees, a relatively early retirement age seems to be quite acceptable to most workers. Thus

far, the long term early retirement trend has been only slightly
affected by cyclical economic trends. Adverse economic conditions
may, of course, continue to force some individuals to retire
involuntarily.

In many developed countries, the older worker has become a
policy paradox. Pressures on income-support systems point to the
need for policies to encourage worklife extension and many older
people express a preference for working past normal retirement age,
usually on a part-time basis. Yet, faced with high unemployment
among youth, many European countries have developed policies to
encourage even earlier retirement. In the light of population
aging, these policies have serious long-range cost implications. In
raising the age of eligibility for full social security retirement
benefits for future retirees, the United States appears to have
taken a different policy approach. It is unclear, however, whether
older persons in the future will, in fact, continue to work longer
to obtain full benefits, or whether employers will "make up the
difference" for individuals who retire early.

Technological Change and the Older Worker

It is extremely difficult to accurately determine the effects
of technological change on the employment of older workers. Part of
the problem stems from the fact that it is difficult both to define
and measure technological change and to obtain reliable information
from industry about the rate and magnitude of change. Technological
changes continue to take place irrespective of cyclical economic
adjustments and distinguishing their specific contribution to
unemployment, productivity shifts and job creation is a complex
research task. For example, it is difficult to determine how much
of the job loss in the automobile industry is due to technological
advances and how much is due to reduced economic performance and
international competitive factors.

Considerable disagreement exists about the future impact of
technological change on older workers. Generally, two views are
advanced. One holds that the labor market has adjusted to
technological advances in the past and will continue to do so in the
future with new jobs and occupations developing to gradually absorb
workers displaced by technology. Since future cohorts will be
better educated, more continuously trained, and perhaps healthier,
older workers may not be as relatively disadvantaged by
technological change in the future as they have been in the past.
Indeed, the aged may benefit from technologies which make jobs less
physically demanding.

An opposing view is that technological changes will result in a
decline in the overall level of employment, underemployment, job
instability, permanent displacement, and the deskilling of many

existing jobs. Displaced workers are not likely to be re-employed
in sectors which provide pre-displacement wage levels, suggesting
that the standard of living of many workers will fall. If
technologies such as computers disproportionately erode middle-level
positions, skill levels may become polarized, possibly leading to
the development of tensions between highly skilled and low skilled
workers. Older women in the labor force, who are concentrated in
clerical and other positions likely to be affected by computer
technologies, may be particularly negatively impacted on by
technological change. These negative consequences may be
exacerbated if employers continue to hold the attitude that,
contrary to the evidence from recent research, older workers fear
change and will have trouble adapting to new technologies.

Aside from the macro-level impacts that changing workplace
technology may have on disemploying or employing older workers,
technology has the potential of being used at a micro-level to
measure individual performance against physical standards or norms,
eliminating the need to make employment decisions on the basis of
age. Such technology based performance appraisal systems could
result in narrowing current legal exceptions to the U. S. Age
Discrimination in Employment Act which, for certain occupations,
permits employers to retire workers before age 70 or refuse to hire
workers above a given age.

FUTURE RESEARCH NEEDS

It is clear that additional research is needed before the
impacts of technological advances on older persons in the labor
force can be fully understood. Basic data are needed about the
extent to which older workers have been displaced or, alternatively,
employed as a result of technological change and about the extent to
which technological change has positively or negatively affected
labor force attachment and mobility of older workers. Personnel
policies concerned with the introduction of new workplace
technologies and the provision of training need to be explored.
Older persons' responses to new technologies also warrant further
examination.

Given the difficulties in defining and measuring technology,
the methodologies for studying these research issues are limited.
Indeed, since the effects of technology are mediated through other
labor market events, it may be that, at best, they can be only
indirectly observed through changes in labor supply and demand,
occupational and industrial composition and union membership
patterns. Retrospective studies of substitution, displacement and
transfer and longitudinal studies of employment experience in
industries affected by technological change may be helpful in
clarifying the effects of technology on older workers.

ECOLOGY, AGING AND HEALTH IN A MEDICAL PERSPECTIVE

Alvar Svanborg

Department of Geriatric and Long-Term Care Medicine
University of Gothenburg, Sweden

INTRODUCTION

The aim of this chapter is to present:

- A brief summary of the biomedical aspects on aging as far as we
  can judge, mainly from our own research findings. We have studied
  a representative group of 70-year-olds longitudinally from age 70
  to, at the present time, age 82 (for reviews see Rinder et al.,
  1975; Svanborg, 1977; Svanborg, Landahl, and Mellstroem, 1982).
  Two other studies at younger ages in the same population
  (Bengtsson et al., 1973; Tibblin, 1967) in which certain
  methodological factors have been identical have also allowed
  certain retrospective longitudinal conclusions.

- The question, to what extent are the elderly still vital and
  healthy enough to be able to contribute in a productive way in our
  society? Advancing technology would imply less demand on heavy
  physical load but simultaneously often increased demand on
  adequate psychomotoric speed and reaction time.

- The possible influence of environmental factors both on aging and
  on state of health. To what extent might environmental factors,
  our life style, our profession, influence performance and health
  when we are old? In this context, what are the possibilities for
  preventive/postponing activities and measures? In a
  technologically advancing world, the rapidly occurring changes
  might of course improve both the environment and the possibilities
  for vital aging, but also add new risk factors for aging and state
  of health that we at the present time can hardly predict.

- The phenomenon in most countries in the industrialized world of an <u>ongoing</u> <u>increasing</u> <u>difference</u> <u>in</u> <u>life</u> <u>expectancy</u> <u>between</u> <u>males</u> <u>and</u> <u>females</u>. To what extent might it be possible, in a technologically advancing world, to improve the vitality at old age also of males also, to prolong their longevity and generally to approach, both as far as aging and state of health are concerned, the situation that females experience at the present time?

In medicine there has been, and still is, a tendency to comprehend aging of old people as something pathological. The first phase of the life span, implying growing and functional improvement, we all consider as the consequence of normal processes. Why should not other phases of the life span with proceeding morphological and functional decline also be natural?

In everyday clinical work we use the term health versus different forms of more or less well-defined, clinically observable disorders. We know, however, that a real clear delineation between health and disease is not possible. The older the patient is, the more difficult such a delineation will become, due to the fact that aging might cause impairment, disability and handicaps similar to those caused by disease. The border zone of overlapping becomes broader at higher ages. The problem of distinguishing between physiological aging and definable disorders has in recent time become even more difficult but also more challenging. The populations in many counties seem just now to live in a period of human history when both the rate of and/or manifestations of aging and state of health are undergoing marked and rapid change. The vitality of the elderly of today in Sweden seems to be different from what it was only five to ten years ago. These age cohort differences obviously imply effects mainly of a changing environment. Indirectly we can, thus, begin to define possible active preventive/postponing measures.

The populations in the Nordic countries are at the present time the oldest in the world, although male Japanese already now live longest and Japanese females obviously will do so in 1983. In Sweden, the percentage of those age 65 and above is at the present time approaching 17% of a population of about 8.3 million. The predictions for the future 40 years indicate certain variations from time to time but generally a further increase of the 65+ both absolutely and relatively up to at least 20% in the year 2025. As in many other populations the increase of the oldest of the elderly will be the most pronounced. During the period from 1970 to the year 2000 the absolute number of the 85+ population will have doubled and that of the 95+ population tripled, implying a dramatic increase in the need of social support and medical service to the elderly in Sweden. The trends are similar in the United States but with somewhat lower percentage figures.

It should, however, be emphasized that such a composition of our population must also mean that the future greater than one-fifth of the population will have great impact not only on care programs but also on society planning in general. The advances in technology will obviously, from many points of view, markedly improve the possibilities to adjust in a better way, not only to the need of support of the elderly, but also to their demand for meaningful, productive and responsible contribution for as long as possible in their lives.

My presentations will to a great extent be based upon results obtained in our gerontological and geriatric longitudinal study of the 70-year-olds in Gothenburg, Sweden. Gothenburg, situated on the west coast of Sweden, is the second largest city in the country, with 450,000 inhabitants in the downtown area and approximately 1 million people in the metropolitan area. It is rather heavily industrialized, has the biggest harbour in the country and also has a well developed school and university life.

The objectives, design, representativeness and results of the longitudinal study of 70-year-olds, which started in 1971/72, have been described previously (Rinder et al., 1975; Svanborg, 1977). Follow up studies have been performed in 1976/77 at the age of 75, in 1980/81 at the age 79 along with another group of 70-year-olds, and later on followed up in 1981/82 at the age of 75. Because of the rather marked cohort differences observed at age 70 and 75, a third age cohort of 70-year-olds is at the present time under investigation.

ASPECTS OF PHYSIOLOGICAL AGING

It seems reasonable to state that the general ideas about aging have been, and still are, that life implies only two phases, either growing and functional improvement, or somatic atrophy and functional decline. There are in fact certain functional parameters that show almost such a two-phased curve between functional ability and chronological age. A typical example is the perceptual speed (psychomotoric speed) that starts to go down as early as in the age interval 20-30. Due to reasons that also are not wholly understood the basal oxygen consumption also starts to decline in an almost linear pattern early in life.

But for many other functions the relationship between functional performance and chronological age shows curves that can generally be considered to imply at least four phases (Figure 1). Following the first phase of growing and functional improvement, there is often a shorter or longer period of life when the functional ability seems to be rather constant or declining only slightly. This second phase is then followed by a third phase with

a functional decline, commonly at the rate of 1% per year (Svanborg,
Landahl, and Mellstroem, 1982). Finally, our present experiences
indicate that many of these longitudinally followed people will
experience something I would like to describe as a terminal phase, a
phase when the rate of appearance of, as well as aggravation of
manifestations of aging increase and the vitality rapidly goes
down.

Most organs, presumably all of them, have a marked reserve
capacity implying that at younger ages only about 20% of the total
organ function capacity is used ordinarily. It seems obvious that
this reserve capacity goes down with increasing age. The rate of
decline of this reserve capacity of different organ functions is,
however, impossible or extremely difficult to analyze, especially in
men. It is difficult to state to what extent this final and
terminal aging phase, (predicted by our present experiences from the
longitudinal study of 70-year-olds as well as from clinical
experiences) is an indication that we then have reached a phase of
life when we no longer have any reserve capacity available.

The length of the second phase with a rather stationary
functional period is different for different functions. The
composition of muscle fibres in the striated muscles and the enzyme
composition have been reported to be rather unchanged, indicating an
unchanged qualitative functional ability, up to age 60. Studies of
the immune response in our own samples of 70-year-olds indicate no
measurable decline in immune response until after age 70. Studies of
body composition show only very small changes in the Gothenburg
population up to about age 65 in males and 70 in females (Steen et
al., 1977; 1979).

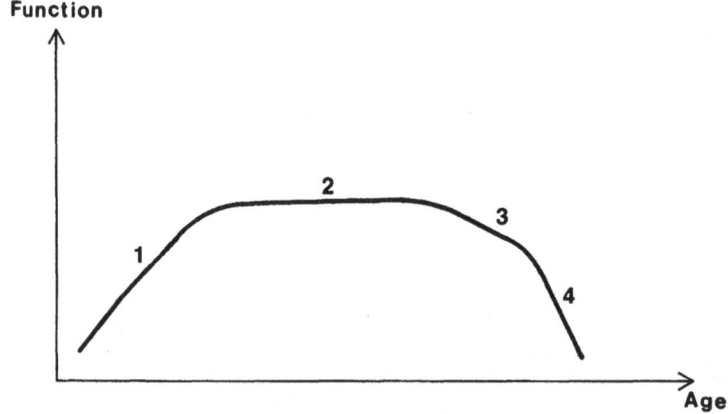

Figure 1.   Four phases in the relationship between functional
            performance and chronological age.

Many of these differences in levels and curve forms for the different organ functions have important clinical consequences. It should, however, be emphasized that many of these facts, based upon recently obtained research data, have significant consequences also in the perspective of the productivity of the elderly. One example is that many studies definitely show that, with the exception of psychomotoric speed, other measurable cognitive abilities are well preserved in healthy elderly up to at least age 70-75 (Schaie, Labouvie-Vief, 1974; Botwinick, 1977; Berg, 1980). Furthermore, in the 70 year-old study, Berg (1980) found a positive cohort difference implying that the 70-year-olds of 1976/77 were even more intellectually able than those who became 70 years old in 1971/72, i.e., five years earlier. To a certain extent it is a conflict in our societies when we find that the intellectual function, for example, of the elderly improves but the economical situation, as well as technical advances, minimizes the availability of meaningful professional activities in many modern societies, with ongoing increase in life expectancy also at those ages commonly used for mandatory retirement.

To what extent, furthermore, are the elderly unable to contribute in a meaningful way in our societies because of bad health? And to what extent do they have disabilities or handicaps that need advancement in aid resources? Is it really true that the elderly, if we examine them carefully enough, will be found to be suffering from definable disease? That they need to avoid an active life because of bad health? There is no doubt that a considerable proportion of elderly are suffering from diagnosed disorders and another proportion are suffering from a group of undiagnosed disorders, i.e., underdiagnosis. But according to our experiences the real problem is , on the contrary, the tendency to overdiagnosis. This is caused by the fact that the limited knowledge of how aging is manifesting itself has resulted in an inadequate interpretation of manifestations of aging as being symptoms of disease. We are overdiagnosing to a considerable extent! In the study of 70-year-olds in Gothenburg, Sweden, we can definitely say that at least 30-40% of the elderly are totally lacking symptoms of any definable disorder (Svanborg et al., 1982). Among those with definable disorders many have diseases that cause only slight impairment and no real disability or handicap.

Within medicine there are already today essentially improved possibilities to measure functional performance instead of the common classifications according to chronological age. Our studies show for example that neither the graying of the hair, nor the wrinkling of the skin around the eyes nor the psychomotoric speed are reliable indicators on the degree of age-related decline in, e.g., the function of the kidneys, the density of the skeleton or decline of the blood flow in the brain. But there seems to be reason to anticipate that within a few years methods for an

individualized functional analysis will be available, necessary for
clinical purposes and motivated also by economic reasons.

Although we are living in a world that is technologically
rapidly improving, these technical advancements have only to a
limited extent up to now been used for the support of age-related
handicaps and health problems. Examples of these are many. In
Gothenburg, among those elderly at age 70, a majority of the 48% who
declared that their only hobby was reading lacked adequate light
above the chair and the bed. Generally, only 10-15% of the
70-year-olds had adequate illumination above their reading-place.
An advanced visual handicap is uncommon at age 70. At maximal
correction with glasses a reduction of visual capacity to 50-70% of
"normal" capacity was found in only 9%; a reduction to 20-50% of
"normal" capacity in 2%; and down to 5-10% of "normal" capacity only
in 1%. To those who have misleading information concerning the
visual acuity of, e.g., 70-year-olds I would like to mention that
97% were able to read the standardized text on medicine bottles and
other packages of drugs in Sweden when they had adequate glasses and
light.

From many other points of view we found that the elderly
generally have a much better capacity, both intellectual and
physical, than generally thought. At age 70, 95% are lacking
advanced handicap, at age 75, 90% and at age 79, 80%.

Accidents are common in the elderly. When we analyzed where,
when and, if possible, why hip fractures occurred we found that
contrary to what is usually considered, only 5% occurred in the
bathroom (Zetterberg et al., 1983). The majority took place in the
living room, bedroom and kitchen. At the same time as architects
must try to adjust homes in order to prevent/postpone accidents they
should also try to find out to what extent the homes of the elderly
can be adapted to a better level of activity. And housing in a
wider perspective implies also staircases, lifts, traffic
situations, etc. Why is it that many elderly stop driving cars at
ages when they really need this form of transport? We know that the
elderly are no real risk group, actively causing traffic accidents.
In Sweden, traffic accidents with lethal outcome are about four
times more common in 65+ than in the other high risk age-group,
namely age 5-14. Among pedestrians killed in Sweden, 48% are 65
years old or older. In this context I would like to mention also
that we have measured comfortable walking speed (Rundgren, 1982).
At age 70 comfortable walking speed was found to be 1.12±0.29 metres
per second for males and and 1.04±0.34 for females and at age 79 for
males 0.98±0.25 and for females 0.84±0.33 metres per second. This
should be compared to the time allowed for street intersections that
usually is adapted to a walking speed of 1.4 metres per second. At
age 79 at fast walking speed, the average walking speed for males
was found to be 1.21±0.74 and for females 1.05±0.49 metres per

second. I would like to once again emphasize that our studies have been performed on samples shown to be representative of the total population at these ages. These are only a few examples of the fact that at the present time our general societal planning has only to a limited extent been adapted to the demands and needs of the elderly.

Recent research has, to a great extent, altered the rather common idea that at old age we need rest more than activity. The idea has been that when our cells and organs have only a limited storage of energy producing capacity left at old age, this implies a low rate of usage of these resources, and that such a low rate of use is one way to prolong life. There is much research data indicating, however, that within rather wide limits cells, organs and the whole individual need the stimulation of being used. The practical problem is to find a border zone between use and overload in age intervals when the rate of aging is relatively fast. On the other hand, so many facts show that inactivity generally is more negative and dangerous than activity. Physical training will also at age 70 and 75 stimulate both slow and fast reacting muscle fibres and improve muscle function both quantitatively and qualitatively (Aniansson et al., 1980; Aniansson & Gustafsson, 1981). A better use of the productive ability of the elderly might contribute to society, not only by the use of the resting resource in the society, but also as a method to prevent and postpone disability. In this context, I would like to emphasize that there are greater interindividual variations, for example at the age interval 65-75, than at younger ages. I would like to warn against segregation of the elderly as a group. It is necessary to accept certain forms of statistical grouping but more individually related evaluation of ability and productivity is definitely necessary.

To what extent advances in technology will improve the social situation of the elderly in our present society is difficult to say. Recent studies in our samples, and also in the total population in Sweden, show marked differences in longevity between those who at age 50-90 are still living together with a spouse compared to those who lost their spouses at these ages. Some of these differences can be due to life style factors that the couple have in common. Those who lose their spouses in this age interval live on an average three years less if they are widowers and one year less if they are widows, compared to those still living together with their spouses. The marked difference in morbidity and death rate occurs during the first 3-6 months of bereavement. The mortality among widowers during the first three months of loneliness is 48% higher among widows, 22% higher than among those still living together with their spouses. These are the most dramatic examples of how sociological environmental factors influence aging and health in the elderly of our society. Apparently, the sudden decline in intellectual, physical and emotional activity is one of the

situations with high risk of morbidity, especially in age groups
where the reserve capacity is low. We should be aware of the fact
that it is not only the death rate that increases after bereavement
but also morbidity and demand for both medical and social support.
The extent to which advances in technology in our societies will
improve or worsen the situation of the elderly in this perspective
is difficult to foresee.

It is difficult to say to what extent advances in productivity
will change the profession-related risks of aging and morbidity. At
the present time, there are great differences in longevity among
different professions in many countries. In Sweden, those who live
longest are the farmers and some skilled workers like carpenters and
masons. Those who have the shortest life are sailors, restaurant
workers and journalists. I recently studied profession-related
longevity in Japan and found, for example, that those Japanese who
have the lowest life expectancy are unskilled workers and farmers,
and those who live longest are apparently policemen. To what extent
this profession-related difference in longevity is due to the
profession itself or to life style factors and personality is
difficult to say. Our general conclusions are that life style
factors, especially smoking and alcohol habits, are of great
importance.

Smoking has been found not only to have the earlier wellknown
negative influence on state of health but also to negatively
influence certain age-related changes in organ function and
morphology in a way similar to the phenomenon we used to identify as
aging. Smokers at age 70, 75 and 79 have, e.g., a lower muscle
strength and also markedly lower density of the skeleton. All
through our lives males used to have a higher density of the
skeleton than females. Our studies show, however, that smoking
males have lost more bone than non-smokers. Their bone density is
on an average 30% lower than that of non-smokers and about the same
as that of non-smoking females. In Gothenburg, the incidence of hip
fractures (Zetterberg and Andersson, 1982) increases much more than
can be explained by the aging of the population. And the increase
in males is even more pronounced than in females. The extent to
which this increase in the incidence of hip fractures can be caused
also by less physical loading, alcohol abuse, lower calcium intake
and/or exposure to sunshine, etc. is difficult to state.

When we talk about advances in technology we must be aware of
the fact that further research is needed also on the problem of how
much physical loading the human needs, and especially the skeleton,
in order to be adequately stimulated to go on producing bone.
Previous ideas that physical loading at older ages no longer
stimulates bone production - or at least no longer slows down the
rate of bone loss - are definitely shown to be wrong. At least up
to ages 70-80 reasonable loading still seems to stimulate bone
stability.

One interesting observation is that smokers (Mellstrom et al., 1982) seem to have a faster rate of aging of their gonadal function than non-smokers. In Gothenburg, females get their menopause on an average 2 years earlier if they are smokers. Our studies of certain metabolic parameters indicating the balance between male and female gonadal hormone function also showed that smokers seem to have aged faster than non-smokers.

In a technologically advancing world it must be of utmost interest to realize that environmental factors and our life styles thus play a much greater role in aging and health than we usually have been aware of. It is still obviously necessary to choose the right parents in order to have a basic chance to live a long and vital life. But the influence of the ecology is definitely an important co-factor. This is indeed something that has to be taken into consideration, especially in a world that is undergoing such a rapid change as ours, to a great extent due to the rapidly occurring changes in technology.

REFERENCES

Aniansson, A., Grimby, G., Rundgren, A., Svanborg, A. and Orlander, J., 1980, Physical training in old men, Age and Ageing, 9:186.

Aniansson, A. and Gustafsson, E., 1981, Physical training in elderly men with special reference to quadriceps muscle strength and morphology, Clin. Physiol., 1:87.

Bengtsson, C., Blohme, G., Hallberg, L., Hallstrom, T., Isaksson, B., Korsan-Bengtsen, K., Rybo, G., Tibblin, E., Tibblin, G. and Westerberg, H., 1973, The study of women in Gothenburg 1968-1969 - A population study, Acta Med. Scand., 193:311.

Berg, S., 1980, Psychological functioning in 70- and 75-year old people: A study in an industrialized city, Acta Psychiat. Scand., 62, Suppl. 228.

Botwinick, J., 1977, in,: Handbook of the Psychology of Aging, J.E. Birren and K.W. Schaie, Eds., Van Nostrand Reinhold, New York.

Mellstrom, D., Rundgren A., Jagenburg, R., Steen, B., and Svanborg, A. 1982, Tobacco smoking, ageing and health among the elderly: A longitudinal study of 70-year-old men and an age cohort comparison. Age and Ageing, 11: 45-48.

Rinder, L., Roupe, S., Steen, B. and Svanborg, A., 1975, 70-year-old people in Gothenburg. A population study in an industrialized Swedish city. I. General presentation of the study, Acta Med. Scand., 198:397.

Rundgren, A., 1982, Gron gubbe - om aldres ganghastighet, in:
    "Aldreoskyddade i trafiken," eds. Trafiksakerhetsverket.
Schaie, K.W. and Labouvie-Vief, G.V., 1974, Developmental
    Psychology, 10:305.
Steen, B., Bruce, A., Isaksson, B., Lewin, T. and Svanborg, A.,
    1977, Body composition in 70-year-old males and females in
    Gothenburg, Sweden. A population study, Acta Med. Scand.,
    Suppl. 611:87.
Steen, B., Isaksson, B. and Svanborg, A., 1979, Body composition at
    70 and 75 years of age. A longitudinal population study, J.
    Clin. Exper. Gerontol., I:185.
Svanborg, A., 1977, Seventy-year-old people in Gothenburg. A
    population study in an industrialized Swedish city. II.
    General presentation of social and medical conditions, Acta
    Med. Scand., Suppl. 611:5.
Svanborg, A., Bergstrom, G. and Mellstrom, D., 1982,
    "Epidemiological studies on social and medical conditions of
    the elderly. Report on a survey," Regional office for
    Europe, World Health Organization, Euro reports and studies
    62, Copenhagen.
Svanborg, A., Landahl, S. and Mellstroem, D., 1982, Basic issues of
    health care, in: New Perspectives on Old Age. A Message to
    Decision Makers. On Behalf of the International Association
    of Gerontology, H. Thomae, and G.L. Maddox, eds., Springer
    publishing Company, New York.
Tibblin G., 1967, High blood pressure in men aged 50 - A population
    study of men born in 1913. Acta Med. Scand., Suppl. 470.
Zetterberg, C., and Andersson, G., 1982, Fractures of the proximal
    end of the femur in Gothenburg, Sweden, 1940-1979. Acta
    Orth Scand. 53:419-426.
Zetterberg, C., Mellstroem, D., Rundgren, A., Andersson, G.,
    Frennered, K. and Hansson, T., 1983, Hoftfrakturer hos aldre
    - Bakgrundsfaktorer, Lakartidningen, 80:2043.

THE COMPRESSION OF MORBIDITY

James F. Fries

School of Medicine
Stanford University

In this century, we are progressing through three separate eras with dramatically different characteristics of health and illness. We entered the century in an era of infectious disease, with tuberculosis the number one killer of our population, and smallpox, diphtheria, tetanus, and other infectious illnesses extremely prevalent. A reduction in mortality from these diseases of over 99% (Fries and Crapo, 1981; Cooper, 1982), has led to the present era, where the major illness burdens of the United States are the chronic diseases. Atherosclerosis and its complications, neoplasia, emphysema, diabetes, cirrhosis, and osteoarthritis have increased in prevalence even as the infectious illnesses which preceded them declined. It is one thesis of this discussion that this chronic disease era in its turn will slowly decline in significance, leaving a third era in which major health problems of the United States will be directly related to the process of senescence, and where the aging process itself, independent of specific disease, will constitute a major illness burden for the United States.

The ultimate constraint within which we must develop health policy for the future is, of course, the limit for life itself. Man is mortal, and the limits to what presently may be accomplished in decreasing mortality are set by the lifespan of our species. In the following discussion we will examine natural limits to the lifespan, develop an incremental model of chronic disease and of aging which focuses upon the postponement both of disease and of senescent change, and examine the implications of improvement of vitality in a finite world.

It is important to define some terms. The "maximum life potential" is approximated by the older age achieved by any human

169

being. In the United States, this age is 113 years, 214 days
(McWhirter, 1980). This figure represents a point far out on the
"tail" of a distribution of genetically different individuals.
"Life expectancy" is the average length of life which we may expect,
given current age specific death rates, for an infant born today.
This figure is 73 at present, approximately 70 for men and 77 for
women. The "life span," on the other hand, represents the average
longevity in a society without disease or accident. The life
expectancy can rise toward, but cannot exceed, the life span. The
human life span appears to be approximately 85 years, with a broad
distribution in which natural longevity for individuals falls nearly
entirely within the range of from 70 to 100 years. Much
misunderstanding in discussion of this topic comes from confusion of
the mean of a distribution with its extremes.

National mortality figures demonstrate a smooth decline in
number of deaths and a smooth increase in mortality rates as we move
toward higher ages (Fries and Crapo, 1981). The 41,000 deaths per
year in the United States in the eighty-fifth year of age has
decreased to 24 by age 110, in a smooth progression which shows no
exceptions. The absence of exceptions carries important
implications, since it demonstrates that particular lifestyles or
particular food or vitamin intakes which have been promoted as aids
to longevity and which have been used extensively throughout the
culture do not, in fact, prolong the genetically determined life
span. If Vitamin C or lifelong aerobic exercise extended the life
span, we should expect to have seen exceptionally long lives (beyond
115) in at least a few practitioners of such habits. The ultimate
limits appear to apply to the aerobically fit and to the megavitamin
faddist, to the farm or city dweller, and to all societies.

Examination of mortality rates in different societies confirms
the actuarial "law" first proposed by Benjamin Gompertz (Gompertz,
1825). Gompertz noted a linear increase in mortality rates with age
when rates were plotted on a logarithmic scale. That is, the
mortality rate increased exponentially with age, doubling
approximately every eight years of age. Gompertz' law is an
empirical observation which fits closely with the observation of
smoothly and rapidly declining numbers of individuals alive at
successive ages.

Different species, on the other hand, have very different life
spans; the obvious fact that we tend to outlive our pets is not due
to increased disease or accident among the animals, but rather to
the difference in species life span (Rockstein, 1958). Rodents live
at best a few years, while some of the Galapagos tortoises which
formed a part of Charles Darwin's observations about the origin of
species were still alive on the 100th anniversary of Darwin's
death.

While the process of senescence certainly has biochemical and cellular underpinnings, it is presently best understood by decline of maximal function of vital organs (Finch, 1976; Shock, 1960; Strehler, 1960). The organ reserve potential, greatest in early life, shows a functional decline which is essentially linear and which is roughly parallel for all major organs. The decline is in the "reserve power" of the particular organ, and thus is apparent on measurement of maximum performance well before it is clinically visible as a limitation to activity of the organism. Studies of this physiological deline, about 1-1/2% per year, uniformly indicate that the decline begins early in life in healthy individuals--well before it is reasonable to postulate any specific chronic disease effects (Figure 1).

Decline in the function of multiple organs may be considered in the context of preservation of homeostasis. Reserve function is required when the organism is stressed in order to restore the normal homeostatic equilibrium. As the reserve of individual organs declines in a linear fashion, the ability to maintain homeostasis in the face of a threat of a given magnitude declines exponentially (Strehler, 1960), hence the observations of physiology and the actuarial observations of Gompertz are reconciled (Figure 2). Natural death must ensue, without disease, when the reserve function has declined below that point, probably about 20% above basal levels, at which routine daily perturbations cannot be weathered. A transition from premature death to natural death occurs as the characteristics of the host resistance (homeostatic reserve) become

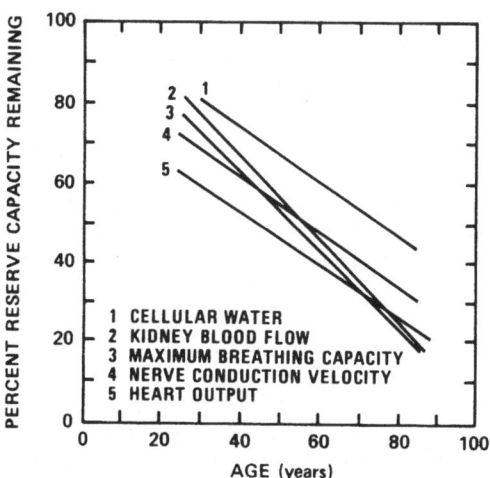

Figure 1. Linear decline in organ function with age. Based on Shock, 1960 and Strehler, 1960. Reprinted with permission from Vitality and Aging, W. H. Freeman and Company, San Francisco, 1981.

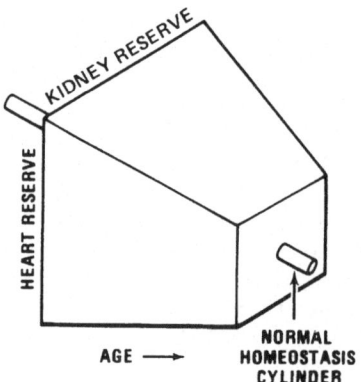

Figure 2. Linear decline in multiple organs exponentially decreases
           the area within which perturbations may be restored as
           shown here for two organs. Reprinted with permission
           from Vitality and Aging, W. H. Freeman and Company,
           San Francisco, 1981.

more important than the specific nature of the insult to the
equilibrium. The concepts of premature death (due to disease or
accident) and natural death (due to senescent frailty) are
complementary rather than antagonistic, and any dividing line must
be an arbitrary one.

     Life expectancy from birth in this century in the United States
has increased from 47 to 73 years (Fries, 1980); an increase of 26
years (Figure 3). The rise has been reasonably constant throughout
the century, with some periods of plateau and some periods of
acceleration. This striking advance is not as apparent when one
considers life expectancy from age 20, which has increased only some
13 years; life expectancy from age 40, which has increased eight
years; life expectancy from age 60, which has increased 5 years;
from age 80, 2.5 years; or from age 100, 0.7 years (Faber, 1982).
The greater slope of the curve representing life expectancy from
birth reflects the great improvement in infant mortality over this
period. In contrast, improvement in chronic disease control will
result in a more nearly parallel slope to all lines, since these
benefits accrue to individuals later in life. We are beginning to
see, in terms of rate of change, some of these effects. To avoid
misinterpretation of these "rate of change" data as indicative of
galloping longevity, it is essential to look at absolute changes in
life expectancy at the same time. Absolute changes in life
expectancy, as above, show a progressive decline at higher ages.

     Mortality statistics graphed as the percent surviving versus
age are perhaps the most dramatic and decisive way to view the
mortality rate events of this century (Figure 4). In 1900,

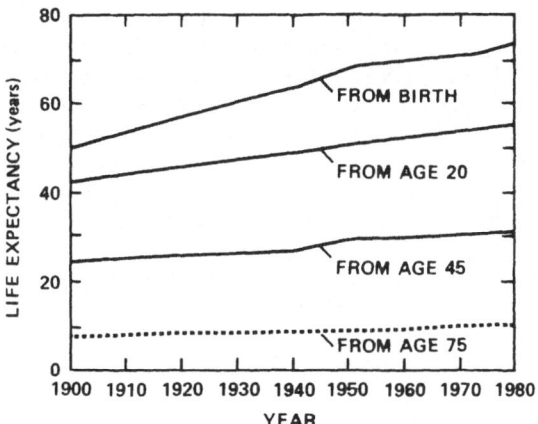

Figure 3. Changes in life expectancy from different ages in the
          twentieth century. Reprinted with permission from
          Vitality and Aging, W. H. Freeman and Company, San
          Francisco, 1981.

mortality occurred at a relatively steady rate throughout the
lifespan. In successive decades, the curves have begun to bend
upwards and to the right, each considerably different from the
last. The form of the curve is increasingly rectangular, having an
increasingly flat top and an increasingly sharp downslope. This
observation is frequently referred to as "rectangularization of the
survival curve." The point on the age axis at which the curves
intersect has remained approximately the same, the differences lying

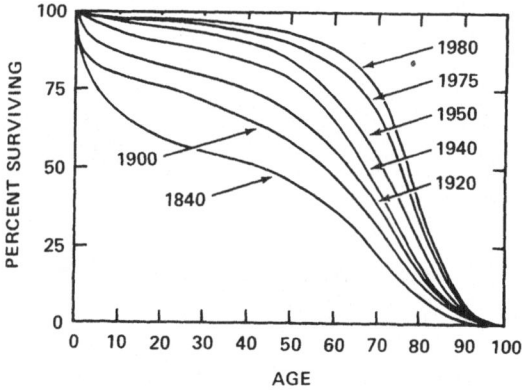

Figure 4. Changes in survivorship curves in the United States in
          the twentieth century. Reprinted with permission from
          Vitality and Aging, W. H. Freeman and Company, San
          Francisco, 1981.

within the width of the lines used to plot such curves. The
progressive shape of these curves allows visual prediction of future
trends. The ideal curve must lie in its initial 60 years within a
very narrow flat zone, and then must plunge quite precipitously if
it is to meet the historical lines at the bottom of the graph
(Figure 5). The increasingly rectangular survival curve, with the
clear convergence of curves from different decades, demonstrates
visually the limits of the human life span.

Thus, nine general lines of evidence confirm the existence of a
finite human life span. (1) There are no exceptions to the
declining numbers of individuals present at successive ages. (2)
Gompertz' law appears to hold in all populations and assures an
exponentially increasing mortality rate, and therefore death for the
entire population within a decade or two past the age of 100. (3)
There has been no historical change over several centuries of
observation with regard to maximum life potential, as underscored by
studies of centenarians. This observation has been repeatedly made
in the United States with good data since 1939 or earlier (Bowerman,
1939; Faber, 1982). Life expectancy at age 100 has changed at most
0.7 years over 80 years (Faber, 1982), and much of this improvement
must have been due to reduction of premature death, not change in
life span.

Moreover, (4) there is no biological reason to assume that any
change in genetic longevity characteristics should have occurred
merely because we have improved infant mortality, cleaned up water
supplies, or discovered penicillin. (5) The difference in species

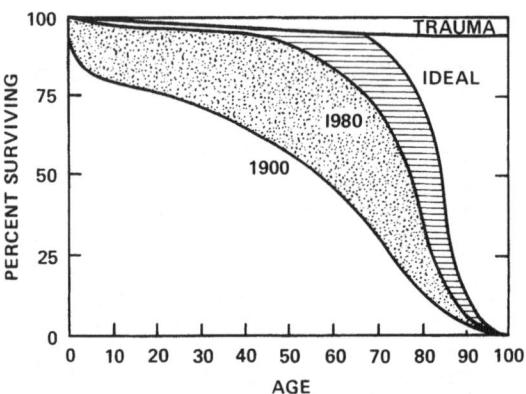

Figure 5. Ideal Survivorship Curve. Trauma plays a large and
          potentially reversible role. Chronic disease accounts
          for almost all of the approximately ten year-wide area of
          premature death remaining over ages 60-90. Reprinted
          with permission from Vitality and Aging, W. H. Freeman
          and Company, San Francisco, 1981.

life span among animals is a commonplace daily observation. (6) Anthropological analyses (Cutler, 1979) suggest a formula by which mammalian life spans may be predicted by the brain size/body weight ratio; such models suggest an approximately constant life span for the human species for the past 100,000 years. (7) The linear decline in organ reserve, repeatedly the subject of physiologic observation, mandates a point at which function must be inadequate to support life; that point apparently being when organ reserve is reduced to approximately 20% over that function required for the maintenance of basic life processes; reserve of this magnitude is required for daily functions outside of bed. (8) The increasingly rectangular curve demonstrates the barrier to immortality.

Finally, (9) we have the important phenomenon of a priori aging, the daily evidence of our senses. People do grow older, with changes which are apparent to all of us, as we age. And these changes--from hair color to hearing--are not the result of disease as we usually define it. A new group of Americans has become a subject of increasing concern: the "frail elderly." This term is a new one; it refers to individuals, often without demonstrable disease, who have manifestly limited organ reserve and increased frailty to external perturbation.

There are several methods of estimating the human life span. One may use the anthropological formulas, reconstruct an ideal survival curve from the tail of the present curve using the assumption that these individuals have been essentially free of disease, make extrapolations from the rectangularizing survival curve, or use estimates based on observed decline in organ reserve. All suggest an average life span of approximately 85 years, with a distribution which includes 99% of individuals between the ages of 70 and 100 (Figure 5). It is not clear whether this distribution is "normal," based on the Gompertz function (which gives a slightly sharper drop-off) or some other distribution. For policy purposes, these distinctions are minor.

There is some controversy about whether the life span is totally fixed, and about the precise projection of life span, but these disagreements fall within a narrow range. Advocates of a slowly increasing life span tend to cite continued gains by white females (Manton, 1982), and note increased percentage gains past age 50 in recent years; these observations are fully consistent with those discussed here. Present gains, accruing by postponement of chronic disease, are reflected in life expectancy increases at all ages; hence, the historically more slowly rising life expectancy from advanced ages will show a larger percentage, although smaller absolute, gain. The model presented here predicts that the male-female gap may eventually decrease, but does not anticipate dramatic change to occur in the next few years; such change requires that premature death in males decreases more rapidly than in females, and may occur if the present cardiovascular disease decline broadens, lung cancer rates follow anticipated trends, and traumatic deaths of males in early life decrease. The Hayflick phenomenon (Hayflick, 1980) of cellular senescence is sometimes argued to be allegorical or even to represent a laboratory artifact (Manton, 1982). This may or may not be true, but the discussions presented here are not based on the Hayflick phenomenon.

When we ask medical students to draw a set of curves which represent changes in life expectancy in this century, they usually draw curves closely similar to a set of advancing sigmoid curves, equidistant along their entire length. Such are the curves that would be represented if the life span, as well as the life expectancy, were increasing (Figure 6). As noted (Figure 4), the actual curves show an increasingly rectangular character (Fries and Crapo, 1981). One does not have to be sophisticated in interpretation of mortality curves to have a reasonable feeling for where future progress may be made, and it is curious when demographic projections (Faber 1982, Manton, 1982) show characteristics not present in the historical record. A problem with elaborate demographic projections may arise if a faulty model is used, a model which has built into it the shape of the equidistant curves of Figure 6. The competing risk model, used without a hazard function representing natural death, leads to such a result. This model assumes that if there were no disease, there would be no death, and thus underestimates future mortality at the higher ages by not accounting for the mortal effects of physiologic frailty. A similar error can arise if recent "rates of change,"

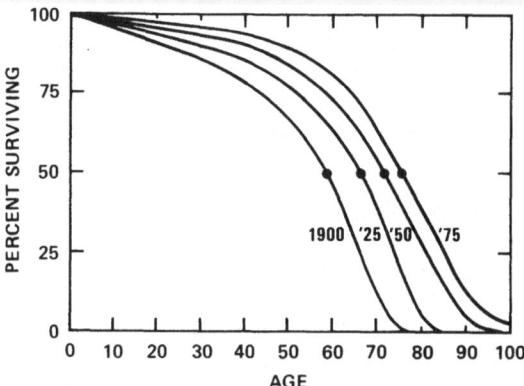

Figure 6. The Curves That Are Not. These survival curves would suggest an increasing life span, but only the median values and the 1950 curve are the true ones. The actual data are shown in Figure 4. Reprinted with permission from Vitality and Aging, W. H. Freeman and Company, San Francisco, 1981.

rather than absolute change, are used in projections. Since we have
moved from an acute disease era to a chronic disease one, premature
death now occurs at higher ages, and percentage gains with further
improvement will accordingly be greater at higher ages.

Clinical observation suggests that a significant number of
deaths, perhaps as many as one-fourth, presently occur in
individuals with minimal organ reserve, and hence are essentially
natural deaths, occurring within a few months of ultimate
physiologic limits. The elderly individual who gradually begins to
"fail"; the quiet death at home; or the terminal "multiple
castrophe" hospital course, characterized by failure of several
organs, are examples of nearly natural death. This phenomenon is
obscured by our social customs, which prevent tabulation of natural
death. "Natural" deaths are hidden in the statistics for
bronchopneumonia, heart failure, generalized atherosclerosis, and
other categories, since there is no death certificate category for
natural death, and everyone must be assigned. When the number of
natural deaths is relatively low, projections ignoring these
classification errors are reasonably accurate, but as the frequency
increases, such models increasingly underestimate future mortality
at advanced ages. If there is a major force for mortality, natural
death, with a hazard function increasing rapidly from essentially
zero at age 70, to nearly 99% at age 100, and the demographic model
does not include this term (Faber, 1982), then the projections will
be wrong.

However, quibbling about the precise numbers obscures the much
greater area of agreement in all projections. Even the most
optimistic calculations (Faber, 1982), project that life expectancy
at age 85 will increase by only two years by the year 2020. Life
expectancy at age 100 is projected to increase from 2.45 years in
1980 to only 3.01 years in 2000, and 3.35 years in 2020. Reduction
of premature death, as opposed to change in life span, must account
for at least part of such projected change. But such an increase of
eleven months over the next forty years, even if due entirely to a
change in the genetic life span, does not distort the policy
implications of the "compression of morbidity" (Fries, 1980).

The compression of morbidity has a central thesis: the age at
first appearance of aging manifestations and chronic disease
symptoms can increase more rapidly than life expectancy. This
statement of the thesis recognizes that increases in life
expectancy, whether or not associated with minor changes in the life
span, are likely over the next 25 years. The question of whether
the period of morbidity may be shortened depends upon whether the
average onset point of a marker of morbidity can increase more
rapidly than does life expectancy from the same age. If it does,
then the period between that marker and the end of life is
shortened. Absolute compression of morbidity occurs if we decrease

age-specific morbidity rates more rapidly than age-specific
mortality rates. Relative compression of morbidity occurs if the
amount of life after first chronic morbidity decreases as a
percentage of life expectancy.

The acute infectious diseases have ceased to be statistically
major causes of mortality in the United States. Tuberculosis, small
pox, diphtheria, tetanus, polio, typhoid fever, and others have
declined by 99% in this century (Fries and Ehrlich, 1980). In turn,
the major medical problems are now well known to be chronic
illnesses: atherosclerosis in all of its guises, cancer in its many
forms, emphysema, diabetes, cirrhosis, osteoarthritis. These
illnesses are not well conceptualized under the old model of
diseases with single causes and specific cures. The present health
problems are characterized by "risk factors" which accelerate their
course or which increase the probability of their occurrence. Their
"cause" is thus multifactorial, and no single cause is essential.
Even more importantly, these illnesses have other characteristics
which are not those of the acute diseases. They are, to one degree
or another, universal. Every individual has, to a greater or lesser
degree, the potential for increasing atherosclerosis, an increasing
statistical possibility of malignant change, and slow degeneration
of the articular cartilage. Moreover, the chronic illnesses have
their onset early in life; signs of such problems may be found in
autopsy studies of individuals in their twenties. The severity of
the conditions increases progressively with age (Figure 7).

Thus, our presently most important illnesses are universal,
have early onset, are progressive, are generally characterized by a
symptom threshold at which time they become clinically obvious, and
are multifactorial in cause (Fries and Crapo, 1981). The
differences between individuals are manifested not as much by the
presence or absence of the condition as by the rate at which the
condition progresses. The rate may be very low (as in atherosclerosis in native Japanese on native diets) or much higher (as in
Japanese on American diets).

As a caveat, there are a number of major chronic diseases, less
important statistically, which do not have these characteristics,
and which are not the subject of this discussion. Such illnesses
include rheumatoid arthritis, Hodgkin's disease, systemic lupus,
ulcerative colitis, and multiple sclerosis. These illnesses
ultimately may fit the traditional medical model more closely than a
universal chronic disease model. Alzheimer's disease, a major
problem currently increasing in prevalence, is difficult to
classify, and clearly deserves both preventive and curative study.
This condition is heavily age dependent, and has specific
pathology.

The multiple risk factor, universal susceptibility model fits

our most prevalent health problems, with important implications. In
this model, as risk factors are modified, the slope of the
progression is decreased. As the slope decreases, the date of
crossing the symptomatic threshold (Figure 7) can be postponed,
death due to the disease can be postponed or even prevented, and the
severity of symptoms experienced can be decreased. If the slope is
sufficiently reduced, the disease may be said to be "prevented,"
since the symptom threshold may not be passed during life.

A model of universal progressive disease with a symptom
threshold allows one to divide life into a "firm" portion, occurring
before the threshold is passed, and an "infirm" portion following
passage of that threshold. As the slope is decreased, the "firm"
period of adult vigor is prolonged and the "infirm" period of
disease or senescence is compressed against the natural barrier at
the end of life. Both the absolute amount and the percentage of
life spent in less than good health may thus be decreased.

Another significant attribute of this model is that any
reduction in the average slope of lines representing individuals in
a population will result in a decrease in age-specific mortality
rates. And, it will also result in an increase in the average age
at which the first symptom is experienced. Thus, as improvement in
the rate of accretion of chronic disease occurs, an effect on
morbidity is linked to the effect on mortality. Importantly, this
model may be used to describe the senescent changes of aging in
multiple organ systems as well as the accelerated decrepitude in a

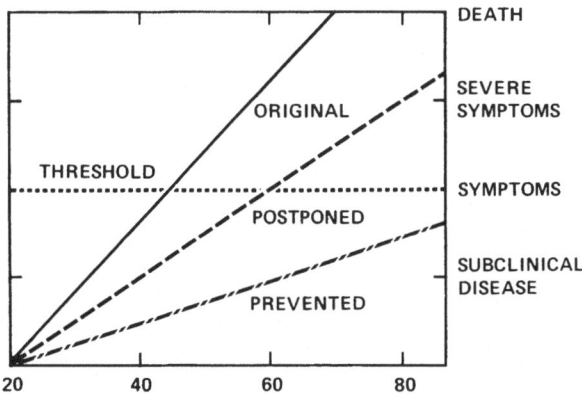

Figure 7. An Incremental Model of Chronic Disease. The model is
          characterized by early age of onset, progression at
          various rates, and passage of a symptomatic threshold
          at which time a clinical diagnosis may be made. Reprinted
          with permission from Vitality and Aging, W. H. Freeman
          and Company, San Francisco, 1981.

particular organ associated with a chronic disease.

Consider two brothers (Figure 8), one of whom smokes three
packages of cigarettes a day. The top line represents the life of
the heavy-smoking brother. Moving life expectancy toward the right
along a life history like this provides insights into many of the
phenomena of contemporary medicine and its interactions with
society. In 1900, perhaps this individual would have encountered
pneumonia at age 30 and have died, after a life of 30 years and an
illness of three days. Premature death, to be sure, but inexpensive
(at least in terms of direct medical costs), with relatively little
illness burden upon the society, and with a high proportion of
vigorous life to sickness. Now with penicillin, this man survives
to begin to develop cough, wheezing, and shortness of breath at age
40. If he continues to smoke, he will be increasingly short of
breath for the remainder of his life. In his fifties he has a heart
attack; perhaps prior to modern management he might have died at
this point. Now his arrythmia is controlled and he goes on to
encounter a stroke a few years later, requiring intensive
rehabilitation efforts. Throughout he remains short of breath.
Finally, a lung cancer develops and he dies, in a crescendo of
chronic disease. Such patients appear to require, not surprisingly,

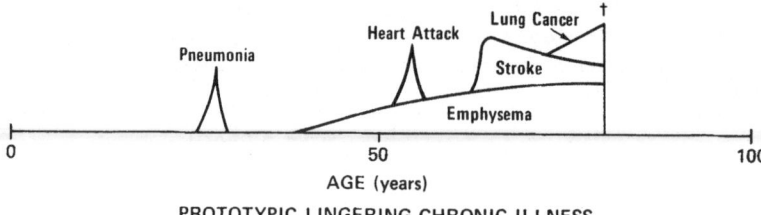

PROTOTYPIC LINGERING CHRONIC ILLNESS

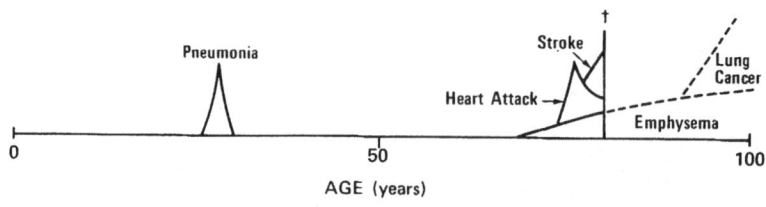

EFFECTS OF THE POSTPONEMENT OF CHRONIC DISEASE

Figure 8. The Compression of Morbidity. Two health-lives are
          diagrammed, the upper with poor health habits and the
          lower with better health habits. The period of adult
          vigor represented of infirmity is reduced in the lower
          example. Reprinted with permission from Vitality and
          Aging, W. H. Freeman and Company, San Francisco, 1981.

up to four times the medical resources of the average individual. Moreover, the more that the life of such individuals is extended toward the right, the greater illness burden they represent to society and the greater amount of their life which is spent in less than good health. Such individuals may "linger" for actually half or more of their lifetime, at enormous personal and social cost.

In contrast, the light-smoking brother does not develop symptomatic emphysema until perhaps 70. The heart attack is postponed a few years, as is the stroke. The lung cancer is postponed all the way out of his lifetime and does not occur. This individual is more vigorous, with a higher quality of life, for a longer period of time, and represents socially a much smaller burden on society. The change in the point of first breathlessness represents, in this common example, as much as thirty years of improved quality of life for the individual without a particularly noxious habit.

The same linear senescence observed by Shock in cross-sectional studies of organ function with age is seen with human optimal performance as, for example, with the world age group records for men in the marathon (Fries, 1980; Fries and Crapo, 1981). World-class performance is optimal in the twenties and early thirties and shows a linear decline after that point (Fries, 1980), up to the point where sample size is inadequate for estimation. Similar linear decline is evident in age group records for other athletic endeavors. Is is also present in longitudinal data of the same marathon runner (e.g., Clarence DeMar) over 50 years. However, the decrement associated with age is relatively small; in the marathon it approximates two minutes per year. Variation within individuals of the same age is very much larger than this. The individual not performing at maximum level may readily improve marathon times with age, and this phenomenon is commonplace. This ability of the individual to swim against the current of senescence holds as a general truth when one considers the modifiability (or lack thereof) of the physical and psychological markers of aging. Training in a particular faculty results in improvement in performance in that faculty, at any age.

Some aspects of the aging process appear to be non-modifiable. For the most part, these have in common the slow accumulation of fibrous tissue, replacing tissue which previously functioned. Thus, developing rigidity of the arterial wall, cataract formation, the graying of hair, the gradual loss of glomeruli in the kidney, thinning of hair, and the elasticity of skin. These capacities appear to be insidiously and irreversibly lost with age, according to present data (Fries and Crapo, 1981). Given present knowledge, there is little reason to expect that lifestyle or therapeutic interventions will reduce decrements in these areas.

When questioned about fears of growing old, individuals over
the age of 50 usually do not cite fear of death (Neugarten 1977;
Baltes 1982). As more significant concerns, they first describe a
dread of approaching chronic illness, pain, and inability to
physically get around. Second, they report fears of approaching
senility and loss of memory. And third, they describe a fear of
total dependence upon others.

In these areas, there is good reason to expect improvement from
preventive and lifestyle approaches. There are convincing studies
which indicate that cardiac reserve, dental decay, glucose
tolerance, intelligence tests and memory, osteoporosis, physical
endurance, physical strength, pulmonary reserve, reaction time,
social ability and blood pressure, among other variables, are
modifiable by the individual at any age. This modifiability is
sometimes termed the "plasticity of aging" (Baltes, 1982).
Modification consists in most instances of training and practice in
the specific faculty (Bortz, 1982; Dehn, 1972; Farquhar, 1978;
Labouvie-Vief, 1982; Langer, 1976; Paffenbarger, 1978; Rodin, 1977;
Spirduso, 1980; Valliant, 1979; Baltes, 1980; Fries and Crapo,
1981). In many instances, there is relatively little cross-over
from training in one attribute to success at another.

The dual ability to postpone chronic disease (Walker, 1977) and
to utilize the plasticity of aging affords an effective approach to
the major fears of chronic disease and loss of mobility, loss of
intellect and memory, and total dependence upon others. The
compression of morbidity and the plasticity of aging are related
concepts, and are applicable both to the problems of chronic disease
and to the problems of senescence.

Emerging data suggest that the compression of morbidity may be
a present, as well as a future, phenomenon. Conclusive data are
difficult to come by because data on the incidence of markers of
morbidity have not been systematically collected, nor have we
attempted even in cross-sectional studies to establish prevalence
figures for the "quality of life." Indeed, it is not even clear what
measures of morbidity might be used. If we accept as such measures
the age at first heart attack or stroke, or the age at development
of identifiable lung cancer, or age at admission to nursing homes,
then available data suggest that, in the United States, the onset
age may be increasing more rapidly than is the life expectancy from
age 40. For example, the average age at first heart attack for men
appears to have increased approximately four years in the past
sixteen, while life expectancy from age 40 increased but two years
over the past 20. Despite the decreasing prevalence of heart attack
over the period of 1968-1978, the percentage of hospital discharges
for this condition over the age of 65 has risen from 47 to 52% of
the total (National Center, 1965, 1968, 1974, 1978; Faber, 1982;
Elvebach, 1981; Connolly, 1981). The ratio of fatal to non-fatal

heart attacks appears to have remained constant during this period, suggesting that morbidity and mortality from this cause have both been reduced (Paffenbarger, 1979). The stroke data are more equivocal, but appear to show the same thing (Kramer, 1982; Robins, 1981). Lung cancer age-specific incidence rates appear to have similarly changed, and the risk factor model based heavily on cigarette smoking (pack/years) suggest that as smoking decreases the effect will first be postponement of onset; that is, the requisite number of pack/years will be reached later in life. Lifetime medical costs of heavy smokers and/or heavy drinkers, inferred from studies of high-cost hospital users (Schroeder, 1980), appear to be as much as four times those of individuals with moderate habits, even though the duration of life is shorter. The rate of admission to nursing homes has remained essentially constant for those over 65 or over 85 (Neugarten, 1977), despite empty beds and an increasing age as the population drifts gradually older. As noted above, physical ability is well demonstrated to be inherently trainable.

Human potential may be conceptualized within the following paradigm. There is a level of optimal performance in each faculty, approximated by world-class performance at each age. (Athletic performance is only the most measurable endeavor; improvement is possible for almost any activity, from shopping to playing chess.) The decline in optimal performance is linear. The mean performance of a population, on the other hand, has also been measured to show a linear decline, at a level of performance markedly below that of the best human performance.

By inference, each individual has his or her own particular curve of optimum performance. That is, with maximum training and effort, there is a theoretically optimum performance for each individual, and this curve should be expected to decline with age as do the measurable curves. The same individual, untrained and expending less than maximal effort, will show a similar curve but at a lower level of performance. For each individual, the area between the line of present performance and that of potential performance enables the plasticity of aging. With time and age, one may improve, regress, or stay the same within surprisingly broad limits of performance. If one is not performing near an individually optimal level with regard to a particular faculty, then improvement, despite increasing age, remains possible.

As an additional factor tending to reduce the effects of biological senescense, it should be observed that many psychologists favor a life span concept with regard to cognitive and social functions (Labouvie-Vief, 1982). In these models, a variety of stages occur, one following the other. With regard to certain functions, such as the accumulation of wisdom and judgment, an individual may move from "immature" thought processes through successive stages of "maturity" as he or she ages. Studies showing

declines in cognitive function with age are felt by these
researchers as due largely to the use of youth-oriented measurements
of factual cognitive function which do not account for the more
complex and less easily measurable manifestations of mature wisdom.

Human variability, the compression of morbidity, and the
recognition of untapped human potential at advanced age contain
important implications for research policy and for public policy.
They suggest shifts in research direction. First, they suggest that
a broad biopsychosocial model of health must be utilized, since many
of the interventions important in affecting senescent phenomena are
certain to be outside of a traditional biological model. Second,
there is a lack of good data on the "quality of life," and efforts
at systematic accumulations of such data must be accelerated.
Initially, decisions on those specific measures of "quality of life"
need to be collected, and cross-sectional data on these measures
need to be gathered. Third, at the same time, longitudinal studies
need to be initiated,and these need to include biologic,
psychological, and social variables in the same studies. Fourth,
the means for effecting positive change in an aging population are
to be found in the variability of the population, as well as in the
average values. It can be argued that we already have too many data
showing that everything declines with age. Studies need to identify
and quantitate inter-individual variation for specific marker
valuables at particular ages; identify the factors which predict the
variability; and design rigorous prospective intervention trials
based upon this identification of associated "risk" factors.

On the public policy side, there are several obvious
implications of this model. First, there should be no mandatory
retirement age. Studies of plasticity suggest strongly the health
and vitality benefits of continuing challenge, problem-solving,
perception of productivity, continued activity, and more money; for
some, these features will be best obtained by continued employment.
Second, creative vocational opportunities should be available, with
multiple and flexible pathways toward optimal use of the later
years. Third, health enhancement programs must begin early in adult
life and be continued throughout. Aging programs directed only at
the aged will have less impact than those addressing the same
problems earlier. Fourth, to the extent possible, we should seek
de-institutionalization of long term care programs which in their
most severe forms prevent individual initiative. Usually this will
consist of seeking the most independent living alternative possible
for the individual. It will require better development of home
health care services, a redevelopment of the role of the family, and
a more peripheral distribution of needed care away from large
impersonal institutions. Fifth, programs should stimulate the
independence of individuals. A false dichotomy is sometimes raised
between "caring" and "curing," and there is some risk that our
elderly may be smothered by good intentions. The elderly have a

right to an independent life, and to the execution of personal choices within the broadest possible framework. Finally, solutions are hampered by the existence of certain adverse incentives within our society. We subsidize bad health habits and we encourage disuse of the mental and physical faculties of older individuals. Our health insurance programs take money from those with good health habits and give it to those with poor health habits. We need to look closely at how our laws and our customs may affect the independent expression of vitality in the older individual.

The rectangular morbidity curve represents, in many ways, a social ideal. A long, vigorous life culminating, as in Oliver Wendell Holmes' One-Hoss Shay, with a sudden terminal collapse (Holmes, 1908). Vitality until the end, and death coming without fear or fury at the natural end of the individual life span. It will not happen this way, of course. Utopias may be envisioned, but not totally achieved. Increasing birth cohorts will continue to discharge ever larger numbers of individuals into the older age groups until equilibrium is reached after some 50 years, and the problems we have been experiencing will grow worse before they grow better. Clinical observation of our most vital citizens suggests a usual terminal decline of months to even a few years, not an abrupt collapse as that of the one-hoss shay. But, the compression of morbidity is an achievable phenomenon; it is already occurring in some areas, and it can be made to grow importantly. Projection of health needs under the scenario presented are more favorable than sometimes supposed. But the many problems of our increasingly elderly population continue to exist and to require vigorous attack from many directions. The paradigm of increasing vitality and finite life, with the consequent compression of morbidity into a shorter period prior to the end of life, offers a framework within which to view the problems, and within which we may begin to develop some constructive solutions.

REFERENCES

Baltes, P.B., Baltes, M.M.: Plasticity and Variability in Psychological Aging: Methodological and Theoretical Issues. In: Determining the Effects of Aging on the Central Nervous System. Berlin: Schering, 1980.
Baltes, M.M.: Environmental Factors in Dependency Among Nursing Home Residents: A Social Ecology Analysis. In: T. A. Willis (Ed), Basic Processes in Helping Relationships. New York: Academic Press, 1982.
Bortz, W.M.: Disuse and Aging. JAMA. 1982;248:1203-1208.
Bowerman, W.G.: Centenarians. Tr Acturarial Soc Am. 1939;40:360-378.
Connolly, D.C., Oxman, H.A., Nobrega, F.T., Kurland, L.T., Kennedy, M.A., Elvebach, L.R.: Coronary Heart Disease in Residents of

Rochester, Minnesota 1950 to 1975. I. Background and Study
Design. Mayo Clinic Proc. 1981;56:661-664.

Cooper, R.: Recent Health Gains for Adults. NEJM. 1982;307:631.

Cutler, R.G.: Evolution of Human Longevity: A Critical Overview.
Mechanisms of Aging and Development. 1979;9:337-354.

Dehn, M.M., Bruce, R.A.: Longitudinal Variations in Maximal Oxygen
Uptake With Age and Activity. J. Applied Physiology.
1972;33:805-807.

Elvebach, L.R., Connolly, D.C., Kurland, L.T.: Coronary Heart
Disease in Residents of Rochester, Minnesota. II. Mortality,
Incidence, and Survivorship 1950-1975. Mayo Clinic Proc.
1981; 56:665-671.

Faber, J.F.: Life Tables for the United States, 1900-2050.
Actuarial Study No. 87. U.S. Department of Health and Human
Services, SSA Pub. No. 11-11534, Office of the Actuary, 1982.

Farquhar, J.W.: The American Way of Life Need Not be Hazardous to
Your Health. New York: Norton, 1978.

Finch, C.E.: The Regulation of Physiological Changes During
Mammalian Aging. Quarterly Review of Biology. 1976;51:49-83.

Fries, J.F.: Aging, Natural Death, and the Compression of
Morbidity. NEJM. 1980;303:130-135.

Fries, J.F., Ehrlich, G.E.: Prognosis: Contemporary Outcomes of
Disease. Bowie, Maryland: Charles Press, 1980.

Fries, J.F., Crapo, L.M.: Vitality and Aging: Implications of the
Rectangular Curve. San Francisco, CA: W.H. Freeman, 1981.

Gompertz, B.: On the Nature of the Function Expressive of the Law of
Human Mortality. Philosophical Transactions of the Royal
Society of London. 1825;I:513-585.

Hayflick, L.: The Cell Biology of Human Aging. Scientific American.
1980;242:58-65.

Holmes, O.W.: The Deacon's Masterpiece; or, The Wonderful 'One-Hoss
Shay'. From: The Autocrat of the Breakfast Table, 1857-1858.
In: The Complete Poetical Works of Oliver Wendell Holmes.
Boston: Houghton Mifflin, 1908.

Kramer, S., Diamond, E.L., Lilienfeld, A.M.: Patterns of Incidence
and Trends in Diagnostic Classification of Cerebrovascular
Disease in Washington County, Maryland 1969-1971 to 1974-1976.
Am. J. Epidem. 1982;115:398-410.

Labouvie-Vief, G., Blanchard-Fields, F.: Cognitive Aging and
Psychological Growth. Ageing and Society. 1982. (Spring)

Langer, E.J., Rodin, J.: The Effects of Choice and Enhanced Personal
Responsibility for the Aged: A Field Experiment in an
Institutional Setting. J. Personality and Social Psych.
1976;34:191-198.

Manton, K.G.: Changing Concepts of Morbidity and Mortality in the
Elderly Population. Milbank Memorial Fund Quarterly.
1982;60:183-244.

McWhirter, N.: Guiness Book of World Records. 1980. Banton Books,
New York.

National Center for Health Statistics: Inpatient Utilization of
    Short- Stay Hospitals by Diagnosis: United States, 1965, 1968,
    1974, 1978. Series 13, Numbers 6, 12, 26, 46.
Neugarten, B.L., and Havighurst, R.J.: Extending the Human Life
    Span: Social Policy and Social Ethics. (National Science
    Foundation.) Washington, D.C.: U.S. Government Printing
    Office, 1977. (Stock No. 038-000-00337-2)
Paffenbarger, R.S.: Proceedings of the Conference on the Decline in
    Coronary Heart Disease Mortality. NIH Publication, No.
    79-1610, 1979; 298-311.
Robins, M., Baum, H.M.: Incidence of Stroke. In: Weinfield, F.D.,
    The National Survey of Stroke. Stroke. 1981;12:145-157.
Rockstein, M.: Heredity and Longevity in the Animal Kingdom. J.
    Gerontology. 1958;13:7-12.
Rodin, J., Langer, E.J.: Long-Term Effects of a Control-Relevant
    Intervention with the Institutionalized Aged. J. Personality
    and Social Psych. 1977;35:897-902.
Schroeder, S. A., Showstack, J.A., Roberts, H.E.: Frequency and
    Clinical Description of High Cost Patients in 17 Acute Care
    Hospitals. NEJM 1979;300:1306-1311.
Shock, N.W.: Discussion on Mortality and Measurement. In B.L.
    Strehler et al, eds.: The Biology of Aging, A Symposium.
    Washington, D.C.: American Institute of Biological Sciences,
    1960.
Spirduso, W. W.: Physical Fitness, Aging, and Psychomotor Speed: A
    Review. J Gerontology. 1980;35:850-865.
Strehler, B.L., Mildvan, A.S.: General Theory of Mortality and
    Aging. Science. 1960;132:14-21.
Valliant, G.E.: Natural History of Male Psychological Health:
    Effects of Mental Health on Physical Health. NEJM. 1979;
    301:1249-1254.
Walker, W.J.: Changing United States Life-Style and Declining
    Vascular Mortality: Cause or Coincidence. NEJM. 1977;
    297:163-165.

NOTE

This paper was first delivered as a keynote address to the
Institute of Medicine, National Academy of Sciences, Washington,
D.C., October 20, 1982. It is adapted from 'The Compression of
Morbidity,' "Milbank Memorial Fund Quarterly," 1983, and is
presented here by permission.

Address reprint requests to Dr. Fries at Department of
Medicine, Stanford University School of Medicine, Stanford,
California 94305.

AGING, HEALTH, STRESS AND TECHNOLOGY IN THE

WORK CONTEXT: CONCEPTS AND ISSUES

Ilene C. Siegler and
Linda K. George

School of Medicine
Duke University
Durham, North Carolina

In this paper we briefly review diverse but related literatures in order to speculate on the role that technology currently plays and will play in the future in defining the intersections of age, health and stress in the workplace.

Technology

Gomory (1983) in discussing technology development compared and contrasted scientific versus technical dimensions of problem solving. Science was seen as being practiced primarily in academic settings, university oriented, with the results of research publicly distributed. These results add to a large pool of knowledge that is fed by basic research, where a main motivator is curiosity, and the main labor force is staffed by Ph.D.'s in various disciplines. Technology, on the other hand, is often developed in industrial settings, driven by applications of knowledge and product development needs. The results are proprietary and the workforce mostly engineers. By this definition, technology is a priori neutral and most commonly includes products and devices offered for sale. Technological development can thus have an impact in terms of products developed specifically for older persons, or for those with specific deficits (e.g., sensory or motor deficits), elderly or not, or through research (design, data collection, analysis). At least for those currently alive, technological developments related to genetic engineering may indeed change the consequence of known health problems and provide for additional avenues of cure; but probably will not change the basic nature of aging itself except possibly for future generations. As changes related to devices and research strategies are the most likely for the near future, we will confine our hypotheses to those technological innovations.

Health Concerns

One of the most expected correlates to aging in our times is a decline in physical capacity reflected in poorer health. In general, in the workplace, remember that the older worker is anyone over forty. This is the age at which legal protection from age discrimination in employment becomes available. Thus we are talking about fairly severe illnesses which are disabling to relatively younger workers and subtle diseases not severe enough for disability pensions in older workers. However, our concerns with stress suggest that even without major health problems, stress itself can lead to decrements in performance which then may lead to the development of various disorders, at least in those individuals who have not developed adequate coping mechanisms to handle the stress (see Siegler, Nowlin & Blumenthal, 1980; Siegler & Costa, in press). Verbrugge (1983) has looked at data over the past 30 or so years from the National Health Interview Study in a most interesting way--by looking at patterns of morbidity and mortality as they have changed in middle aged (45-64) and older (65+) persons in the United States. She asks some very provocative questions in her paper with the informative title "Longer Life but Worsening Health?" In the 1950s and 1960s in the United States, mortality rates were essentially stable. We all learned about the squaring of the mortality curve and that that squaring implied that further public health measures would be useless and we would see no gains in life expectancy unless we understood the very basic mechanisms of life itself. While this is not to deny the need for understanding aging at the cellular level, since 1970 there have been increasing declines in mortality at all ages and in both sexes for a variety of conditions (see also Manton, 1982).

Verbrugge argues that we are seeing worsening health in middle aged and older persons because there is a greater awareness of health and disease, therefore earlier diagnoses and earlier and more complete reporting in the health interview surveys as well as lower mortality resulting in increased survival of those with nonetheless compromised health; and that we are seeing longer life or decreased mortality because of a lowered incidence of chronic disease. This lowered incidence appears to be the result of safer and cleaner environments, better habits, earlier and better medical care and better self-care. She suggests that these trends will continue in the future. Thus technological change can be expected to have an impact on the age distribution of the population in the near future and the health status of the various subgroups of the population of older persons. For the near term we can then predict that we will see continuations of the trends she has described--increased morbidity coupled with decreased mortality. For the longer term, if individuals choose to modify their lifestyles, and/or medical science develops some cures for current chronic diseases, we could predict decreases in both morbidity and mortality.

Aging and Environmental Change

An extremely useful perspective on aging and its relationship to technology is the perspective that James L. Fozard has been developing over the years in terms of his view of optimizing environments for the elderly (Fozard, 1981; Fozard & Popkin, 1978). Three major principles for designing environments for the elderly are given in Table 1 and provide a basis for discussion of the ways in which age can be expected to interact with technological change. Fozard's principles are useful in that they reinforce the dynamic nature of the processes we are dealing with, and see change as essentially neutral. These propositions remind us that technology may have an impact in changing the environment in which the older person operates; but also in changing implications of the age changes internal to persons. Various systems that have been developed to deal with issues in job discrimination for aged and handicapped workers give an excellent place to start thinking about these issues (see Edelman & Siegler, 1978 for a review).

Occupational Stress

Holt (1982) has provided an excellent review of the occupational stress literature where stress is seen as a major etiologic factor in disease. In a series of detailed tables, he reviewed the literature organized by objective stressors (e.g., noise in decibels, heat, other physical stressors defined independently of the individual; stressors dealing with time, for example, shift-work, time-zones, time pressure; social and

Table 1. Principles for Designing Environments for the Elderly

1.   PLAN ENVIRONMENTS THAT ALLOW FOR THE DEVELOPMENTAL CHANGES OCCURRING OVER THE ADULT LIFESPAN.

2.   RECONSIDER HOW PEOPLE REACT TO THE PSYCHOLOGICAL, PHYSICAL, AND SOCIAL ENVIRONMENT SO AS TO SUPPORT HUMAN DEVELOPMENT IN THE ADULT YEARS.

3.   RECOGNIZE THAT THE ENVIRONMENT IN WHICH PEOPLE AGE IS ITSELF ALSO CHANGING RAPIDLY OVER TIME, AND THAT GENERALIZATIONS ARE NOT NECESSARILY GOING TO REMAIN VALID.

Source: Fozard, James L. Person-Environment Relationships in Adulthood: Implications for human factors engineering. Human Factors, 1981, 23 (1), p. 8-9.

organizational properties of the work setting; and changes in the
job including unemployment and transfers). Subjectively defined
stressors include those that are role related and other
miscellaneous factors such as job complexity, relationships with
co-workers, person-environment fit, and off the job stresses such as
family problems. Major dependent variables in studies of
occupational stress include strains defined as symptoms and
variables predictive of upset, such as pulse rate, blood pressure,
catecholamine and other physiological indices of stress;
psychological indices including anxiety, and dissatisfactions;
behavioral and social responses by workers such as absenteeism, and
disruption of behavior; mental and physical illness and increased
mortality. In reviewing the set of moderator variables used, age is
cited for two studies, and stage of life for one additional study.
While all of this work reviewed is on adults, the role of age as a
contributing variable is modest, as is true in studies in industrial
psychology in general. If we put this literature together with the
data that we do have on the role of age in the workplace (see
Edelman & Siegler, 1978; pp. 22-27) then it is more likely that what
is required is retraining strategies that are designed for older
workers (Belbin & Belbin, 1972). Studies of the impact of stress
itself are exploding (see Elliot & Eisdorfer, 1982; Goldberger &
Breznitz, 1982 for two recent collections of review papers) where
the role of age plays a similar role. Models of social stress such
as the one given in George (1980;1982) relate stress to models of
coping.

Coping

The literature on coping is large and expanding. . A few
insights from our own work (George & Siegler, 1982; Siegler &
George, 1983) as well as similar findings by others suggest some of
the following ideas: (1) Data from our own work provides an
interesting unobtrusive measure of the role of technological change
in an open ended life history interview. We looked at the extent to
which events where work was the major context provided stress for
individuals. One hundred individuals were asked about 3 positive
and 3 negative events in the present, the past and the past 10
years. We had a maximum of 600 person/events. Only one individual
mentioned a technological change in work--a newspaper/advertising
man who reported that technology significantly increased the ease of
his job. This is a most interesting phenomenon. Our respondents
were born between 1899 and 1922. At the time of interview in the
late 1970s most were somewhere in the age range of 55-80 and had
lived through a time of significant technological change. (2) Men
and women, at least older men and women, live very different lives,
they have different stresses to cope with and use different coping
mechanisms. However, both sexes use coping mechanisms that are
appropriate to the event requiring coping. Thus the differences in
coping mechanisms can be accounted for by the differences in life

events experienced. (3) The role of expectation, and the individuals' defining of the event was critical in understanding the processes observed, and (4) Age per se was not important in understanding coping.

PROJECTIONS OF TECHNOLOGY NOW AND IN THE FUTURE

One obvious role for technology is in the types of research we can do to understand the aging process. We are developing the technology of monitoring physiological parameters, on line, in individuals while they go about their normal activities. Data in the neuroendocrine measures of stress and their correlates to known experimental tasks are also in a phase of rapid development. Prospective studies which included data on personal characteristics have yielded important insights. Secondly, various jobs are changing rapidly as technological advances replace certain types of workers and change the day to day stresses and strains. It is possible to thus consider a future in which technology has positive, negative or mixed impacts and that these impacts may or may not interact with age. It is more difficult to imagine the impact that technology may have on the aging process per se.

REFERENCES

Belbin, E. & Belbin, R.M. Problems in adult retraining. London: Heineman, 1972.

Edelman, C.D. & Siegler, I.C. Federal Age Discrimination in Employment Law: Slowing Down the Gold Watch. Charlottesville: Michie Company, 1978 (and 1980 Cumulative Supplement).

Elliot, G.R. & Eisdorfer, C. (Eds.) Stress and human health: Analysis and implications for research. New York: Springer Publishing Company, 1982.

Fozard, J.L. Person-environment relationships in adulthood: Implications for human factors engineering. Human Factors, 1981, 23(1), 7-27.

Fozard, J.L. & Popkin, S.J. Optimizing adult development: Ends and means of an applied psychology of aging. American Psychologist, 1978, 33, 975-989.

George, L.K. Role Transitions in Later Life. Belmont, CA: Brooks Cole, 1980.

George, L.K. Models of transitions in middle and later life. Annals, 1982, 464, 22-37.

George, L.K. & Siegler, I.C. Stress and coping in later life. Educational Horizons, 1982, 60(4), 147-196.

Goldberger, L. & Breznitz, S. (Eds.) Handbook of stress: Theoretical and clinical aspects. New York: Free Press, 1982.

Gomory, Ralph E. Technology development. Science, 1983, 220: 576-580.

Holt, R.R. Occupational stress. In L. Goldberger & S. Breznitz,
    (Eds.) Handbook of Stress: Theoretical and Clinical Aspects,
    New York: Free Press, 1982, pp. 419-444.
Manton, K.G. Changing concepts of morbidity and mortality in the
    elderly population. Health and Society (Millbank Memorial Fund
    Quarterly), 1982. 60, 183-244.
Siegler, I.C. & Costa, P.T., Jr. Health behavior relationships. In
    J.E. Birren and K.W. Schaie (Eds.). Handbook of the Psychology
    of Aging. New York: Van Nostrand Reinhold, in press.
Siegler, I.C. & George, L.K. Sex differences in coping and
    perceptions of life events. Journal of Geriatric Psychiatry,
    1983, 16(2), 197-210.
Siegler, I.C., Nowlin, J.B. & Blumenthal, J.A. Health and behavior:
    Methodological considerations for adult development and aging.
    In L.W. Poon (Ed.) Aging in the 1980's, Washington D.C.:
    American Psychological Association, 1980, 559-612.
Verbrugge, L.M. Longer life but worsening health? Trends in health
    and mortality of middle aged and older persons. Paper
    presented at the Population Association of America Meetings,
    Pittsburgh PA, April, 1983.

TECHNOLOGICAL CHANGE AND THE AGING

OF WORKING CAPACITY

Roy J. Shephard

School of Physical & Health Education
University of Toronto
Toronto, Ontario

INTRODUCTION

Unlike many previous alliances, NATO is pledged to joint action not only in war, but also in peace, safeguarding the "freedom, common heritage and civilization of its peoples, founded on the principles of democracy, individual liberty and the rule of law." Aging and technological progress is a particularly apt theme for NATO.

A central tenet of classical Marxist theory was that technical progress would divide western society into two opposing camps, labour and capital; violent conflict between these two groups was postulated as inevitable until the means of production was brought under collective social ownership.

More recently, the Warsaw pact powers have had sufficient faith in their economic system to accept a policy of co-existence, arguing that greater social and technological progress would soon allow them to out-perform the west.

Historically, Marxism has thrived in a climate of social discontent, or "alienation" to use the communist jargon. The nations that comprise NATO currently face a danger of increasing alienation through (a) a rising dependency ratio (the proportion of citizens who are too young or too old to work); (b) a concentration of high technology upon preparations for war rather than fundamental human needs, and (c) rapidly decreasing employment opportunities for those of working age.

The present paper will focus specifically on the need to replace the energy demands of traditional work with voluntary leisure activity if physical and mental health are to be conserved in an aging and technically advanced society.

WORK AND TECHNOLOGICAL CHANGE

Neolithic Culture

Studies of the Eskimo or Inuit[1,2] allow us to follow the problems of western society on a vastly accelerated time scale. In a mere decade, an entire community has moved from a neolithic to an urban Canadian culture. When we first reached the high arctic, many of the Inuit living around Igloolik ($69°40'$ N) were still "primitive" hunters. During both summer and winter, hunting demanded a high 24h energy expenditure, the average for 8 types of hunt being 15.4 $MJ \cdot day^{-1}$. This is comparable with mining (15.3 $MJ \cdot day^{-1}$) and forestry (15.4 $MJ \cdot day^{-1}$).

Physiological consequences of this vigorous lifestyle included a virtual absence of subcutaneous fat, strong leg muscles, and large maximum oxygen intake (Table 1). The entire population had a much higher maximum oxygen transport than their white counterparts in Toronto, although the rate of aging was similar for the two groups.

The Inuit women also worked somewhat harder than the chatelaines of a technically advanced society. There were skins to chew and scrape, and a baby was frequently carried in the traditional "yappa" at a cost of about 18 $kJ \cdot min^{-1}$. With these exceptions, most of the women's tasks had a similar energy cost to light industrial work (24-hour energy expenditure about 10 $MJ \cdot day^{-1}$). Figures were higher at field camps, where more improvisation of domestic equipment was necessary.

Rapid Acculturation

When our initial observations were made, in 1969-1970, many of the people had reached an intermediate stage of technology. Inuit who had accepted permanent employment within the settlement mostly had blue collar occupations, such as water delivery, garbage clearance, and garage mechanic. The average energy cost of 25 such activities[1] was 14.4 $kJ \cdot min^{-1}$. "Work" rarely lasted more than 8 hours per day, so that the 24 hour energy expenditures were less than for the hunters (around 13.7 $MJ \cdot day^{-1}$).

To assess the impact of technological change, we used an acculturation index developed by Ross MacArthur (Table 2). This considered such items as education, housing, geographic mobility, wage income, and household equipment. Acculturation was associated with an increase of subcutaneous fat, a decrease of leg extension force and a decrease of maximum oxygen intake (Table 3).

Over the past decade, Igloolik has undergone a further dramatic change. Kayaks have been replaced by motor launches and seaplanes. Dog teams have given way to snow-mobiles. Powerful equipment such as graders keep the village streets clear of snow. Homes have video television and hi-fi sets to occupy the arctic nights, and a new runway allows a 5-hour jet excursion to the night-life of Montreal. A 10 year follow-up survey (Table 3) has shown a marked decrease of maximum oxygen intake, increase of fat and loss of muscle strength in all groups except the young boys. Data is now similar to that

Table 1. Physiological differences between Inuit following the traditional neolithic hunting culture and those adopting the lifestyle of a western settlement, based on crossectional data of Shephard.[1] Data for summer, with winter in parenthesis.

Physiological variable	Traditional neolithic hunters (n=20)		Transitional group (n=22)		Workers in Settlement (n=18)	
Skinfold thickness (average of 3, mm)	5.0	(6.4)	6.1	(6.7)	6.7	(7.9)
Leg extension force (N)	798	(857)	888	(984)	891	(930)
Maximum oxygen intake $(ml \cdot kg^{-1} min^{-1})$	56.6	(56.2)	54.9	(54.9)	51.2	(50.1)

Table 2. Coefficients of correlation between an arbitrary index of acculturation and physiological variables, based on data of Shephard.[1]

Variable	Men (n=132)	Women (n=93)
Sum of 3 skinfolds	0.29	0.34
Leg extension force	-0.15	n.s.
Maximum oxygen intake	-0.23	n.s.

Table 3. Follow-up Data for Igloolik Community Showing the Physiological Consequences of 10 Years Acculturation to a Modern Technological Society, Based on Data of Rode & Shephard[2]

Group age in 1970/71 and sex (years, M/F)	Height (cm)	Body mass (kg)	Sum of 3 skinfolds (mm)	Aerobic power ($l \cdot min^{-1}$)	Aerobic power ($ml \cdot kg^{-1} \cdot min^{-1}$)	Leg extension force (N)	Handgrip force (N)
20 - 29							
M	-1.7*** ±1.7	+2.3 ±6.2	+6.7* ±14.4	-0.63*** ±0.68	-11.2*** ±9.8	-55 ±183	+22* ±39
F	-2.1*** ±1.1	+4.4** ±4.8	+17.5*** ±15.7	-0.34*** ±0.32	-9.2*** ±6.4	-76** ±95	-3 ±39
30 - 39							
M	-2.2 ±1.8	+3.9* ±7.7	+11.5** ±15.9	-0.85*** ±0.48	-13.2*** ±6.4	-151** ±202	+4 ±41
F	-1.8*** ±1.2	+6.4* ±8.7	+14.5* ±15.6	-0.37* ±0.43	-6.7* ±6.9	-58 ±106	+2 ±38
40 - 49							
M	-2.0 ±3.2	+3.9 ±9.2	+11.6 ±17.3	-0.65*** ±0.20	-12.1*** ±5.7	-175 ±300	0 ±35
F	-2.1 ±1.8	+1.5 ±6.7	+14.5 ±22.9	-0.53*** ±0.16	-8.3*** ±1.5	-124 ±119	-33*** ±15

Significance of differences   *p < 0.05   **p < 0.01   ***p < 0.001

observed in "white" society.[2] One particularly interesting
finding is that all age groups have become 1-2 cm shorter over the
ten year interval. Again, the change far exceeds the normal rate of
aging, and we suspect that spinal damage has been caused by driving
snowmobiles long distances at high speed over the rough pack-ice.

Intermediate Technologies

Studies of primitive populations by indirect calorimetry show
quite high energy expenditures (Table 4). Viteri et al.[4] found values of 13.2-17.9 MJ·day^{-1} in Guatemalan highlanders. Mowing, cutting wood and hoeing were all carried out using simple hand tools.
In Israel, farming is somewhat more advanced,[21] but there is still
much heavy work--walking in mud (33.5 kJ·min^{-1}), moving irrigation pipes (32.2 kJ·min^{-1}), forking and scything grass
(24.7-25.1 kJ·min^{-1}) and spreading manure (26.4 kJ·min^{-1}).

Advanced Societies

Edholm[8] made the intriguing suggestion that mechanization
caused people to work faster, without reducing their daily energy
expenditure. This may be true for intermediate levels of technology.
For example, the maneuvering of a rotary cultivator in heavy soil
leaves little opportunity for rest, whereas when digging a person can
adjust the speed of movement of a spade or fork to ground conditions.

However, Edholm's view is not true for our "post-industrial"
society. The energy demands of operating a modern air-conditioned
tractor or advanced forestry equipment are low relative to traditional and intermediate technologies (Table 5). The need for human
power has declined in most industries, and it will continue to
decline as robots begin to control production lines. Physical
effects of inactivity are now compounded by the social consequences
of reduced employment opportunities in most daily tasks.

In Britain, the number of full-time agricultural workers
dropped by two thirds between 1921 and 1975, while productivity per
worker jumped 30 fold from 1826 to 1971.[9] Of particular interest
to NATO, the Russian farm-productivity of 1966 was still inferior to
that of Britain in 1826. The labour force of the Canadian forest
industry decreased 11.4% between 1965 and 1978, but productivity per
worker jumped by 71% over the same interval.[10]

Power equipment has also reduced the energy cost of operating a
home. The list of "domestic tasks" proposed by Durnin & Passmore[5]
is now almost entirely obsolete. Carpets are cleaned by vacuum
cleaner rather than beaten, floors are no longer scrubbed, furniture
is no longer wax polished, clothes are washed by machine, shopping
is carried in a car, and few householders chop wood or carry buckets
of coal.

Table 4. Daily energy expenditures associated with intermediate
          technology. Based on data of Viteri et al.,[4] Edholm
          et al.,[5] Miller et al.,[6] and Norgan et al.[7]

Population	24-hour energy expenditure (MJ)	Population	24-hour energy expenditure (MJ)
Guatemalan highlanders:[4]		Israeli farmers:[5]	
		Kurds	12.9 (9.7F)
agricultural workers	16.2	Yemeni	12.7 (9.8F)
horseman	17.9	Jamaican farmers[6]	14.6
carpenter/mason	14.8		
dairymen/herdsmen	13.2	New Guinea	
foreman	11.6	peasants (scaled to 65 kg, 55 kg)[7]	10.6 (7.9F)

Table 5. The impact of technological change on the energy cost of
          some common tasks. Based on data collected by Durnin &
          Passmore.[3]

Activity	Traditional technology (kJ·min^{-1})		Intermediate or modern technology (kJ·min$_{-1}$)	
Mowing	Hand	23-43	Tractor	7.5-18.8
Grain harvest	Binding & stooking	21-36	Combined harvester	8.4-13.0
Milking	Hand	9.2-21.3	Machine	6.3
Tree planting	Hand	27	Machine	11.7
Horizontal sawing	Hand	30	Power saw	22.6
Digging	Pick & shovel	20-42	Mechanical digger	15.5-30.5

TECHNOLOGICAL CHANGE AND COMPULSORY RETIREMENT

Physical Limitation

Until recently, compulsory retirement was justified on the grounds that workers who had passed a certain age would be physically over-taxed.

The permissible energy expenditure for an 8-hour day was set at 33-50% of aerobic power,[11-13] depending upon the environmental temperature and humidity, along with such adverse factors as an awkward posture, high peak loads and use of small muscle groups. In support of this limit, the self-paced energy output is often about 40% of maximum oxygen intake.[14] A greater demand causes fatigue, and the task must then be rated as too arduous.

By the age of 65, the aerobic power of the average man has dropped to 26.8 $ml \cdot kg^{-1} min^{-1}$, while that of the average women is 24.1 $ml \cdot kg^{-1} min^{-1}$. Assuming also respective body masses of 70 and 55 kg, the 40% loading thus corresponds to an energy expenditure of about 15.7 $kJ \cdot min^{-1}$ in the men, and 11.1 $kJ \cdot min^{-1}$ in the women. Given also a 20% inter-individual variation of aerobic power, one sixty-five year old man in 40 will have an upper limit of 9.4 $kJ \cdot min^{-1}$, while one sixty-five year old woman in 40 will have an upper limit of 6.7 $kJ \cdot min^{-1}$.

Many tasks that use traditional technology thus seem too strenuous for a person aged 65. It is surprising that complaints are not more frequent! Possible explanations include (i) promotion out of physically demanding jobs, (ii) preservation of fitness by the demands of the job or leisure activities, (iii) delegation of strenuous tasks to younger workers, (iv) a slowing of pace, (v) a reduction of energy cost by control of body mass or efficiency of effort, and (vi) a maximization of available oxygen transport through such measures as avoidance of smoking.

Impact of Technology

Has technology reduced energy demands to the point that physical limitations are no longer important? Certainly, there has been much mechanization in such heavy occupations as forestry, mining, farming, steel work, and the building trades. At the same time, the daily energy demands of some occupations have been increased by shift work and compressed working weeks. Tasks that remain heavy work involve walking, carrying and climbing steps, for example marine surveying[15] and the delivery of mail.[16] Marine surveying is resistant to automation, although as cargo-holds become larger a power-operated inspection-chair becomes a technical possibility. Mail carrying is also threatened by electronic mail systems.

It seems unlikely that technical progress will extend the working span of the present generation, since the oldest employees are usually found in occupations using traditional methodologies. Thus, Sachuk[17] noted a much younger median age for computer operators than for railway trackmen or irrigation workers. Lack of modern skills forces older workers to continue with strenuous manual tasks.

Whether subsequent generations will push back the age of retirement depends in part on the availability of employment, and in part on voluntary maintenance of health and fitness by leisure pursuits. Lack of physical activity can cause a 30% deterioration of oxygen transport,[18] equivalent to a 25-year handicap of physiological function. It is thus disturbing that the older half of many adult populations takes almost no vigorous activity.

TECHNOLOGICAL PROGRESS AND HEALTH

How far has the reduced energy demands of modern society influenced health?

Cardiovascular Health

One of the fascinating phenomena of the 20th Century has been the epidemic of cardiovascular disease. It was tempting to correlate the increase of heart attacks with measures of technical progress such as the number of vehicle registrations, and the use of fossil fuel in industry. A decreased food consumption (Table 6) apparently confirmed declining physical effort, and investigators such as Paffenbarger[19] demonstrated a strong correlation between a low energy expenditure at work and the incidence of ischemic heart disease, even after allowance for such variables a cigarette smoking and blood pressure.

Some exercise enthusiasts have also attributed the waning of the epidemic to an increase of voluntary leisure activity, although other important factors include improved treatment of cardiac patients, a reduced intake of dietary fat, and a diminished proportion of cigarette smokers.[20]

Injuries

Lack of physical activity also enhances the risk of injury by weakening muscles and tendons, slowing reflexes, and exacerbating the normal age-related demineralization of bone.[21,22] The risk is particularly great in situations where the physical demand is normally low, but rises to a substantial level on rare occasions.

Table 6. Daily per capita food sales in the U.S.*

Year	Food energy consumed (MJ)	Year	Food energy consumed (MJ)	Year	Food energy consumed (MJ)
1930	14.4	1960	13.1	1975	13.6
1940	14.0	1965	13.1	1977	14.1
1950	13.6	1970	13.8		

*Note: There is no guarantee that wastage remained constant. More-over, because body mass increased from 1930 to 1970, an increased intake of some 0.6 $MJ \cdot day^{-1}$ would be needed for equal activity.

Thus in South African industry, manual work is normally undertaken by "black" labourers, but "white" supervisors may exert themselves in an emergency. In consequence, back injuries are more frequent in "whites" than in "blacks".[23] Likewise, back injuries are seen in Americans who move from the high technology of the U.S. to the less developed society of an Israeli kibbutz.[24]

Perceived Health

Several recent surveys have shown relatively high levels of job satisfaction in North American workers.[25] There is little evidence of the alienation that Marxist theorists have postulated for a capitalist technocracy. Nevertheless, the groups surveyed have tended to be white-collar workers employed by non-union companies with a long history of good labour relations.

Dissatisfaction as expressed by absenteeism, abuse of alcohol and drugs, poor quality work and strikes is much greater in companies where automation has reduced the worker's responsibility to a dull, repetitive job. In older workers, the need to maintain a home and support a family reduce outward manifestations of dissatisfaction, but covert dissatisfaction may lead to a worsening of perceived health.[26] At the opposite end of the spectrum, modern technology imposes excessive mental demands on some older workers, and this also can lead to a deterioration of health.

POLICY IMPLICATIONS

Restoration of Fitness

Debate continues as to how far the fitness of the older worker
can be restored. Sources of difficulty in interpreting experiments
are methods of matching initial fitness in young and old subjects,
failure to reach a true maximum oxygen intake in an unfit older
individual, and uncertainty whether to express gains as absolute or
percent changes.[21] A well-designed training regimen can boost
maximum oxygen intake by 20%, with parallel gains of muscle strength
and flexibility, dispersal of surplus body fat, and a halting of
bone mineral loss.[21] The rate of subsequent aging is unchanged,
but because the person now starts from a higher baseline, it takes
many more years to reach the situation where working capacity is
inadequate to meet the demands of occupation or personal care.

This has important social consequences. Currently, we provide
some $2,000 of institutional care per year for every senior citizen
in the U.S. and Canada. There are many reasons for institutionalization, but one factor is undoubtedly that domestic tasks exceed 40%
of the individuals's working capacity. An unfit person would probably reach this stage at about 79 years of age, but a fit individual
could remain independent to 87-88 years of age. Assuming that the
exercise has no effect on longevity, some two thirds of the population would die between 79 an 88 years of age, offering the attractive prospect of reducing institutional costs by up to two thirds.[21]

The Age of Leisure

The old-age dependency ratio for Canada and the U.S. will
increase from a current figure of 0.18 to 0.33 by the year 2031.
Moreover, there will be a disproportionate increase in the numbers
of the very old needing institutional care.[27]

Despite these demands, technological advances may lead to a
loss of 10-14% of existing jobs. In the future, people will need to
accept leisure not as a weekend luxury or an inter-job episode, but
rather as a permanent way of life. While the younger generation may
accept this change readily, the protestant "work ethic" is strongly
inculcated into the older employee. Much patient education will thus
be needed if the displaced worker is to find self-fulfillment rather
than disgrace in years of enforced leisure. Recreational programmes
for the older individual[28] can play an important part in meeting
this need, reversing the self-fulfilling expectation that active
leisure declines with age.

Implications for Industry

For the next two or three decades, the work site will remain the most effective place for programmes to restore fitness lost through sedentary living. Time is not lost in travel. Costs are low because parking space is already available and careful scheduling can assure high-volume use of facilities. Recruitment is also facilitated by established channels of communication, and interest is sustained by peer encouragement.

The expense to a company depends upon the luxury of facilities, the size of any employee contributions, and the completeness of accounting. Canadian estimates range from $100-$350 per participant-year,[29] while U.S. Corporations regard a figure of $500 per participant as more realistic. Participation rates are currently about 20%, so that the cost per worker-year is in the range $20-$100.

Among a long list of benefits to both the company and the individual,[29] we may note gains of productivity (through enhanced working capacity and mental arousal), decreased absenteeism and turn-over (presumably an expression of greater job satisfaction), and reduced health care costs (presumably an indication of greater per-ceived health).

CONCLUSIONS

We may conclude that technology has led to a diminution of energy expenditures in both industry and the home, with an associated decrease of physical working capacity. However, programmes of voluntary physical activity can do much to counter adverse health effects, providing self-fulfillment for those with boring jobs or increased leisure, and generating a substantial volume of new employment. Voluntary activity cannot slow the inherent rate of biological aging, but it does set back the time when a person is unable to work and unable to care for him or herself.

REFERENCES

1. R. J. Shephard, Work physiology and activity patterns, pp. 305-338, in: "The human biology of circumpolar populations," F.A. Milan, ed., London: Cambridge University Press (1980).
2. A. Rode and R. J. Shephard, Ten years of "civilization" - fitness of the Canadian Inuit, J. Appl. Physiol., in press (1984).

3. J. V. G. A. Durnin and R. Passmore, "Energy, work and leisure,"
    London: Heinemann (1967).

4. F. E. Viteri, B. Torun, J. C. Galicia, and E. Herrera,
    Determining energy costs of agricultural activities by
    respirometer and energy balance techniques, Am. J. Clin.
    Nutr. 24: 1418-1430 (1971).

5. O. G. Edholm, S. Humphrey, J. Lourie, B. E. Tredre, and J.
    Brotherhood, Energy expenditure and climatic exposure of
    Yemenite and Kurdish Jews in Israel, Phil. Trans. R. Soc.
    (Lond.) B 266: 127-140 (1973).

6. G. J. Miller, J. E. Cotes, A. M. Hall, C. B. Slavosa, and A.
    Ashworth, Lung function and exercise performance of healthy
    Caribbean men and women of African ethnic origin, Quart. J.
    Exp. Physiol. 57: 325-341 (1972).

7. N. G. Norgan, A. Ferro-Luzzi, and J. V. G. A. Durnin, the energy
    and nutrient intake and the energy expenditure of 204 New
    Guinea adults, Phil. Trans. R. Soc. (Lond.) B 268: 309-348
    (1974).

8. O. Edholm, The changing pattern of human activity, Ergonomics
    13: 625-643 (1970).

9. T. Bayliss-Smith, Energy use, food production and welfare:
    perspectives on the "efficiency" of agricultural systems,
    in: "Energy and effort," G. A. Harrison, ed., London:
    Taylor & Francis (1982).

10. M. Lortie, Human resource development in the Canadian Forest
    industry, pp. 80-85, in: "Proceedings of the Canadian
    Forest Congress, Toronto, Sept. 22-23," (1980).

11. F. H. Bonjer, Relationships between physical working capacity
    and allowable caloric expenditure, in: "Int. Colloquium on
    muscular exercise and training," W. Rohmert, ed.,
    Darmstadt: Gentner Verlag (1968).

12. I. Åstrand, Degree of strain during building work as related to
    individual aerobic work capacity, Ergonomics 10: 293-303
    (1967).

13. I. Åstrand, Aerobic work capacity in men and women with special
    reference to age, Acta Phsiol. Scand. 49: Suppl. 169: 1-92
    (1960).

14. A. L. Hughes and R. F. Goldman, Energy cost of hard work, J.
    Appl. Physiol. 29: 570-572 (1970).

15. R. J. Shephard, Equal opportunity for a geriatric labor force.
    Some observations on marine surveying, J. Occup. Med., in
    press (1983).

16. R. J. Shephard, The daily work-load of the postal carrier, J.
    Hum. Ergol., 11(2), (1983).

17. N. N. Sachuk, The aging worker's abilities and disabilities in
    relation to industrial production, pp. 147-162, in: "Work
    and Aging," J. Huet, ed., Paris: Int. Centre of Social
    Gerontology (1971).

18. D. A. Bailey, R. J. Shephard, and R. L. Mirwald, Validation of
    self-administered home test of cardiorespiratory fitness,

Can. J. Appl. Sports Sci. 1: 67-78 (1976).

19. R. Paffenbarger, Physical activity and fatal heart attack: protection or selection? pp. 35-49, in: "Exercise in cardiovascular health and disease," E. A. Amsterdam, J. H. Wilmore, and A. N. deMaria, Eds,. New York; Yorke Books (1977).

20. R. J. Shephard, "Ischemic heart disease and exercise," Chicago: Year book (1982).

21. R. J. Shephard, "Physical activity and aging," London: Croom Helm (1978).

22. R. J. Shephard, "Textbook of exercise physiology and biochemistry," New York: Praeger (1982).

23. D. I. Guthrie, A new approach to handling in industry. A rational approach to the prevention of low back pain, S. Afr. Med. J. 27: 651-656 (1963).

24. A. Magora and I. Taustein, An investigation of the problem of sick-leave in the patient suffering from low back pain, Industr. Med. 38: 80-90 (1969).

25. M. Cox, R. J. Shephard, and P. Corey, Influence of an employee fitness programme upon fitness, productivity and absenteeism, Ergonomics 24: 795-806 (1981).

26. D. Coburn, Work alienation and well-being, pp. 420-437, in: "Health and Canadian Society," D. Coburn, C. D'Arcy, P. New, and G. Torrance, eds., Toronto: FitzHenry & Whiteside (1981)

27. F. Denton, C. Feaver, and B. Spencer, "The future population and labour force of Canada: projections to the year 2051," Ottawa: Economic Council of Canada (1980).

28. R. J. Shephard, Physiological aspects of recreational activity in the older adult, Recreation Research Rev. 9: 48-65 (1982).

29. R. J. Shephard, "Employee Fitness," Toronto: University of Toronto Press.

SOURCES OF OCCUPATIONAL STRESS

AMONG OLDER WORKERS

Cary L. Cooper

University of Manchester Institute
of Science and Technology
Manchester, United Kingdom

INTRODUCTION

Stress-related illnesses such as coronary heart disease have
shown a steady upward trend over the past couple of decades in the
U.K. and other developed countries, particularly for the middle-
aged. In England and Wales, for example, the death rate for men
between 35 and 44 nearly doubled between 1950 and 1973. By 1973,
41% of all deaths in the age group 25-44 were due to cardiovascular
disease, with nearly 30% due to cardiac heart disease. In fact, in
1976 the American Heart Association estimated the cost of cardiovascular disease in the U.S. at $26.7 billion a year.

In addition to the more extreme forms of stress-related illnesses, there has been an increase in other possible stress
manifestations, such as alcoholism (hospital admissions in the U.K.
increased from roughly under 6,000 in 1966 to over 8,000 in 1974),
industrial accidents, and short-term illnesses, with an estimated
300 million working days lost at a cost of 55 million British pounds
in national insurance and supplementary benefits payments alone.
The total cost to industry of all forms of stress-related illness
and other manifestations, a large slice of which can be attributed
directly or indirectly to the working environment of the middle-aged,
must be enormous, beyond the scope of most cost accountants to begin
to calculate. Some estimate that it may represent in the order of
1% to 3% of GNP in developed countries alone (Cooper, 1983).

Many of these stress-related illnesses obviously affect the
older more than the younger employee, regardless of occupational
grouping (as Table 1 indicates and other morbidity and mortality
data would support). The purpose of this paper will be to highlight

Table 1. Acute Sickness and Consultations with General
Medical Practioners, 1974-75, in Great Britain

| | Average number of restricted activity days per person per year (males) | | | Average number of consultations per person per year (males) | | |
	15-44	45-65	All ages	15-44	45-64	All ages
Professional	9	16	12	2.1	2.7	2.7
Employers and managers	11	13	14	1.8	2.4	2.7
Intermediate and junior non-manual	10	21	15	2.0	4.3	3.1
Skilled manual and own account non-professional	15	24	17	2.8	4.0	3.2
Semi-skilled manual and personal service	16	23	18	2.7	4.5	3.7
Unskilled manual	21	28	20	3.5	4.8	3.6
All persons	13	21	16	2.4	3.8	3.1

Source: U. K. Government's General Household Survey, 1975.

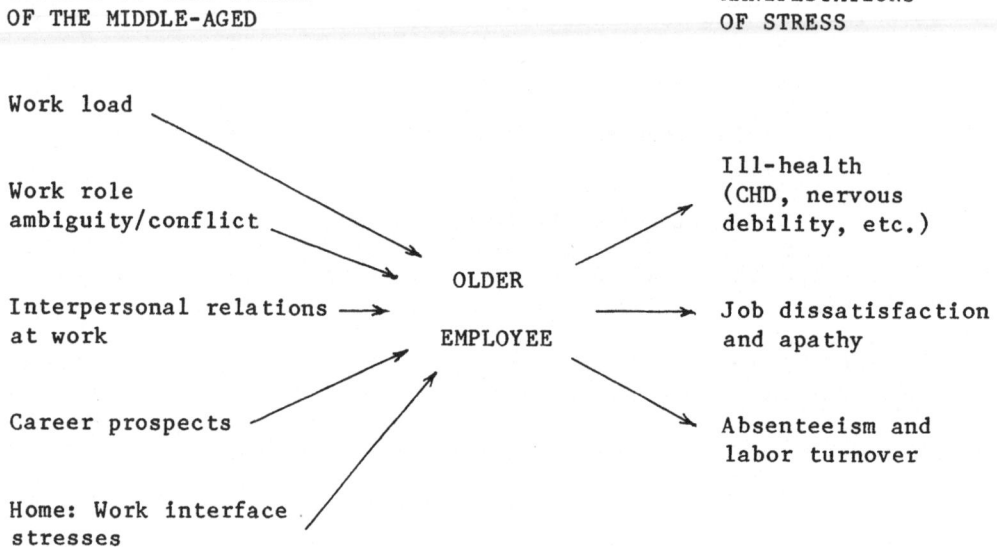

SOURCES OF WORK STRESS
OF THE MIDDLE-AGED

MANIFESTATIONS
OF STRESS

Work load

Work role
ambiguity/conflict

Interpersonal relations
at work

Career prospects

Home: Work interface
stresses

OLDER

EMPLOYEE

Ill-health
(CHD, nervous
debility, etc.)

Job dissatisfaction
and apathy

Absenteeism and
labor turnover

Figure 1. Occupational stressors among older workers.

the _sources_ of stress acting on older workers in work context.

Five major categories of work stressors can be identified from the large volume of work undertaken in the occupational health field (see Figure 1). These are work load, work role ambiguity/conflict, interpersonal relations at work, career prospects (or lack of them) and home:work interface stresses.

WORK LOAD AND THE OLDER EMPLOYEE

One of the most important sources of job stress for middle-aged individuals is their tendency to work long hours and to take on too much work. Research into work overload has been given substantial empirical attention. French and Caplan (1973) have differentiated overload in terms of quantitative and qualitative overload. Quantitative refers to having "too much to do," while qualitative relates to work that is "too difficult." In one of their early studies, French and Caplan (1970) found that objective overload was strongly linked to cigarette smoking (an important risk factor in CHD). Persons with more phone calls, office visits, and meetings per given work time were found to smoke significantly more cigarettes than persons with fewer such engagements. In a study of 100 middle-aged coronary patients, Russek and Zohman (1969) found that 25% had been working at two jobs and an additional 45% had worked at jobs which required (due to work overload) 60 or more hours per week. They add that prolonged emotional strain preceded the attack in 91% of the cases while similar stress was observed in only 20% of a control group. Breslow and Buell (1960) have also reported findings which support a relationship between hours of work and death from coronary disease. In an investigation of mortality rates of men in California, they observed that workers in light industry under the age of 45, who are on the job more than 48 hours a week, have twice the risk of death from CHD as similar workers working 40 or under hours a week. Another substantial investigation on quantitative work load was carried out by Margolis et al. (1974) on a representative national U.S. sample of 1496 employed persons. They found that overload was significantly related to a number of symptoms or indicators of stress: escapist drinking, absenteeism from work, low motivation to work, lowered self-esteem, and an absence of suggestions to employers.

There is also some evidence that "qualitative" overload can be a source of work stress for the middle-aged. French et al. (1965) looked at qualitative work overload in a large university. They used questionnaires, interviews, and medical examinations to obtain data on risk factors associated with CHD for 122 university administrators and professors. They found that one symptom of stress, low self-esteem, was related to work overload but that this was different for the two occupational groups. Qualitative overload was not significantly linked to low self-esteem among the administrators but was significantly correlated for the professors. The greater the "quality" of work expected of the professor, the lower the self-esteem. French and Caplan (1973) summarise this research by suggesting that both qualitative and quantitative overload produce at least nine different symptoms of psychological and physical strain: job dissatisfaction, job tension, lower self-esteem, threat, embarrassment, high cholesterol levels, increased heart rate, skin resistance, and more smoking.

It is also interesting to note that overload is not always externally imposed. Many middle-aged professionals react to overload by working longer hours (Cooper, 1979). For example, in an American study (1975) it was found that 45% of the executives investigated worked all day, in the evenings, and during weekends, and that a further 37% kept weekends free but worked extra hours in the evenings. In many companies this type of behavior has become the norm to which executives feel they must adhere.

WORK ROLE AMBIGUITY/CONFLICT

Another major source of work stress among middle-aged individuals is associated with a person's role at work. Role ambiguity exists when an individual has inadequate information about his work role, that is, when there is a "lack of clarity about work objectives associated with the role, about work colleagues' expectation of the work role and about the scope and responsibility of the job." Kahn et al. (1964) found in their study that men who suffered from role ambiguity experience lower job satisfaction, higher job-related tension, greater futility, and lower self-confidence. French and Caplan (1970) found, at one of NASA's bases, in a sample of 205 middle-aged engineers, scientists, and administrators, that role ambiguity was significantly related to low job satisfaction and to feelings of job-related threat to one's mental and physical well-being. This also related to indicators of physiological strain, such as increased blood pressure and pulse rate. Margolis et al. (1974) also found a number of significant relationships between symptoms or indicators of physical and mental ill health, on the one hand, and role ambiguity, on the other, in their national U.S. sample. The stress indicators related to role ambiguity were depressed mood, lowered self-esteem, life dissatisfaction, job dissatisfaction, low motivation to work, and intention to leave job.

Role conflict exists when an "individual in a particular work role is torn by conflicting job demands or doing things he really does not want to do or does not think are part of the job specification" (Kahn et al., 1964). The most frequent manifestation of this is when an individual is caught between two groups of people

who demand different kinds of behaviour or expect that the job
should entail different functions. Kahn et al. (1964) found that
men who suffered more role conflict had lower job satisfaction and
higher job-related tension. It is interesting to note that they
also found that the greater the power or authority of the people
sending the conflicting roles messages, the more job dissatisfaction
produced by the role conflict. This was related to physiological
strain as well, as the NASA study illustrates (French and Caplan,
1970). The NASA group telemetered and recorded the heart rate of 22
men for a two-hour period while they were at work in their offices.
They found that the mean heart rate for an individual was strongly
related to his report of role conflict. A larger and medically more
sophisticated study by Shirom et al. (1973) found similar results.
Their research is of particular interest, since they tried to look
simultaneously at a wide variety of potential work stresses. They
collected data on 762 male kibbutz members aged 30 and above, drawn
from 13 kibbutzim throughout Israel. They examined the
relationships between CHD, abnormal electrocardiographic readings,
CHD risk factors (systolic blood pressure, pulse rate, serum
cholesterol levels, etc.), and potential sources of job stress (work
overload, role ambiguity, role conflict, lack of physical
activity). Their data were broken down by occupational groups:
agricultural workers, factory employees, craftsmen, and managers.
It was found that there was a significant relationship between role
conflict and CHD (specifically abnormal electrocardiographic
readings), particularly for the middle-aged managers.

Another important potential source of middle-aged stress
associated with work role is responsibility for people, as
differentiated from responsibility for things (equipment, budgets,
etc.). Wardwell et al. (1964) found that responsibility for people
was significantly more likely to lead to CHD than responsibility for
things. Increased responsibility for people frequently means that
one has to spend more time interacting with others, attending
meetings, working alone and, in consequence--as in the NASA study
(French and Caplan, 1970)--more time in trying to meet deadline
pressures and schedules. Pincherle (1972) also found this in his
U.K. study of 2,000 executives attending a medical centre for a
medical checkup. Among the 2,000 managers sent by their companies
for annual examinations, there was evidence of physical stress being
linked to age and level of responsibility; the older and more
responsible the executive, the greater the probability of the
presence of CHD risk factors or symptoms. French and Caplan (1970)
support this in their NASA study of managerial and professional
workers--they found that responsibility for people was significantly
related to heavy smoking, raised diastolic blood pressure, and
increased serum cholesterol levels; however, the more the individual
had responsibility for things as opposed to people, the lower were
each of these CHD risk factors.

Having too little responsibility, lack of participation in decision-making, lack of managerial support, having to keep up with increasing standards of performance and coping with rapid technological change are other potential middle-aged role stressors found at work.

INTERPERSONAL RELATIONS AT WORK

Another major potential source of stress for the middle-aged worker has to do with the nature of the relationships with older people in the work environment. Behavioural scientists have long suggested that good relationships among members of a work group are a central factor in individual and organisational health (Cooper, 1979).

Buck (1972) focused on the attitude and relationship of middle-aged workers and managers to their immediate boss, using Fleishmans's leadership questionnaire on consideration and initiating structure. The consideration factor was associated with behaviour indicative of friendship, mutual trust, respect and a certain warmth between boss and subordinate. He found that those managers who felt that their boss was low on consideration reported feeling more job pressure. Managers who were under pressure reported that their boss did not give them criticism in a helpful way, played favourites with subordinates, "pulled rank" and took advantage of them whenever he got a chance. Buck concludes that the "considerate behaviour of superiors appears to have contributed significantly inversely to feelings of job pressure."

Morris (1975) encompasses this whole area of work relationships in one model, which he terms the "cross of relationships." While he acknowledges the differences between relationships on two continua --one axis extends from colleagues to users and the other intersecting axis from senior to junior staff--he feels that the focal middle-aged manager must bring all four into dynamic balance in order to be able to deal with the stress of his position. Morris's suggestion seems only sensible when we see how much of his work time the middle-aged manager spends with other people. In a research programme to find out exactly what managers do, Minzberg (1973) showed just how much of their time is spent in interaction. In an intensive study of a small sample of chief executives, he found that in a large organisation a mere 22% of time was spent in desk work sessions, the rest being taken up by telephone calls (6%), scheduled meetings (59%), unscheduled meetings (10%), and other activities (3%). In small organisations, basic desk work played a larger part (52%), but nearly 40% was still devoted to face-to-face contacts of one kind or another.

CAREER PROSPECTS

Two major clusters of potential work stressors can be identified in this area: 1. Lack of job security, fear of redundancy, obsolescence or early retirement, etc.; and 2. status incongruity, under- or over-promotion, frustration at having reached one's career ceiling, etc. For many middle-aged workers, career progression is of overriding importance--by promotion they earn not only money but also status and the new job challenges for which they strive. However, career progression is, perhaps, a problem by its very nature. For example, Sofer (1970) found that many of his sample believed that "luck" and "being in the right place at the right time" play a major role in career advancement.

Typically, in the early years at work, ambition and the ability to come to terms quickly with a rapidly changing environment are fostered and suitably rewarded by the company. At middle age, and usually middle management levels, career advancement becomes more problematic and most workers find their progress slowed, if not actually stopped. Job opportunities become fewer, those jobs that are available take longer to master, past (mistaken?) decisions cannot be revoked, old knowledge and methods become obsolete, energies may be flagging or demanded by the family, and there is the press of fresh young recruits to face in competition. Both Levinson (1973) and Constandse (1972)--the latter refers to this phase as "the male menopause"--depict the managers as suffering these fears and disappointments in "silent isolation" from their family and work colleagues.

The fear of demotion or obsolescence can be strong for those who know they have reached their "career ceiling"--and most will inevitably suffer some erosion of status before they finally retire. Goffman (1952), extrapolating from a technique employed in the con-game "cooling the mark out," suggests that the company should bear some of the responsibility for taking the sting out of this (felt) failure experience.

From the company perspective, on the other hand, McMurray (1973) presents the case for not promoting a manager to a higher position if there is doubt that he or she can fill it. In a syndrome he labels "the executive neurosis" he describes the over-promoted manager as grossly overworking to keep down a top job and at the same time hiding his insecurity, and points to the consequences of this for the manager's work performance and the company. Age is no longer revered as it was; it is becoming a "young man's world." The rapidity with which society is developing (technologically, economically, and socially) is likely to mean that individuals will now need to change career during their working life (as companies and products are having to do). Such trends breed undertainty, while research suggest that older workers look for

stability (Sleeper, 1975). Unless the middle-aged adapt their
expectations to suit new circumstances, career development stress,
especially in later life, is likely to become an increasingly common
experience.

HOME/WORK INTERFACE STRESSES

This section covers those interfaces between life outside and
life inside the organisation that might put pressure on the
middle-aged worker: family problems, life crises, financial
difficulties, conflict of personal beliefs with those of the
company, and the conflict of company with family demands.

The area which has received most research interest here is that
of the middle-aged worker/manager's relationship with his wife and
family. This individual has two main problems vis-a-vis his family:
the first is that of time-management and commitment-management. Not
only does his busy life leave him few resources with which to cope
with the details of house management, etc., to relieve stress when
possible, and to maintain contact with the outside world. The
second, often a result of the first, is the spillover of crises or
stresses from one system which affect the other.

MARRIAGE PATTERNS

The arrangement the middle-aged man reaches with his wife will
be of vital importance to both problem areas. Pahl and Pahl (1971)
found that the majority of wives in their middle-class sample saw
their role in relation to their husband's job as a supportive,
domestic one; all said that they derived their sense of security
from their husbands. Gowler and Legge (1975) have dubbed this bond
"the hidden contract," in which the wife agrees to act as a "support
team" so that her husband can fill the demanding job to which he
aspires. Handy (1978) supports the idea that this is "typical" and
that it is the path to career success for the middle-aged male
manager. Based on individual psychometric data, he describes a
number of possible marriage-role combinations. In his sample of top
British executives (in mid-career) and their wives, he found that
the most frequent pattern (about half the 22 couples interviewed)
was the "thrusting male/caring female." This he depicts as a highly
role-segregated combination, with emphasis on separation, silence
and complementary activities. Historically, both the company and
the manager have reaped benefits from maintaining the segregation of
work and home implicit in this pattern. The company thus
legitimates its demand for a constant work performance from its
employee, no matter what his home situation, and the manager is free
to pursue his career but keeps a safe haven to which he can return
to relax and recuperate. The second and most frequent combination

was "involved/involved" - a dual career pattern, with the emphasis on sharing of responsibility. This, while potentially extremely fulfilling for both parties, requires energy inputs which might well prove so excessive that none of the roles involved is fulfilled successfully.

MOBILITY

Home conflicts become particularly critical in relation to work relocation and mobility among the middle-aged. Much of the literature on this topic comes from the United States, where mobility is much more a part of the national character for the middle-aged managers and other workers than in the U.K., but there is reason to believe that in the U.K. (Pierson, 1972), too, it is an increasingly common phenomenon.

At an individual level, the effects of mobility on the manager's wife and family have been studied (Cooper, 1983). Researchers agree that whether she is willing to move or not, the wife bears the brunt of relocations, and they conclude that most husbands do not appreciate what this involves. American writers point to signs that wives are suffering and becoming less cooperative. Immundo (1974) hypothesizes that increasing divorce rates are seen as the upwardly aspiring manager races ahead of his socially unskilled stay-at-home wife. Seidenberg (1973) comments on the rise in the ratio of female to male alcoholics in the United States from 1:5 in 1962 to 1:2 in 1973 and asks, provocatively, "Do corporate wives have souls?" Descriptive accounts of the frustrations and loneliness of being a corporate wife in the U.S. and U.K. proliferate. Increasing teenage delinquency and violence are also laid at the door of the mobile, middle-aged manager and the society which he has created.

Constant moving can have profound effects on the life-style of the people concerned--particularly on their relationships with others. Staying only two years or so in one place, mobile families do not have time to develop close ties with the local community. Immundo (1974) talks of the "mobility syndrome," a way of behaving geared to developing only temporary relationships. Packard (1975) describes ways in which individuals react to the type of fragmenting society this creates, e.g., treating everything as if it were temporary, being indifferent to local community amenities and organizations, living for the present and becoming adept at instant gregariousness. He goes on to point out the likely consequences for local communities, the nation, and the rootless people involved.

Pahl and Pahl (1971) suggest that the British reaction is, characteristically, more reserved and that many middle-aged mobiles retreat into their nuclear families. Managers, particularly, do not become involved in local affairs due both to lack of time and to an appreciation that they are only short-stay inhabitants. Their wives find participation easier (especially in a mobile rather than a static area) and a recent survey suggested that, for some, involvement is a necessity to compensate for their husband's career involvement and frequent absences. From the company's point of view, the way in which a wife adjusts to her new environment can affect her husband's work performance. Guest and Williams (1973) illustrate this by an example of a major international company who, on surveying 1800 of their executives in 70 countries, concluded that the two most important influences on overall satisfaction with the overseas assignment were the job itself and, more importantly, the wife's adjustment to the foreign environment.

CONCLUSION

In summary, the sources for stress among middle-aged workers are many and varied, but an awareness of these is critical if we are to minimise their adverse effects, both for the individual and the organisation. As Wright (1975) so aptly suggests, "the responsibility for maintaining health should be a reflection of the basic relationship between the individual and the organisation for which he works: it is in the best interests of both parties that reasonable steps are taken to live and work sensibly and not too demandingly."

REFERENCES

Breslow, L. and Buell, P. (1960). Mortality from coronary heart disease and physical activity of work in California. J. Chron. Dis., 11, 615.

Buck, V. (1972). Working Under Pressure. London: Staples Press.

Constandse, E.J. (1972). A neglected personnel problem. Personnel J., 51, 129.

Cooper, C.L. (1979). The Executive Gypsy: The Quality of Managerial Life. London: Macmillan. New Jersey: Petrocelli Books.

Cooper, C.L. (1983). Stress Research: Issues for the 80s. New York: John Wiley & Sons.

French, J.R.P. and Caplan, R.D. (1970). Psychosocial factors in coronary heart disease. Indus. Med., 39, 383.

French, J.R.P. and Caplan, R.D. (1973). Organizational stress and individual strain. In: Marrow, A.J. (ed.), The Failure of Success. New York: AMACOM.

French, J.R.P., Tupper, C.J. and Mueller, E.I. (1965). Workload of University Professors. Unpublished Research Report. Ann Arbor, Michigan: The University of Michigan.

Goffman, E. (1952). On cooling the mark out. Psychiat., 15, 451.

Gowler, D. and Legge, K. (1975). Stress and external relationships - the 'hidden contract.' In: Gowler, D. and Legge, K. (eds.). Managerial Stress. Epping: Gower Press.

Guest, D. and Williams, R. (1973). How home affects work. New Society. January.

Handy, C. (1978). The fmmily: help or hindrance. In: Cooper, C.L. and Payne, R. (eds.)., Stress at Work. London: John Wiley & Sons.

Immundo, L.V. (1974). Problems associated with managerial mobility. Personnel J., 53, 910.

Kahn, R.L., Wolfe, D.M., Quinn, R.P., Snoek, J.D. and Rosenthal, R.A. (1964). Organizational Stress. New York: John Wiley & Sons.

Levinson, H. (1973). Problems that worry our executives. In: Marrow, A.J. (ed.), The Failure of Success. New York: AMACOM.

McMurray, R.N. (1973). "The executive neurosis." In: Noland, R. Industrial Mental Health and Employee Counselling. New York: Behavioral Publications.

Margolis, B.L., Kroes, W.H. and Quinn, R.P. (1974). Job Stress: an unlisted occupational hazard. J. Occup. Med., 16, 654.

Minzberg, H. (1973). The Nature of Managerial Work. New York: Harper and Row.

Morris, J. (1975). Managerial stress and 'the cross of relationships.' In: Gowler, D. and Legge, K. (eds.), Managerial Stress. Epping: Gower Press.

Packard, V. (1975). A Nation of Strangers. New York: McKay.

Pahl, J.M. and Pahl, R.E. (1971). Managers and Their Wives. London: Allen Lane.

Pierson, G.W. (1972). The Moving Americans. New York: Knopf.

Pincherle, G. (1972). Fitness for work. Proc. Roy. Soc. Med., 65, 321.

Russek, H.I. and Zohman, B.L. (1969). Relative significance of heredity, diet and occupational stress in CHD of young adults. Am. J. Med. Sci., 235, 266.

Siedenberg, R. (1973). Corporate Wives - Corporate Casualties. New York: American Management Association.

Shirom, A., Eden, D., Silberwasser, S. and Kellerman, J.J. (1973). Job stress and risk factors in coronary heart disease among occupational categories in kibbutzim. Soc. Sci. Med., 7, 875.

Sleeper, R.D. (1975). Labour mobility over the life cycle. Brit. J. Indus. Rel., 13.

Sofer, C. (1970). Men in Mid-Career. Cambridge: Cambridge University Press.

Uris, A. (1972). How managers ease job pressures. Int. Mgt., June, 45.

Wardwell, W.I., Hyman, M. and Bahnson, C.B. (1964). Stress and coronary disease in three field studies. J. Chron. Dis., 17,73.

Wright, H.B. (1975). Executive Ease and Dis-ease. Epping: Gower Press.

SOURCES OF STRESS AMONG OLDER PERSONS

ANTECEDENTS OF EMOTIONAL HEALTH STATUS AMONG THE ELDERLY

Laurence G. Branch, Linda J. Evans, and
Pamela J. Perun

Harvard Medical School,
Department of Social Medicine and Health Policy
Boston, Massachusetts

PROLOGUE

As the conveners and participants of the NATO Symposium on
Aging and Technological Advances are all well aware, technological
advances are a two-edged sword for whoever wields it. From the
perspective of older people themselves, many specific technological
advances occasion both benefits and liabilities depending upon the
criterion. As others have noted, for example, the mechanization of
the workplace has enabled us to produce more goods and services with
decreasing levels of physical fitness.

New technologies present a less frequently recognized
double-edged sword for professionals trying to understand the
underlying principles characterizing the later decades of the life
span. Many concur that the discipline of gerontology has a minimum
of unifying principles, models, or theories. The few previous
attempts have enjoyed only limited success. The disengagement
theory offered by Cumming and Henry (1961) has not achieved its
promise as a unifying principle. The health service utilization
model offered by Aday, Andersen, and their colleagues (1974) is a
unifying conceptual model for researchers, but has had only limited
success as a cohesive principle to explain the utilization practices
of U.S. elders in the post-Medicare era.

Our thesis is that the technology of high-speed computers is a
two-edged sword for gerontological professionals, on the one side
contributing more information, but on the other side not contributing greater understanding. Those who remember the methods of
statistical analyses prior to computerized data processing recall

221

that the hum of mechanical calculators lasted hours, and usually reflected an equal amount of careful consideration about the purpose of the undertaking. In the days of the Monroe and Frieden calculators, one did not relish the recalculation of a factor analysis because the basis of the analysis was not conceptualized adequately. In this era of high-speed computers, too often the amount of forethought about how our understanding will be advanced by the findings equals the computer time required for the data analysis -- about 1.5 seconds.

To change metaphors briefly, high-speed computers have facilitated the identification of numerous trees (i.e., regression analyses which specify the relationship of variables taken two at a time while controlling for others), but not necessarily our understanding of the forest (i.e., the unifying principles underpinning the relationships).

In the following sections, we will summarize overlapping areas of gerontological research, and attempt to show how alternative quantitative methods can assist in the quest for unifying principles.

BACKGROUND

Increased longevity and declining fertility rates contribute to America's emergence as an aging society. Advanced age has been associated with increased risks for a broad array of negative outcomes for the individual, including increased physical and emotional health problems, and increased utilization of costly health care services. The anticipated numbers of older people, coupled with their increased risk of poor health and increased use of medical care, have given rise to fiscal concerns, and spawned research on alternatives to institutional health care, various health payment plans, and possible redefinitions of service eligibility from chronological to functional criteria (Hodgson and Quinn, 1980; Winn and McCafree, 1979; Branch, 1980). Substantial efforts have been made to identify individual predictors of health service utilization patterns (Berki and Ashcraft, 1979; Verbrugge, 1976; Goldberg et al., 1980; Branch et al., 1981)

In addition to physical health risks, old age has been postulated to facilitate emotional health risks as well. Indeed, the assertion that old age inherently brings role losses, increased isolation, and less expansive and pleasant emotions among the elderly has underpinned more than 20 years of research on their adjustment to presumably reduced circumstances (Chown, 1977; Cumming and Henry, 1961; Neugarten et al., 1961). Despite considerable research, it is still not clear whether emotional health status is a stable personality trait, a developmental phenomenon possibly based in biological changes, or a situational adaptation (George, 1981; Shock, 1962; Stenbeck, 1980; Mechanic, 1976; Kessler et al., 1981).

Our analysis is an attempt to bridge the gap between two major research traditions -- one that emphasizes health service utilization and its correlates, and the other which examines emotional health status and its demographic and social correlates.

PREVIOUS RESEARCH

Three bodies of literature contribute to the conceptual design of our analyses. The first group of studies has focused upon individual factors associated with health service utilization among the non-aged and the aged, and has considered the predictive value of what are termed predisposing, enabling, and needs variables (Aday and Andersen, 1974; Andersen and Newman, 1973). Emotional health status has generally been excluded from these investigations. A second group of studies has emphasized factors related to emotional health among the old. The dominant outcome variable, emotional health status, has been variously defined in terms of distress, stress, life satisfaction, depression, morale, well-being, and adaptation. A third and fairly scanty volume of literature has attempted to link emotional health status with health service utilization patterns among the non-old and old.

Health service utilization research has tended to emphasize socio-cultural and needs predictors of utilization, typically relying upon Aday and Andersen's (1974) individual determinants: predisposing, enabling, and need variables. Need usually has been measured by symptoms of illness, diagnoses, evaluations by medical personnel, functional or disability levels, and/or perceived health status. Predisposing factors are social and demographic characteristics that affect a person's general stance toward his or her body, health, and medical care. Frequently used indicators of this concept are age, sex, race ethnicity, family size-composition, social class, residence, and attitudes and beliefs about health conditions and medical care providers (Andersen and Newman, 1973; Rosenstock, 1974). Enabling factors refer to resources such as income, education, health insurance, friends, and transportation that enhance or diminish the likelihood of seeking medical service, assuming a predisposition to do so has been established. In general, studies of utilization which employed large-scale multivariate techniques have explained more of the variance in health service utilization with needs variables than with predisposing or enabling predictors (Wolinsky, 1978; Branch and Jette, 1982; Branch et al., 1981). Nonetheless, some researchers have identified age, gender, residence status, education level, and belief in the effectiveness of medical care as predisposing variables related to health service utilization (Branch et al.,

1981; Kulka et al., 1979; Marcus and Seeman, 1981; Crandall and Duncan, 1981).

One of the most important yet difficult independent variables studied with respect to elderly emotional health status is age. Separating age-based changes from those that are socially-mediated by situational circumstances is difficult and perhaps artificial. Nonetheless, age is reported in the literature as having a direct and inverse association with emotional health (Rabbitt, 1977; Palmore and Cleveland, 1976; Beck, 1967, 1974; Pfeiffer, 1977). Other demographic variables and social correlates of elders' emotional health include gender, marital status, income, education level, type of residence, ethnic background, rural-urban location, race, and geographical options (Coulton and Frost, 1982; Gerber et al., 1975; Gubrium, 1976; Kohen, 1983; Albert and Zarit, 1977; Woehrer, 1978; Adams, 1971; Liang and Warfel, 1983; Usui et al., 1983; Cutler, 1975; Clemente and Kleiman, 1976). An implicit though rarely stated unifying principle of this literature is that life stresses (for example, threat of crime commonly associated with low income) can lead to depression and somatic symptoms which, if unalleviated, in turn can result in low morale or life satisfaction (Rahe et al., 1964; Tessler and Mechanic, 1978).

A third body of literature linking emotional health with health service utilization presupposes emotional health as a determinant of health service use. Very little research has been done in this area, and findings vary. Some researchers, for example, have found that depressed/distressed persons are more likely to list physical illnesses, contact health professionals, and report hospital admissions and physician utilization than nondepressed/nondistressed individuals (Frerichs et al., 1982; Tessler et al., 1976). In contrast to these studies reporting a strong inverse association between emotional health and service utilization among non-aged samples, Coulton and Frost (1982) found that, within an elderly sample, distress showed no direct effect on medical care utilization when perceived need was controlled.

CONCEPTUAL MODEL

Our approach is based upon these research endeavors, but we altered our conceptual model to reflect our judgment of which variables precede others in their influences on the outcome. Our conceptual model as suggested in Figure 1 incorporates internal characteristics such as the demographic and social correlates that have been associated with both emotional health status and health service utilization. In the tradition of health service utilization research, we subsume these variables into two categories -- those which usually are treated as predisposing such as age and gender and those considered as enabling factors for utilization such as insurance coverage or income.

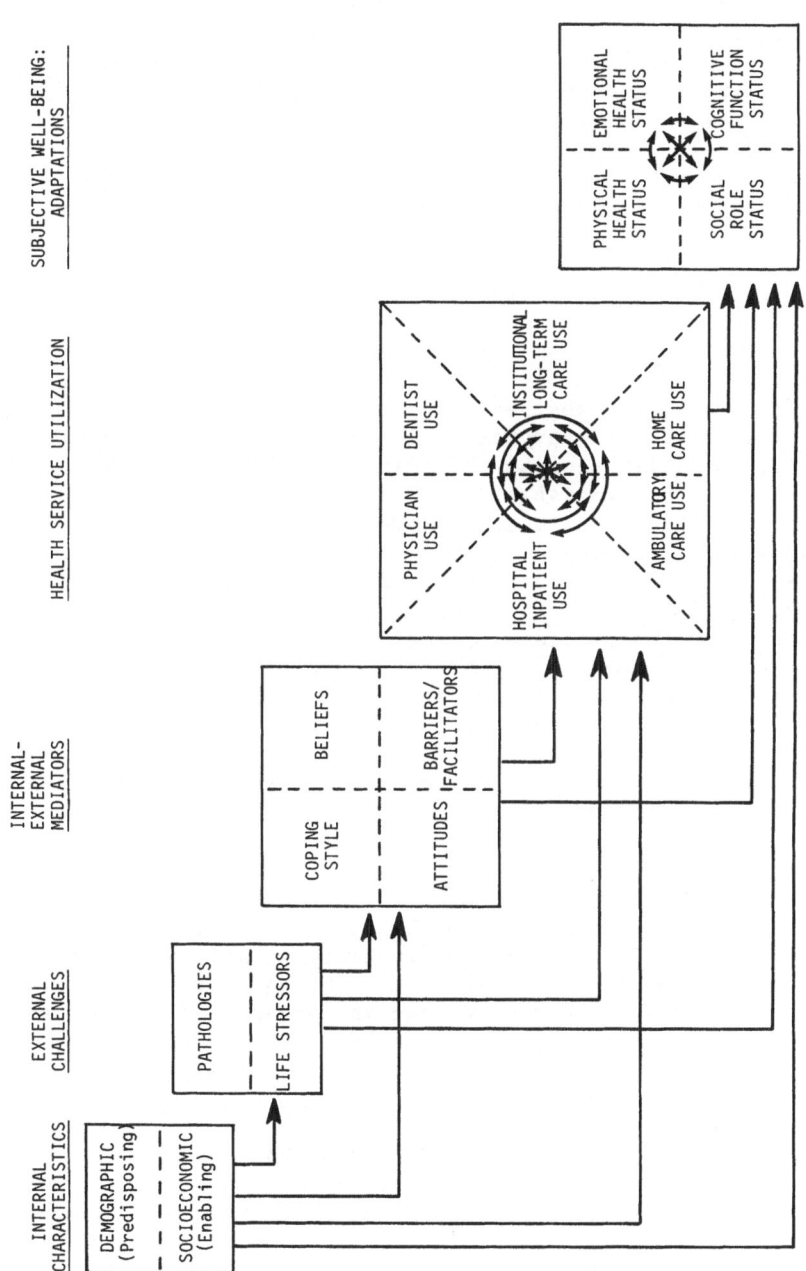

Figure 1. Conceptual model. (Recursive Loops Presumed But Not Indicated)

Our conceptual model also includes the influence of <u>external challenges</u> such as life stress events and specific medical conditions and pathologies or subsequent adaptations that might be expected to affect both health service utilization and emotional health status. Interaction between the internal characteristics of the individual and the external challenges conceptually can influence another set of variables labelled the <u>internal-external mediators</u>. These variables conceptually would encompass the <u>attitudes</u> and beliefs of the individual, coping styles including proclivities to seek assistance, and perceptions of barriers to assistance, expectancies for care, health beliefs, and the like. While health belief variables are usually subsumed under predisposing variables within the health service utilization tradition, we have placed these with other beliefs, attitudes, and expectancies on the assumption that they reflect a synthesis of internal characteristics and external challenges.

Our departure from most existing research which considers both health status and health service utilization is our consideration of health status or well-being as an indicator of adaptation and therefore the outcome variable, and health service utilization as an intervening variable. Our model is predicated on the assumption that seeking and receiving ameliorative medical treatments substantially affects the components of an elder's health status or well-being. <u>Health service utilization</u> conceptually refers to use of services such as physicians, hospitals, dentists, long-term care institutions, ambulatory care, and home care. <u>Subjective well-being</u> reflects enduring adaptations on various dimensions, including emotional health status, physical health status, cognitive functional status, and social role status. <u>Emotional health status</u>, the outcome variable of focus in this analysis, incorporates the concepts of morale and depression.

While this five-category conceptual model emphasizes four dimensions of adaptation or subjective well-being as the outcomes, we recognize that the dimensions of health service utilization and the dimensions of well-being may be interactive and recursive (Vaillant, 1979; Tessler and Mechanic, 1978; Kahn et al, 1975). Accordingly, the model would encourage the investigation of the pathways of influence on any of the dimensions of subjective well-being or health service utilization as outcomes.

TRADITIONAL MULTIVARIATE TESTS OF THE CONCEPTUAL MODEL

Multiple regression analysis can estimate the simultaneous independent influence of numerous potential predictor variables on a dependent variable within a linear model. For all of its considerable strengths, regression analysis does not suggest which variables might precede others in a chain of influence on an outcome variable. It is limited to a sophisticated estimate of the linear association of two variables while holding the influence of other variables constant. This can be seen in the series of multiple regression analyses which we performed for men and women on the influence of 12 possible predictors on two dimensions of health care utilization (Home Care Use and Number of Physician Visits) and one dimension of subjective well-being (Emotional Health Status: Morale). The data for these analyses come from the Year 6 (Fall of 1980) interviews from a statewide sample of 776 noninstitutionalized, continuing participants of the Massachusetts Health Care Panel Study (Branch and Jette, 1982). All participants were over 70 years of age at the time of the interview -- 37 percent were 71-74 years of age; 52 percent were 75-84 years; 11 percent were 85 years or older. The sample is predominantly female (65 percent) and almost exclusively Caucasian. At the time of these interviews, 38 percent lived alone. The sample's median income in 1979 was between $6,000 and $7,000. Table 1 presents the results of the regression. The predictor variables combine to account for a limited amount of the variance of two types of health service use. (R^2 can be interpreted as the percentage of the dependent variable which has been statistically accounted for by the unique contributions of the predictors.) The regression analyses explained 24 percent of women's use of home care, 20 percent of men's use of home care, 10 percent of women's visits to physicians, and 20 percent of men's visits to physicians. The predictor variables accounted for considerable more of the variance of morale -- 41 percent for women and 38 percent for men. The dominant predictor of morale for both women and men was depression.

LISREL ANALYSES

As useful as the regression analyses are, the significant influences of previous life events, physical health status, and depression do not disclose the pathways of influence among these variables on morale which in turn might suggest the underlying principles. Recent advances in computer-generated structural equation models enable us to take a step in the direction of identifying such underlying principles. Based on the research previously reviewed and on the criterion that it is better in this case to postulate a path or a relationship between variables and then fail to confirm it rather than fail to postulate it and then observe an empirical justification for a path, we modified our conceptual model and specified a limited theoretical model for a Lisrel analysis (Joreskog and Sorbom, 1981) (Figure 2). The decision to specify a limited model with eight variables for the Lisrel analysis rather than the approximately 25 variables specified in the conceptual model was made for two reasons. First, the

Table 1. Multiple Regression Analyses for All Predictor Variables for Women and for Men on Home Care Use and Physician Visits

	Home Care Use[a]		Physician Visits		Morale	
	Women	Men	Women	Men	Women	Men
	R^2=0.240 n=433	R^2=0.200 n=259	R^2=0.099 n=433	R^2=0.201 n=259	R^2=0.406 n=433	R^2=0.380 n=259
	beta	beta	beta	beta	beta	beta
INTERNAL CHARACTERISTICS						
Age	0.09					
Income		-0.20				
EXTERNAL CHALLENGES						
Sum of life events[b]	0.10	0.13			-0.09	
HEALTH SERVICES USE						
Number of physician visits		0.16	INAP	INAP		
Use of home care services	INAP	INAP		0.16		
SUBJECTIVE WELL-BEING: PHYSICAL HEALTH STATUS						
Subjective health	-0.11				0.13	0.27
Use walker or wheelchair	0.25	0.13				
Adapted Rosow-Breslau Scale						
Assistance in Basic ADL	0.17		0.17	0.14		
Adapted Nagi reported difficulties						
Adapted Keitel impairments						
SUBJECTIVE WELL-BEING: EMOTIONAL HEALTH STATUS						
Adapted Lawton Morale Scale					INAP	INAP
Adapted Zung Depression Scale					-0.47	-0.42

a. Standardized beta coefficients are presented only for those variables which were significant predictors at the .05 level.

b. Sum of the occurrence within five years of these seven life stressors: death of spouse; death in family or of close friend; family member had major illness or injury; family member entered nursing home; change in household composition; change in household income; person or family member robbed or attacked.

interrelationships and pathways of influence among eight variables
is complex enough in its own right. Second, some of the variables
in the conceptual model such as physical health status are not
measured directly, and therefore require the specification of a
measurement model before inclusion into a Lisrel path analysis. For
our purposes we chose the simpler alternative of including eight
variables, all directly measured.

Based on preliminary analyses, we opted to estimate our models
separately for both females and males, rather than consider gender
as an exogenous variable in the Lisrel analysis. Among the many
things which can be observed from Figure 2 are the numerous paths
specified in the model to be estimated. In this theoretical model
with eight variables, 22 potential paths were specified.

The analyses reported below are therefore based on the maximum
likelihood estimation of two models -- one for women (n=434) and one
for men (n=260). For the two variables which had missing data
(income in 53 cases and depression in 2 cases), the mean value was
substituted in the preparation of the covariance matrix to be
analyzed by Lisrel. Initial work with the theoretical model
indicated that for both women and men, the variable LIVING
ARRANGEMENTS was not a significant contributor to the model and was
eliminated from the final model as presented in Figures 3 and 4.

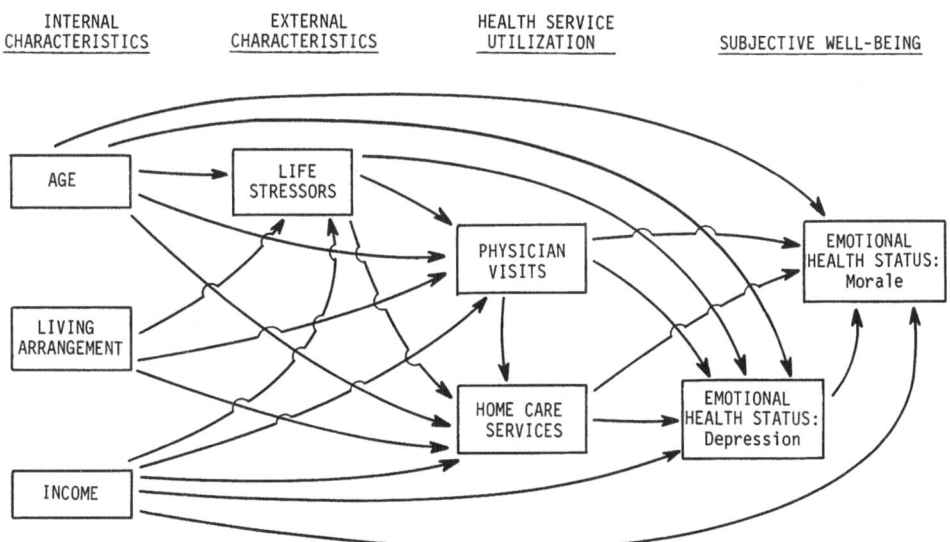

Figure 2. Theoretical model specified for Lisrel path-modeling
analysis.

The Lisrel estimates provided in the tables are the standardized
coefficients.

For elderly women living in the community, eight pathways were
significant in the overall model (Figure 3). Three of the
significant paths were associated with the use of home care services
which in turn was not significantly relate to our outcome variable
of emotional health status as measured by morale.

AGE was significantly related to the experience of life
stressors and to the receipt of home care services among female
participants, but not to physician contact, depression as measured
by an adaptation of the Zung Scale, nor to morale as measured by an
adapted version of the Lawton Morale Scale.

Among these women, INCOME was significantly inversely related
to the receipt of home care services, inversely related to
depression, and positively related to our outcome variable of
morale. Income was not significantly related to the experience of
life stressors or physician visits.

Experience of LIFE STRESSORS among elderly women living in the
community was significantly and negatively influenced by their age,
which in turn significantly influenced their use of home care

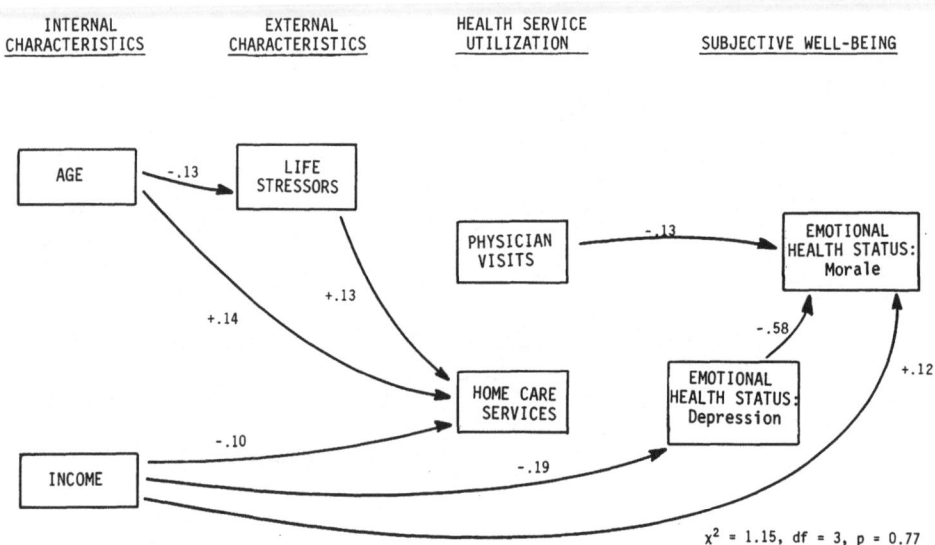

Figure 3. Lisrel model for females over age 70 living in the
communities of Massachusetts.

services. The experience of life stressors was not significantly related to physician visits, depression, or morale.

Women's PHYSICIAN VISITS were significantly and negatively related to morale. The use of HOME CARE SERVICES among elderly women in the community was influenced by their age, income, and experience of life stressors as previously indicated, but the use of home care services in turn was not subsequently related to either depression or morale. In addition, there was no significant path from physician visits to home care services.

DEPRESSION in these women, which we saw was influenced by their income, in turn is the factor most strongly associated with morale. The nature of the relationship supports the theoretical contention that depression as measured is the more transient dimension and morale as measured is the more enduring characteristic.

Summarizing the relationships depicted in Figure 3, we observe that our outcome variable of emotional health status as measured by morale was directly influenced by depression, income, and frequency of physician visits among elderly women in the community.

The adequacy of a Lisrel model is gauged by the criterion of a goodness-of-fit chi-square measure which indicates how well the model as specified fits the data. In this instance, the chi-square statistic was 1.15 (3df), p=.77, indicating an excellent fit.

Figure 4 depicts the nine significant paths of influence on morale among elderly males living in the community. This model also has an excellent fit (x^2=.21,1df,p=.65).

For these men AGE significantly influenced depression, but exhibited no other significant influence on any other variable in the model.

INCOME exerted a direct influence on the experience of life stressors, as well as significant inverse influences on both the use of home care services ad the experience of depression. Income was not directly related to our outcome measure of morale.

While LIFE STRESSORS were influenced by income among these men, they in turn significantly influenced their use of home care services. Again as was the case with elderly women, life stressors did not significantly influence the level of depression reported by these elderly men in the community.

For male participants, PHYSICIAN VISITS significantly influenced the use of home care services and the level of depression reported. In addition, their use of HOME CARE SERVICES was significantly associated with their level of reported depression.

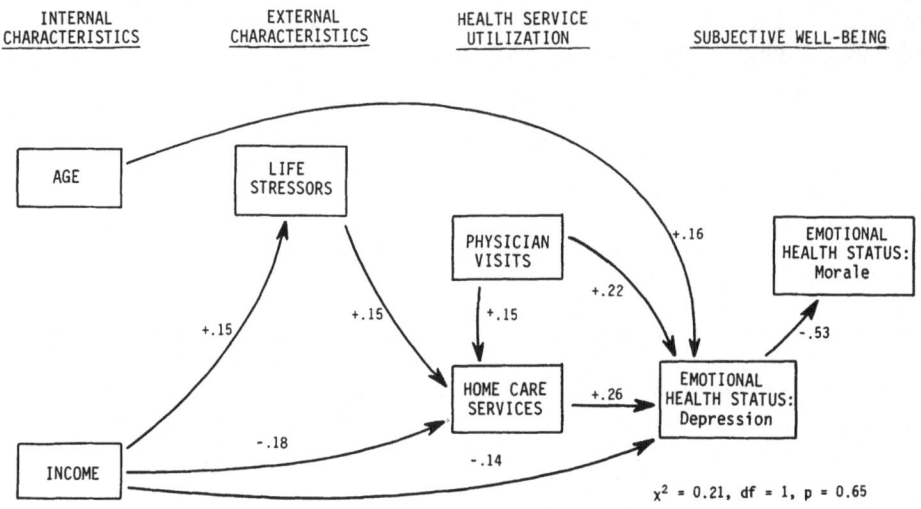

Figure 4. Lisrel model for males over age 70 living in the
communities of Massachusetts.

     In sum, we observed that the reported level of DEPRESSION among
elderly men in the community was influenced significantly by their
age, income, number of physician visits, and use of home care
services; furthermore, depression in turn is the sole significant
path to reported morale.

     Figures 3 and 4 depict noticeably different models for elderly
men in the community compared to elderly women in the community.
For these men, the only significant influence on their level of
morale is their level of depression which in turn was influenced by
four variables (internal characteristics of age and income, and
health service utilization variables of physician visits and home
care services). In contrast, elderly women in the community had
three significant direct influences on their morale, although the
influence of depression was more substantial than the influence on
number of physician visits or income. Interestingly, the level of
depression among community elderly women was influenced
significantly by one variable only (income), not by the four
significant influences observed for community elderly men.

     One interpretation of these gender differences in the pathways
leading to depression and morale is that depression as measured is
indeed an intermediate step to morale for men. But among these
elderly women, level of depression is not the primary path for their
subsequent adaptation in emotional health status as defined by
morale.

We also observed that for elderly men in the community, utilization of health care services was indeed important and significant in their subsequent emotional health status, but the association between health service utilization and emotional health status was much less pronounced for elderly women.

The minimal influence of the experience of life stressors during the previous five years, including loss of spouse, is particularly noteworthy. The direct influence of life stressors with health service utilization and emotional health status has frequently been postulated. It appears that the relationship is limited to the use of home care services.

IMPLICATIONS

The income level of elderly residents in the community has a significant influence on their use of home care services (not surprising because such services in Massachusetts are usually provided through social service agencies which administer an income means test) and to emotional health status as one dimension of subjective well-being. Other characteristics of elderly residents had no independent influence on health service utilization or subjective well-being (e.g., living arrangement). Among men and women over aged 70, age had only modest influence. Age did influence reported depression among elderly men; while among elderly women age influenced their experience of life stressors and their use of home care services.

The experience of life stressors within the previous five years had no direct influence on emotional health status. For both elderly men and elderly women, however, the experience of more life stressors was related to the receipt of home care services. It is possible that the impact of the experience of life stressors would have its influence in a much shorter period of time than five years. Nevertheless these data suggest that an enduring influence of life stressors on emotional status is not supported.

The relationship between health service utilization and emotional health status appears to be different for elderly men than for elderly women. For elderly men in the community, utilization of both physician visits and home care services significantly influences their reported level of depression, which in turn is significantly related to morale. (Presumably all these variables are also influenced by antecedent variables which influence the utilization.) Among elderly women in the community, however, only utilization of physician visits is related to emotional health status, and this at a lesser (although still significant) level than for elderly men.

The analysis reported in this chapter has identified the paths
of influence among variables specified in a theoretical model. This
method of structural modeling and path analysis is to regression
analysis what multiple and partial correlations were to pairwise
correlations -- a sizable step forward in our ability to understand
interrelationships among numerous variables. The Lisrel method can
be considered a technological innovation. Although its costs and
benefits remain to be ennumerated, its potential benefits include a
pressure to consider the underlying principles which underpin the
theoretical model specified for the analysis.

REFERENCES

Adams, D. L., 1971, Correlates of life satisfaction, The
        Gerontologist, 11:64-68.
Aday, L. A., and Andersen, R., 1974, A framework for the study of
        access to medical care, Health Service Research, 9:208-220.
Albert, W. C., and Zarit, S. H., 1977, Income and health care of the
        aging, in: "Readings in Aging and Death: Contemporary
        Perspectives," S. H. Zarit, ed., Harper & Row, New York.
Andersen, R., and Newman, J. F., 1973, Societal and individual
        determinants of medical care utilization in the United
        States, Milbank Memorial Fund Quarterly, 51:95-124.
Beck, A. T., 1967, "Depression," Hoeber Medical Division, Harper &
        Row, New York.
Beck, A. T., 1974, The development of depression: A cognitive model,
        in: "The Psychology of Depression: Contemporary Theory and
        Research," R. J. Friedman and M. M. Katz, eds., Winston &
        Sons, New York.
Berki, S. E. and Ashcraft, M., 1979, On the analysis of ambulatory
        utilization, Medical Care, 17:1163-79.
Branch, L. G., 1980, "Vulnerable Elders," Gerontological Society of
        America Monograph No. 6, Washington, D.C.
Branch, L. G., Jette, A., Evashwick, C., Polansky, M., Rowe, G., and
        Diehr, P., 1981, Toward understanding elders' health service
        utilization, Journal of Community Health, 7:80-91.
Branch, L. G., and Jette, A. M., 1982, A prospective study of
        long-term care institutionalization among the aged, American
        Journal of Public Health, 72:1373-79.
Chown, S. M., 1977, Morale, careers and personal potentials, in:
        "Handbook on the Psychology of Aging," J. E. Birren and K.
        W. Schaie, eds., Van Nostrand Reinhold, New York.
Clemente, F., and Kleiman, M. B., 1976, Fear of crime among the
        aged, The Gerontologist, 16:207-210.
Coulton, C., and Frost, A. K., 1982, Use of social and health
        services by the elderly, Journal of Health and Social
        Behavior, 23:330-339.
Crandall, L. A., and Duncan, R. P., 1981, Attitudinal and
        situational factors in the use of physician services by

low-income persons, Journal of Health and Social Behavior, 22:64-77.

Cumming, E., and Henry, W. E., 1961, "Growing Old: The Process of Disengagement," Basic Books, New York.

Cutler, S. J., 1975, Transportation and changes in life satisfaction, The Gerontologist, 15:155-159.

Frerichs, R. R., Aneshensel, C.S., Yokopenic, P. A., and Clark, V. A., 1982, Physical health and depression: An epidemiologic survey, Preventive Medicine, 11:639-646.

George, L. K., 1981, Subjective well-being: Conceptual and methodological issues, in: "Annual Review of Gerontology and Geriatrics Vol. II," C. Eisdorfer, ed., Springer, New York.

Gerber, I., Rusalem, R., Hammon, N., Battin, D., and Arkin, A., 1975, Anticipatory grief and aged widows and widowers, Journal of Gerontology, 30:225-229.

Goldberg, I., Regier, D., and Burns, B., 1980, "Use of Health and Mental Health Services in Four Organized Care Settings," U.S. Government Printing Office, Washington, D.C.

Gubrium, J. F., 1976, Being single in old age, in: "Time, Roles & Self in Old Age," J. F. Gubrium, ed., Human Sciences Press, New York.

Hodgson, J. H., and Quinn, J. L., 1980, The impact of the triage health care delivery system upon client morale, independent living and cost of care, The Gerontologist, 20:364-371.

Joreskog, K. G., and Sorbom, D., 1981, "Lisrel V," International Educational Services, Chicago.

Kahn, R. L., Zarit, S. H., Hilbert, N. M., and Niederehe, G., 1975, Memory complaint and impairment in the aged. The effect of depression and altered brain function, Archives of General Psychiatry, 32:1569-79.

Kessler, R.C., Brown, R. L., and Broman, C. L., 1981, Sex differences in psychiatric help-seeking: Evidence from four large-scale surveys, Journal of Health and Social Behavior, 22:49-64.

Kohen, J. A., 1983, Old but not alone: Informal social supports among the elderly by marital status and sex, The Gerontologist, 23:57-63.

Kulka, R. A., Veroff, J., and Douvan, E., 1979, Social class and use of professional help for personal problems: 1957 and 1976, Journal of Health and Social Behavior, 20:2-17.

Liang, J., and Warfel, B. L., 1983, Urbanism and life satisfaction among the aged, Journal of Gerontology, 38:97-106.

Marcus, A. C., and Seeman, T. E., 1981, Sex differences in reports of illness and disability: A preliminary test of the "fixed role obligations" hypothesis, Journal of Health and Social Behavior, 22:174-182.

Mechanic, D., 1976, Sex, illness, illness behavior, and the use of health services, Journal of Human Stress, 2:29-40.

Neugarten, B. L., Havighurst, R. J., and Tobin, S. S., 1961,
    Measures of life satisfaction, Journal of Gerontology,
    6:134-143.
Palmore, E., and Cleveland, W., 1976, Aging, terminal decline and
    terminal drop, Journal of Gerontology, 31:76-81.
Pfeiffer, E., 1977, Psychopathology and social pathology., in:
    "Handbook of the Psychology of Aging," J. E. Birren and K.
    W. Schaie, eds., Van Nostrand Reinhold, New York.
Rabbitt, P., 1977, Changes in problem solving ability in old age,
    in: "Handbook on the Psychology of Aging," J. E. Birren and
    K. W. Schaie, eds., Van Nostrand Reinhold, New York.
Rahe, R. H., Meyer, M., Smith, M., Kjaer, G., and Holmes, T. H.,
    1964, Social stress and illness onset, Journal of
    Psychosomatic Research, 8:35-44.
Rosenstock, I. M., 1974, Historical Origins of the Health Belief
    Model, Health Education Monograph No. 2328-335.
Shock, N.W., 1962, The physiology of aging, Scientific American,
    206:100-135.
Stenbeck, A., 1980, Depression and suicidal behavior in old age,
    in: "Handbook of Mental Health and Aging," J. E. Birren and
    R. B. Sloane, eds., Prentice-Hall, Englewood Cliffs, N.J.
Tessler, R. and Mechanic, D., 1978, Psychological distress and
    perceived health status, Journal of Health and Social
    Behavior, 19:254-262.
Tessler, R., Mechanic, D., and Dimond, M., 1976, The effect of
    psychological distress on physician utilization: A
    prospective study, Journal of Health and Social Behavior,
    17:353-364.
Usui, W. M., Keil, T. J., and Phillips, D. C., 1983, Determinants of
    life satisfaction: A note on a race-interaction hypothesis,
    Journal of Gerontology, 38:107-110.
Vaillant, G. E., 1979, Natural history of male psychologic health,
    New England Journal of Medicine, 301:1249-54.
Verbrugge, L., 1976, Sex differences in morbidity and mortality in
    the United States, Social Biology, 23:275-296.
Winn, S., and McCafree, K. M., 1979, Issues involved in the
    development of a prepaid capitation plan for long-term care
    services, The Gerontologist, 19:184-190.
Woehrer, C. E., 1978, Cultural pluralism in American families: The
    influence of ethnicity on social aspects of aging, Family
    Coordinator, 27:329-349.
Wolinsky, F.D., 1978, Assessing the effects of predisposing,
    enabling, and illness-morbidity characteristics on health
    service utilization, Journal of Health and Social Behavior,
    19:384-396.

THE AFFECTIVE CONSEQUENCES OF

TECHNOLOGICAL CHANGE FOR OLDER PERSONS

Johannes J.F. Schroots

Netherlands Institute for Preventive Health Care - TNO
Leyden, The Netherlands

INTRODUCTION

A paradox of our modern technological society is that more and
more technological products and goods are being produced, while at
the same time some crucial emotional ingredients of life are missing
or fading away. Traditionally, religion has played an important
role in people's lives. However, the secularization of society
together with dramatic technological changes has caused not only a
kind of emptiness in human existence, but also an increasing need
for affection and emotional relationships (cf. Wagar, 1971).
Somehow, modern western society does not seem to meet these
affective demands adequately.

Given the foregoing, a number of questions may be asked. For
example: "How is it that technological changes are often followed
by an emotional malaise?" and "How is this impact related to age in
the sense that younger individuals seemingly adapt more rapidly than
older individuals to these changes?" Also, "Are there any scientific
instruments available for the measurement of the emotional
consequences of technological change?"

It is surprising how optimistic scientists sound if they talk
about the use of technology for the elderly and the disabled. In
Bray and Wright (1980), for example, no one was concerned with the
psychological implications of the use of aids and tools for living,
except Wolff (1980), who expressed his concern as follows:

Imagine an old gentleman living by himself who is
physically relatively well, but whose short-term memory is
failing; he can remember with crystal clarity what happened

237

when he was a little boy of 5 years old, but does not
remember that he has put a kettle on the stove. I believe
that even today it would be completely feasible to organize
his living accommodation in such a way that every time he
opened his front door it said, "Have you taken your key
with you?" and ten minutes after he switched on the
electric stove it said, "You know you have got something on
the stove; have you done anything about it?" . . . I am
not suggesting that we install such things tomorrow. We
have not the remotest idea what might be the psychological
repercussions, whether people might become totally
dependent upon it, perhaps vegetables who have to be
programmed through every action of their day. (pp.
115-116) (underlining added)

Technological change and the increasing use of technological
products may not only have unforeseen psychological implications and
repercussions, but also have invaded the methods of science
themselves with many consequences. Some of the negative
consequences for the social sciences have been pointed out
indirectly by Johnson (1976) when he said that it is not enough to
list "barren responses to swiftly delivered questions about 'What do
you need?,' delivered by clipboard interviewers eager to press their
instant replies into a computer." Given the tempo of the questions
asked and their precoded responses it is hardly open to doubt
whether the real needs and demands of the elderly can be assessed in
this way.

Later in this paper I will describe a new assessment method,
called the LIM (Life-line Interview Method), which is
not only self pacing, but also covers the affective aspects of the
most important events in a human life; these events may reflect
technological changes. First, however, I will try to answer the two
questions which I asked earlier: "How is it that technological
change seemingly can have a negative impact on people nowadays?" and
"How is this impact related to age?"

COENOGENESIS AND ONTOGENESIS

Our times may be characterized by at least two important
phenomena: a dramatic increase of average life expectancy and
divergent changes in all sorts of institutions. These institutions
can vary from well-known organizations like the church, family and
state to less familiar, more diffuse conglomerates of institutional
thought such as norms, values and knowledge.

At this point I need to introduce the term coenogenesis which
literally means "coming-into-being of an institution." Normally
defined, however, it refers to the life cycle of a single

institution, i.e., its development over the course of its life. Coenogenesis is distinguished both from ontogenesis, which refers to the life cycle of a single organism, and from phylogenesis, which refers to the evolutionary development of species.

Human ontogenesis can be viewed as a cycle with one period. The length of this period has been increasing dramatically since the last century, especially in developed countries. Birren and Schroots (1980) reported to the World Health Organization that average life expectancy increased so much in a century (from 60 to 100 percent--in some developed countries) that modern men and women almost appear as members of a different species from their forebears. Given the recent rapid increase in average length of life, together with rapid technological changes, individuals undergo more changes in their lives than their ancestors ever did; our generation may be destined to be a very unstable generation.

There has been only a slight, if any, increase in the maximum length of life for the longest lived persons; indeed, the upper limit of length of life seems to be relatively fixed. Projections of life expectancy data show a mean age at death of approximately 85 years in the year 2045 (Fries, 1980). It seems reasonable to expect that human ontogenesis will be stabilized within a few generations when viewed from the vantage points of both average and maximum length of life.

Coenogenesis can also be viewed as a cycle with one period. However, the length of the period is dependent on the types of coenogenesis to be distinguished. We have short term cycles with regard to knowledge, information and communication at the one hand, technological coenogenesis, and we have longer cycles as can be seen in social organizations or institutions (social coenogenesis) at the other hand. In the WHO report quoted before, Birren and Schroots observed the following:

> In view of the many centuries needed for the development of
> a mature culture it is little surprising that a lag, or
> perhaps a void, is created if in a short period of less
> than a century the average life expectancy of a population
> is doubled in length. The matter of defining the proper
> care of the health and social well-being of the elderly
> often finds social institutions historically ill-prepared.
> For the most part they have been concerned with the
> developing child and the nuclear family, and not with the
> adult beyond the age of family rearing or those in very old
> age. (p. 11 and 12)

Part of the tension, uncertainty and stress among the older (and younger) people can be attributed to a phase shift since the last century between the relatively stable, social coenogenesis at

the one side and the temporarily unstable, human ontogenesis at the
other side. Birren and Schroots (op.cit.) gave an example of this
phase shift by pointing out the following dilemma:

> It is most unlikely that those who formalized prescriptive
> rules about "honoring thy father and thy mother" were faced
> with the dilemma of contemporary developed societies in
> which five generations can be living at the same time.
> Members of the middle generation have responsibilities for
> parents and grandparents as well as for children and
> grandchildren. If all of these roles are equally
> prescriptive there isn't sufficient time to allocate to the
> members of each of the different generations. Societies of
> the coming century will indeed have frequent occurrences of
> five generation families. (p. 11)

From the foregoing we may conclude that a change in phase
relationships between social coenogenesis and human ontogenesis
would resolve part of the uncertainty and tension which people are
experiencing. Given the relatively fixed upper limit of the human
life span it seems that church leaders, social policy makers and
other leaders of humanitarian institutions--rather than ordinary
people--have an important task with regard to this change by
accelerating social coenogenesis so that a new equilibrium can be
established. The slow development of many social institutions is in
vivid contrast with the shorter and shorter cycles of technology and
science. The keywords for technological coenogenesis are
acceleration and a decreasing life span. Technological coenogenesis
with its decreasing life span and increasing speed of information
processing constitutes the opposite of human ontogenesis with its
increasing length of life and decreasing speed of information
processing with age (Birren and Schaie, 1977). This opposite phase
relationship is therefore another potential source of stress for the
elderly. It may take more time for the older scientist to work
effectively with a new piece of equipment, say the next generation
of computers, than for a younger colleague.

Apart from its function as a stressor, there are at least three
other negative consequences of technological coenogenesis. Firstly,
modern technology contributes to the obsolescence of skills
possessed by older people, forcing them into dependent roles (Ward,
1981-1982). With the introduction of hi-tech equipment, however,
this dependency also applies to younger persons as they often feel
like an extension of machines, without any control. The
introduction of robots will even accelerate this development with
negative consequences for the self-esteem of skilled and non-skilled
workers.

Also, modernization and informationalization of society can
lead to an atrophy of human communication skills as people

communicate more and more with machines and computers instead of human beings. The term "dyslexithymia" is suggested for the emotional blindness people can suffer. This term refers to the fact that individuals who communicate mainly with or via machines can become blind to the affective qualities of human communication and cannot read the non-verbal cues people give at the affective level of inter-individual relationships.

Technological change with its increasing use of hi-tech machines and equipment will also reduce the opportunities for intimate attachments to other human beings which are, according to Bowlby (1980),

> . . . the hub around which a person's life revolves, not only when he is an infant or toddler or a school child but throughout his adolescence and his years of maturity, and on into old age. (p. 442)

In concluding this section I would like to say that lack of affect due to technological change can not only be anticipated, but also compensated for by active participation in intimate, affective networks of people. Preferably, we should start these networks at a very young age and renew them regularly according to our needs. During the current transition period of informationalization we are in right now, we should keep an open eye for the yet hardly noticed, but vital menace to our mental health: one-dimensional information.

LIFE-LINE INTERVIEW METHOD

The LIM (Life-line Interview Method) is a powerful method for eliciting biographical information with affective value. As one of the interviewees said at the end of the LIM-session (75 minutes): "I never told this before to one single person in my life. I feel entirely exhausted and I really don't know what else to say!" This reaction is not only typical of the emotional involvement of interviewees during a LIM-session, but also in vivid contrast with the behavior of individuals at the end of a semi-structured interview: they start talking about their feelings and emotions when the interview is over, even when the subject of the interview is related to feelings of loneliness, intimate relationships with friends and other emotionally-colored topics (cf. Schroots, 1982). Apparently, questionnaires do not elicit much affect during the interview-session which seeks one-dimensional information.

For most social scientists the evaluation of the affective consequences of technological change does not seem to be a very difficult task at first sight. If no convenient questionnaire is available for that purpose, they usually start to construct one or more scales themselves by formulating pre-determined questions about the subject. This procedure is pretty much routine in the social sciences and almost no one really discusses such well-established practice. However, in spite of their obvious merits, questionnaires do have some serious drawbacks (cf. Johnson, 1982).

One drawback is that, due to a very strong tradition of neo-behaviorism, the dimension of time as represented in the life history of an individual has been completely left out in psychological inquiries. In discussing the concept of life satisfaction Back (1981) refers to this serious omission as follows:

> Social scientists . . . have tended to measure satisfaction in current time, partly because it is technically easier to do so. It is easier to make sure that the respondent is concentrating on current feelings than on the whole life span, perhaps even including expectations for the future. Thus most interviewing scales on life satisfaction consist mainly of questions about current feelings . . . What is needed, therefore, is a convenient measure of assessing the whole life. (p. 1)

Back (1981) and Johnson (1976) are not the only advocates of the reintroduction of a temporal framework within which the behavior of the respondents ought to be interpreted. Many others are also of the opinion that it is necessary to use the life history of the individual as the major framework for their inquiries (Chiriboga, 1978; Cooper et al., 1981; Rakowsky & Hickey, 1981; Rosenmayr, 1981). Raynor and Entin (1982) even declare that the motivation of individuals cannot be understood when the temporal dimension has been omitted.

In addition, questionnaires seldom take into account the affective tones and feelings of personal life events. Questions are asked and interpreted at the cognitive level of organization of behavior, even to the extent that respondents are asked to check a list of words, indicative of personal feelings and emotions. It seems to be extremely difficult to gather information about the affective qualities of events by means of semi-structured interviews and questionnaires. Apparently, these instruments appeal to the more cognitively and verbally oriented qualities of the individual's behavioral organization.

From the foregoing it follows that it will be a difficult task to evaluate the affective consequences of technological change from conventional methods and instruments. Although the free flowing or focussed interview avoids many of the drawbacks of semi-structured interviews and questionnaires, this approach does not seem to be very attractive either, because problems of gathering specific information and of scoring and analysis of the responses are still very complicated (cf. Rosengren, 1981). Therefore I developed a new

method, called the LIM. Essentially, the LIM has been developed on
the basis of metaphors which people use to describe their life
histories and expectations for the future. As I have discussed
before in detail (Schroots, in press):

> . . . metaphor seems to be the key to understanding both
> the methodological and creative aspects of scientific
> discovery and progress. Metaphor is an important way by
> which we create new meanings, making sense and alternate
> sense where there is little or none before. Metaphor, in
> short, is a means of entering the unknown through the
> gateway of the known. The metaphor allows us to map what
> we know onto what we vaguely know and gives rise to new
> hypotheses and integration.

When older persons are asked to describe their life, they
frequently use metaphors like the "river" or "footpath" (cf.
Vischer, 1961). The river symbolizes the stream of life, and the
footpath stands for the journey one makes from birth to death, when
one alternately crosses the mountains and valleys of life. Both
metaphors enclose the forementioned temporal dimension, but only the
"footpath" metaphor refers explicitly to the dimension of affect.
For example, when people say "I'm feeling up" or "I'm really low
these days," they are using a spatial metaphor, i.e., hilly country,
to express the positive and negative feelings they had in life.

From the "footpath" metaphor to the LIM is only one step. Once
I realized that the graphical, two-dimensional representation of a
footpath -- with time on the horizontal dimension and affect on the
vertical dimension -- symbolized the course of human life, I had
found the key to a new method. With the help of this method one can
elicit biographical information about important life-events in a
non-verbal, visual way. Consequently, one avoids the faults of more
cognitively-oriented techniques like questionnaires, open interviews
and auto-biographies which appeal primarily to the verbal capacities
of the individual. Before discussing the LIM in more detail, I will
describe briefly the course of a typical LIM-session.

The interviewer first introduces the general plan of the
session by saying that he is interested in the human life course
with its ups and downs, level periods, rises and declines, etc.,
which are all completely different from one person to another. He
then explains that he would like to hear his subjects's life story
in a special way, which he will illustrate by giving some typical
examples. However, before doing this, he hands a piece of paper
that looks very much like a blank-grid (Figure 1) and says "You will
notice that the bottom and top line represent chronological age
(0-100 years). Please indicate your age at the bottom line."
Drawing a vertical line from the indicated age mark to the top, the

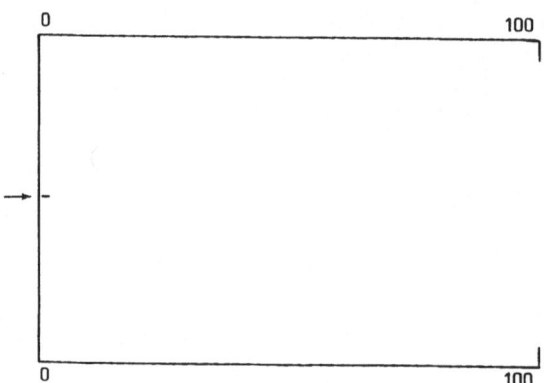

Figure 1.   Blank-grid used for the Life-line Interview Method
(LIM); dimensions: 300 x 185 mm.

interviewer continues in saying "Look, this is your life space. Can
you now visualize how your life could be put into a graph, starting
at this birth dot? First I will give you some examples of the
life-lines other people have drawn." These examples are further
illustrated with explanations like "The higher the life-line, the
happier this particular person feels himself" or "This dip shows
that this person felt very depressed at this age." The subject is
then asked to draw his or her life-line from birth dot to the
present age-line.

As soon as the life-line has been drawn, the subject is asked
to label each peak and each dip by chronological age and to tell
what happened at a certain moment or during an indicated period. At
the same time the interviewer makes a verbatim report of what the
subject sees as the most important events in his or her life.

After the subject's past (and present) have been visualized and
described in detail the future can be explored in the same manner.
Starting from the point where the life-line has stopped, the subject
is asked to continue the line until the age-line of death is
reached. Then the whole procedure of explaining the future
life-line and making a verbatim report is repeated.

When the LIM-session is over, the final result is a visual and
verbal life history of the individual, and his or her visual and
verbal future perspective as well, in terms of what he or she thinks
has been or will be important in life. The life history and future
perspective of the individual can be transposed, then, very easily
into a numbered list of events provided with various labels:
objective time (chronological age), subjective time (age estimation,

horizontal dimension), affect (height estimation, vertical
dimension), and cognition (content-categories). Other measures may
also be computed.

The starting-point for the development of the LIM was the idea
that this method allowed a person to place perceptions of his or her
life visually in a temporal framework. As has been shown in a
pilot-study of mostly older persons (Schroots, 1983) this idea
turned out to be very important for several reasons. Firstly, as
most people had been familiar with the graphical representation of
time by a straight line, they did not need much thinking before
drawing their life-line. However, almost all interviewees from my
study fully realised only after the drawing of the life-line what
they were revealing to the interviewer. Some of them even went so
far that they tried to "correct" the original line. All this is a
positive sign of the LIM's capacity to elicit biographical
information at the affective level of the behavioral organization of
the individual.

A second, innovative aspect of the LIM is its self-pacing
quality. The life-line is drawn and the personal story is told at
the older person's own pace. So, this method not only meets the
objections I raised with regard to modern questionnaires, but also
synchronized perfectly well with the tempo of each individual
person. In practice, the non-directive atmosphere of the interview
facilitates that someone's real opinions, beliefs and attitudes are
expressed.

From the foregoing one may conclude that the LIM can serve
diagnostic, process-facilitating and evaluative functions. The
function of evaluation of technological change needs a few comments
as well. First of all, I want to note that technological change or
its impact on older persons will be seldom mentioned spontaneously
during the interview. The reason for this is rather obvious:
individuals do not think about their own life in terms of the
negative consequences of the introduction of electricity or
micro-computers for the labor-market. They only say that they
became unemployed at a certain age. In other words, the
relationship between the effects of technological change and
someone's life history or expectations for the future is a very
loose one. As such the LIM in its present form can hardly serve an
evaluative function.

However, accompanied by special sheets for the assessment of
specific life themes, such as family, friendships and occupations,
the LIM can be easily extended in the desired direction (Figure 2).
These special sheets look almost identical to the original life-line
sheet, except for the solid middle line and the dotted lines above
and beneath the middle line.

Figure 2. Grid used for the extended LIM; dimensions: 300 x 185 mm.

The middle line represents chronological age (0-100 yrs) and the dotted lines can be conceived as vertical Likert-scales with three plusses and minusses as positive and negative extremes, respectively. The procedure of the extended LIM corresponds in a large measure with the usual LIM, provided that a list of events, i.e., technological events, has been generated by the interviewer or interviewee beforehand. Each event has further to be marked on the middle line in behalf of an age-indication, after which the affective impact can be marked vertically on the dotted lines. The final combination of LIM-data plus extended LIM-data, then, can be used for the evaluation of the affective consequences of technological change.

In conclusion it might be said that modern, technological society has hardly noticed yet the negative consequences of its informationalization. These consequences manifest themselves in an increasing need for affect--especially with older persons--which cannot be satisfied adequately by new devices. It may be noted that the same technological development which created a hunger for affect interfered with the construction of instruments to measure affective needs. With the introduction of the LIM and its extension hopefully a fresh start is given.

In the long term cohort changes with regard to individuals and leaders of humanitarian institutions will solve in all probability most of the problems due to technological change. However, at short notice personal care will be a good substitute for the lack of affect that is felt increasingly by older persons.

REFERENCES

Back, K.W. Types of life course and gerontology. Paper presented at the 12th Int. Congress of Gerontology, Hamburg, July 17, 1981.

Birren, J.E., & K.W. Schaie (eds). Handbook of the Psychology of
     Aging. New York, Van Nostrand Reinhold, 1977.
Birren, J.E. & J.J.F. Schroots. Aging: from cells to society; a
     search for new metaphors. Prepared for the WHO Global Program
     for the Care of the Aged, July 1980.
Bourque, L., & K.W. Back. Life graphs and lifeevents. J. Geront.
     32 (1977) 669-674.
Bowlby, J. Attachment and Loss. 3: Loss-sadness and Depression.
     London, Hogarth Press/Institute of Psycho-Analysis, 1980.
Bray, J., & S. Wright (eds). The Use of Technology in the Care of
     the Elderly and the Disabled. London, Frances Pinter, 1980.
Chiriboga, D.A. Evaluated time: a life course perspective. J.
     Geront. 33 (1978) 388-393
Cooper, P.A., L.E. Thomas, S.J. Stevens & D. Suscovich. Subjective
     time experience in an intergenerational sample. Int.J.Aging
     hum.Dev. 13 (1981) 183-193.
Fries, J.F. Aging, natural death, and the compression of
     morbidity. New Engl.J.Med. 303 (1980) 130-135.
Johnson, M.L. That was your life: a biographical approach to later
     life. In: J.M.A. Munnichs & W. van den Heuvel (eds). Depen-
     dency or Interdependency in Old Age. Den Haag, Nijhoff, 1976.
Rakowski, W., & T. Hickey. A brief life-graph technique for work
     with geriatric patients. J.Am.Geriatrics Soc. 29 (1981)
     373-378.
Raynor, J.E., & E.E. Entin (eds). Motivation, career striving, and
     aging. New York, Hemisphere, 1982.
Rosengren, K.E. Advance in content analysis. Beverly Hill
     (Calif.), Sage, 1981.
Rosenmayr, L. Objective and subjective perspectives of life span
     research. Ageing & Soc. 1 (1981) (part I) 29-49.
Schroots, J.J.F. Metaphors of aging: reactions, implications and
     extensions. Presented at the Invitational Research Symposium,
     "Metaphors in the Study of Aging," University of British
     Columbia, June 1982 (in press).
Schroots, J.J.F. Loneliness from an ontogenetic perspective.
     Presented at the 'Psychologencongres 1982,' Amsterdam, Free
     University, October 1982.
Schroots, J.J.F. Zelfstandigheid van Ouderen; verslag van een
     vooronderzoek. Leiden, NIPG-TNO, 1983. (interne publicatie).
Vischer, A.L. Seelische Wandlungen beim alternden Menschen.
     Bazel/Stuttgart, Benno Schwabe, 1961.
Wagar, W.W. Religion, ideology, and the idea of mankind in
     contemporary history. In: W.W. Wagar (ed.). History and the
     Idea of Mankind. Albuquerque, University of Mexico Press,
     1971. Pp. 196-221.
Wolff, H.S. A cautious forecast on intelligent aids. In: J. Bray &
     S. Wright (eds). The Use of Technology in the Care of the
     Elderly and the Disabled. London, Frances Pinter, 1980. Pp.
     113-118.

IMPACT OF WORK AND TECHNOLOGY ON HEALTH STATUS OF THE OLDER WORKER

Ruth B. Weg

Andrus Gerontology Center
University of Southern California, Los Angeles

Work itself, when appropriately tuned to human capacities and limitations, can be as potent a factor for health promotion as for health abuse. Occupational therapy and vocational rehabilitation attest to the value of work in overcoming physical and emotional disabilities. One study of people who had lost or were about to lose their jobs in the New York garment industry concluded that "for many workers ... we saw their participation in the world of work as oxygen--sustaining them in the face of emotional disorder" (Weiner, Akabas and Sommer, 1973). On the other hand, work characterized by a poor person-job fit, fear of failure, lack of control, or fear of economic or status loss creates stressful situations not easily resolved unilaterally by the worker, resulting in feelings of impotence and hopelessness.

Studies of stress responses in human health status have correlated physical/psychological stressors with particular mediators of stress response and biochemical, structural changes leading to various physical and affective changes/disorders. Age (time) has been addressed frequently in terms of "exhaustion of adaptive capacities," and as time for damage to accumulate and surface. Since aging physiology grows less efficient, stressful experiences may represent an increasing burden. Data suggest that change is frequently perceived as threatening, stimulating varied responses which have been correlated with illness and disease. Early empirical research on life change events (stressors) and health has been criticized methodologically and conceptually. Nevertheless, data have been instructive. Inclusion of several more variables (social environment, life cycle perspective, the notion of no change) could improve usefulness of the data, insure their validity, and potentiate intervention.

In the management of change to automated technology there exists potential for enhancement or deterioration of the quality of worklife and health. Automation and technology can be imposed, mechanically and initially cost effectively--leaving the working individual as a disregarded, dispensable cipher with resulting serious impact on mental and physical health. Or management can choose a participatory role for workers in the production changes into the 21st century. Middle aged and older workers can then be beneficiaries of a humanized, though automated worksite-- characterized by health/fitness promotion programs, worker participation in decisions, appropriate training and retraining with attention to variety and skill demands in keeping with experience and abilities.

Research in the impact of work and technology on the health of the older worker is minimal, reflecting the research design difficulties and complexities, and the recency of the graying of the workplace. A "definitive work" on automation was the report from the National Commission on Technology, Automation and Economic Progress (1966). There was relatively little examination of the effect on the individual of the changing quality of an automated, technological worklife in the report or elsewhere until the 1970's. An important question was asked--"What kind of assemblyline technology provides work dimensions that enhance rather than depress morale and mental and physical health of workers?" That question must be raised again, and reframed to fit current and near future technological picture. Throughout the industrial revolution, human pace of work was made to fit the machine pace--a distinctly nonhuman rhythm and purpose leading to occupational stress.

Responsible research, government and corporate communities must consider the human affective, intellectual and social qualities in the design of automata, and simulate alternative future technology/human interactions so that the most destructive choices (to efficiency and human needs) may be avoided. Longitudinal investigations and interventions under controlled conditions will be necessary to test hypotheses regarding the interrelationships among work, age and health. One-time observations ignore the capacity for learning and becoming--the plasticity of the human being. The challenge remains to break down stereotypes about older workers, to make measurements in real work situations, of capacities, benefits and disadvantages to health for enough time to enable valid conclusions and useful generalizations.

REFERENCE

Weiner, H.J. Akabas, S.H., and Sommer, J.J., 1973, "Mental Health Care in the World of Work," Association Press, New York.

PSYCHOLOGICAL FUNCTIONING IN OLD AGE

AND THE INTRODUCTION OF NEW TECHNOLOGY

Jan-Erik Ruth

Gerontological Research Unit, Kuntokallio-Institute
Finland

A central feature pertaining to an autonomous use of new
technology in the everyday life of the elderly concerns their
cognitive readiness to try out new inventions. Are the elderly
capable of using advanced technology and, above all, are they even
motivated to try? How should the use of technological devices be
presented, based on what we know about the functioning of the aged?
Modes of thinking in the aged have been studied recently from the
perspective of "age and creativity": in Finland (Ruth, 1980;
Mattlar, Ruth and Knuts, 1980-81) and in the United States (Alpaugh,
1975; Alpaugh, Renner and Birren, 1976). The studies were based on
divergent thinking tests, and the cognitive processes of different
age-groups were compared using rather extensive samples.

The results show that age differences to the disadvantage of
the old appear in the fluency, flexibility and originality of
abstract thinking. A rapid production of new ideas reflecting
different categories of thought is not a typical feature of information processing in old age. In concentration, in activation of
information from the memory store, and in shifting lines of thought,
some problems evidently occur. Further, technologically creative
thinking is less typical for old women than old men. The studies
mentioned are all cross-sectional and the results are thus tied to
the cognitive functioning of the aged of today. The results nevertheless can be useful in teaching the use of new technology to the
aged in our modern society. A model encompassing physiological,
cognitive and socializational variables has been constructed to
explain the results presented above (Ruth and Birren, in press).
Making use of the results from the studies above, and additional
knowledge about the behavior of the aged (Ruth, 1983) the following
recommendations can be made:
- The use of new technology must be perceived as meaningful by the

aged users. Efforts must be made to link the use into the life-
style of old persons. If no substantial needs of the aged are met
by the new product, one should abstain from introducing it to them
at all.
- Cautiousness and disbelief in one's own capability may be an
  obstacle in accepting technological advances. Advice and support
  needed in handling the new technology should be provided to the
  aged.
- The old must be given a generous amount of time, as well as repeated short training sessions, when new technology is introduced.
- In the presentation of technological features, more stress should
  be put on practical application than on technical features.
  Selective central facts should be presented, less essential aspects left out.
- Mnemonics and cues will favorably effect self-efficacy in handling
  new technical products. Mnemonical devices should be presented
  when the product is introduced.
- The home and the natural meeting places of the aged should be used
  whenever possible new technology is introduced and training sessions arranged.
- The instructors must be well-known by the aged, or their appearance must be publically announced well in advance, through formal
  and informal channels.
- The attitudes of the instructors towards the aged must be positive, but realistic. The main issue for the instructor should not
  be can the old learn to use the new technology, but how can the
  instruction be arranged so as to meet the special needs and characteristics of the aged.

REFERENCES

Alpaugh, P.: Variables affecting creativity in adulthood: a descrip-
      tive study. MA-thesis. University of Southern California,
      1975.
Alpaugh, P.; Renner, V. J. and Birren, J. E.: Age and creativity.
      Its application for education and teachers. Educational
      Gerontology, 1976, 1, 17-40.
Mattlar, C.-E. Ruth, J.-E. and Knuts, L.-R.: Creativity measured by
      the Rorschach test in a random sample of Finns. Yearbook
      Geron, 1980-81, 23, 15-26.
Ruth, J.-E.: Creativity as a cognitive construct: The effects of
      age, sex and testing practice. Dissertation. University of
      Southern California, 1980.
Ruth, J.-E.: Finnish research on life-style and psychological
      functions in the latter part of the life-span. Adult Education
      in Finland, 1983, 1, 11-17.
Ruth, J.-E. and Birren, J. E.: Creativity in adulthood and old age.
      International Journal of Behavioral Development, 1984.

INTERPERSONAL RELATIONSHIPS AND PSYCHOLOGICAL WELL-BEING

IN MIDDLE-AGED PERSONS*

Joachim Wittkowski

Department of Psychology
University of Wuerzburg
Wuerzburg, Germany

INTRODUCTION

This study aims at discovering the impact of integration in interpersonal relationships as regards an individual's satisfaction with his past and present life, on the one hand, and the extension and evaluation of his future time perspective on the other hand. In accordance with findings reported in the literature, it is hypothesized that there is a positive correlation between variables of interpersonal relationships and variables of life satisfaction, and also a positive correlation between variables of interpersonal relationships and variables of future time perspective.

METHOD

Subjects were 48 men and 47 women, aged 50.0 to 55.0 years (M=52.4; SD=1.4). They were of middle to upper socio-economic class and the majority of them were living in a German town of approximately 130,000 inhabitants. Subjects took part in a semi-structured interview which, among other topics, dealt with "integration in interpersonal relationships" (communication and face-to-face contact with relatives, friends, colleagues, etc., including significant others); with "life satisfaction" (evaluation of one's past and present life); and with "future time perspective" (outlook on one's future). By means of content analysis, the interview material was coded in the dimensions "frequency of interpersonal relationships" (IR-F), "contentedness with frequency of interpersonal relationships" (IR-CF), "satisfaction with one's present life" (LS-PR), "satisfaction with one's past life" (LS-PA), "extension of future time perspective" (FTP-FX), and "evaluation of

future time perspective" (FTP-EV) by two raters independently.
Interrater reliability proved to be above .70.

RESULTS AND DISCUSSION

In the total sample there is a positive association between
IR-F and FTP-EV (r=.30, p<.01). Within the men there are
significant positive correlations between IR-F and LS-PR (r=.30,
p<.05), on the one hand, and FTP-EV (r=.41, p<.01), on the other
hand. Additionally, within the men there is a positive correlation
between IR-CF and LS-PA (r=.29, p<.05). There are no significant
correlations within the female sub-sample. For the present data the
appropriateness of product moment correlation techniques may be
questioned. Therefore, a non-parametric approach was carried out
additionally. Kendall's tau replicates those correlations provided
by Pearson's r, except the association between IR-F and FTP-EV
within the female sub-sample (tau =.20, p<.05).

On the basis of the present findings it may be supposed that
frequency of interpersonal relationships, including close friends,
determines both satisfaction with the present life and positive
evaluation of the future of an individual. However, self-esteem has .
to be taken into account as an intervening variable. Contacts with
a great number of people may be regarded as a premise for a positive
self-concept or high self-esteem, which, on its part, is likely to
determine satisfaction with one's present life as well as a
favorable evaluation of one's future. If this interpretation holds,
those technological advances which lead to a decrease in close
interpersonal relationships within the family and outside the family
may reduce the psychological well-being of the middle-aged and aged
individual. Former experiences in interpersonal relationships and
whether the individual has a low frequency of interpersonal
relationships (voluntarily or involuntarily) are additional
variables that should be considered in further studies.

*The study was supported by the Deutsche Forschungsgemeinschaft
(DFG).

AGING AND TECHNOLOGICAL ADVANCES:

HEALTH AND STRESS

Eileen M. Crimmins

Andrus Gerontology Center
University of Southern California

INTRODUCTION

A summary of the Symposium discussion on the relationship
between technological change, health and stress is presented in this
chapter. First, we discuss the topic of health in relation to
technological change; then outline a model which suggests the paths
through which technological change can make an impact on health.
The latter section includes a discussion of the role of stress as an
intervening variable between technogical change and health. It
should be noted that, for this discussion, specific medical advances
were not considered.

TECHNOLOGICAL CHANGE AND HEALTH

As technology has advanced, health has improved. People are
living longer on the average and more are surviving to old age.
Over much of the past century we have experienced remarkable
declines in mortality due largely to scientific and technological
advance rather than specific medical interventions. Until recently
most of the decline in mortality was due to reduction in deaths from
a variety of infectious diseases which were eliminated through
understanding of the germ theory of disease and concomitant activity
to reduce the spread of disease through public and private health
measures. In addition a general improvement in the economic
condition of our society, fueled by technological advance led to a
reduction in deaths from infectious disease. Data presented by
Fries show that because past mortality improvement has virtually
eliminated deaths from acute diseases, most deaths are now due to
chronic causes. Significant future reduction in mortality will,

thus, require reduced death rates from the major chronic diseases such as heart disease and cancer. Because the pattern of diseases has changed over time, the relationship of health and technological change observed in the past does not have to hold in the future.

To understand contemporary and future relationships between health and factors affecting health we now have to develop new models and measures of health. Commonly employed models of health and techniques of measuring health have not been adapted to this new dominance of chronic diseases. Chronic diseases differ from acute diseases in a number of ways that require this adaptation. They are not cured and eliminated like acute diseases. In general, when one is saved from death from a chronic disease, one continues to have the disease or condition. Because people are not cured of chronic conditions, when mortality from these causes is reduced, it is not clear that the health of the surviving population must be improved. It is possible that a population with impaired functioning survives. Because of this we need better measures of health; we can no longer use mortality as the major indicator of health.

Chronic diseases do not develop quickly and run their courses like acute diseases; but develop slowly over long periods of time. People do not "have" or "not have" some of the major chronic conditions; but all people "have" the condition to some extent even though it may not be clinically evident. This pattern of progression means that determining the cause of many chronic conditions and the effect of interventions to reduce their progression is more difficult than with some of the major acute diseases of the past. Finally, while the etiology of some of the major chronic diseases is not well understood, chronic diseases are less likely to be caused by micro-organisms to the same extent as acute diseases, but appear to be more rooted in factors such as lifestyle, personality, heredity, and in different aspects of the environment. This leads to the conception of health as a bio-psycho-social phenomenon which joins psychological and social factors to the traditional biological factors as influences on the development and course of chronic diseases.

THE BIO-PSYCHO-SOCIAL MODEL OF HEALTH

The bio-psycho-social conception of health provides us with a useful framework for conceptualizing the effects of technological change on health. We can examine separately how technological change might affect biological factors, social factors, and psychological factors influencing health. Of course, both health and technological change are multidimensional concepts themselves. To simplify our discussion, we will use technological change in the workplace as the example to illustrate our framework. Many other forms of technological change affect health as well. Health, too, can be divided into mental, psychological, and physical health, and ill health of any of these types can be subdivided by condition or disease in order to refine the analysis.

Technological Change and Biological Change

Shephard's presentation provides a good illustration of how technological change in the workplace can cause biological changes that can then produce changes in the health status of a population. Without adaptive changes in diet or leisure activity, the reduced physical demands of jobs in more technologically advanced societies result in physiological changes such as increased fat, loss of muscle strength, and a decrease in maximum oxygen intake. Further research is needed on the links between these changes and health but Shephard suggests that such changes may increase levels of heart disease and injury in a population.

Technological Change and Social and Psychological Change

Changes in technology can also cause changes in the social organization of the workplace which are generally thought to affect health through altering the levels of stressors in the environment. Thus stress enters the model as an intervening variable between social organizational factors and health. Stress, in our discussions, was seen to represent a psychological state generally affected by the social environment.

Technology is a determinant of the occupational mix of the labor force at any point in time and change in the occupational mix between points in time. Research presented to our group by Cooper indicated that characteristics of jobs are related to the level of stress experienced by the worker. These characteristics include those intrinsic to the job, the role of the job in the organization, relations among workers, career development pattern of the job, structure of the organization in which the job is performed, and the home-work interface for the holders of the job. This presentation gave us a useful piece of the model relating technological change and health. As technology changes, the stress of work changes with the changing characteristics of jobs and the mix of jobs.

Higher levels of stress are generally thought to be related to higher levels of some diseases such as coronary heart disease; however, the idea that stress can have positive as well as negative effects was emphasized in our discussions. In addition Olbrich brought out the idea that stress is a function of the interaction between stressors and coping; therefore the effect of a stressor on an individual depends on his coping abilities as well as the strength of the challenge to his equilibrium. Of course, abilities to cope may also change with technology and its effects on social organization. The introduction of coping as another variable

affected by technological change further expands the model that must
be developed to understand the role of technological change in
affecting health.

An important generalization across occupations from Cooper's
work makes clear the important role coping could play in a model
relating technological change, health, and age. Cooper found that
within a given occupation older workers experience more stress. The
hypothesis was introduced that coping ability may be lower among
older workers; so that, when job characteristics are constant, older
workers experience more stress. This hypothesis, generated from
cross-sectional studies, may have implications for adaptations to
technological change in the workplace.

FUTURE RESEARCH NEEDS

Our discussion continually emphasized the need for future
research in the area of factors affecting health. The first
priority, however, is in the operationalization and measurement of
health itself. Currently available measures prove best for
measuring health once it degenerates to the phase of
disintegration. Measures of health need to be applicable across the
life span rather than only after significant degeneration. In
addition, they should be cumulative measures not just point
measures. Most importantly, they should be able to measure change
across cohorts and over time and they should indicate when
intervention has been successful.

Research on the factors affecting health should attempt to
explain the mechanisms through which health is affected. In the
case of technological change, the effects should be divided into the
biological, social and psychological. Because of the recent
introduction of social and psychological factors into the model of
factors affecting health, measurement and conceptualization in these
areas remain even more rudimentary than in the biological area.
Future work should attempt to develop biological, social and
psychological markers. All of these markers should be measurable
across the life span and should be able to be related to the
measures of health.

Our knowledge of the effect of technological change on health
would be greatly increased by intensive study of the recent decline
in heart disease death rates which has occurred in many countries of
the world. This not very well understood change in mortality from
the most important cause of death among adults provides us with an
opportunity to gain unprecedented insight into change in a chronic
disease death rate, the factors affecting it, and how this might be
related to technological advance. This opportunity is especially

valuable in that this is the condition most widely believed to be related to technological change.

Our group concluded with the idea that technological change in the future is inevitable. The form technological change in a society takes, however, is affected by the values and power structure in the society. People interested in health and promoting health are responsible for determining the social and health effects of technological change before its mass adoption in order to encourage the development of technology most conducive to health promotion and retention. The broader conception of health that now must be accepted should aid us in accomplishing this task.

TECHNOLOGY AND THE AGING ADULT:

CAREER DEVELOPMENT AND TRAINING

Harvey L. Sterns and Margaret B. Patchett

Department of Psychology and
Institute for Life-Span Development and Gerontology
The University of Akron, Ohio

Impact of Technology on the Organization

Direct observation of the impact of technology on the
organization may be difficult, particularly when resulting changes
occur slowly and on many levels. Theories exist, however, to serve
as guides as to where changes may be expected.

Skinner (1979), in Work in America, sees technology as having
three main organizational effects which he classifies as primary,
secondary, and tertiary. Primary effects directly affect the "work,
product, worker, basic requirements, working environment,
investments, and costs" (p. 209). For instance, as a result of
technology, a new product or process may change the basic task or
job of the worker, requiring new or different job skills and
experience. In turn, the precision, reliability, and quality of the
work may be affected. Secondary effects of technology on the
organization may be reflected in changes in organizational policies
and practices. Personnel policies, production, wage structures,
schedules, even supervisory hierarchies may be altered as a result
of technology. Skinner's third classification, tertiary effects,
refers to the impact technology has on the organization's ability to
perform well. The changes technology brings may affect individual
effort, morale, and productivity as well as the organization's
ability to maintain high investment returns and a competitive edge
in the marketplace.

Burack and McNichols (1973) propose a similar model of
technological impact on the organization. As a result of scientific
innovation or organizational need, new or modified products may be
developed. The research and development department is instrumental

in adapting these new products or materials to organizational needs
and specifications. The engineering department then modifies or
develops production processes to accommodate the changes in
products. As a result of these changes, the organization may need
to reconsider its policies on manpower and capital expenditures.
With regard to manpower policies, a new or modified product line and
method of processing may require a new set of hiring criteria since
job skills needed to perform the task may be altered. Similarly,
the line of progression and promotion may change, requiring new
promotion criteria. Policies regarding education and training may
need to be re-evaluated. The feasibility of retraining existing
employees rather than hiring new ones to learn and perform the new
job may deserve examination. Burack and McNichols also recognize
the possibility that policies regarding capital expenditures may
change as a result of technological innovation. For instance, the
number of personnel may be reduced or increased to fit the new
needs. Changes also may be made in the organization's structure,
function, and overall processes.

Organizational Responsiveness and Resistance to Technological Change

Some organizations are more responsive to innovation and change
than others. In order to remain competitive and survive, certain
industries experience faster and greater technological growth,
forcing companies within these industries to change rapidly.
Organizations have a choice as to which of the available
technological changes they choose to respond to and incorporate. In
turn, these decisions will at some point in time affect the
individual worker. Some jobs may become increasingly specialized as
a result of technology, others may be affected only by an altered
work environment. Some changes may require the worker to simply
update his/her retraining. Failure to remain up to date may in the
future render certain skills obsolete. Similarly, future
technological changes may render some specialized jobs obsolete.

Hughes (1982) refers to two types of technology, conservative
and radical. Conservative technology refers to innovations which
improve or refine existing procedures of production and processing.
It tends to preserve rather than disrupt the current momentum and
structure of the organization and as such meets little if any
resistance in its incorporation. Radical technology on the other
hand represents a strong challenge to the status quo. These are the
innovations which drastically alter the direction and speed the
production. Adopting these innovations may require major
reorganization of the corporate structure. Involved is a certain
degree of risk whereby the changes needed to accommodate the new
technology are met with resistance.

There appears to be a relationship between technical change,
research and development intensity, and growth rates of production

and productivity (Burack and McNichols, 1973; Freeman, Clark and Solte, 1982). Industries which place heavy emphasis on research and development and technology (e.g., chemicals, plastics, electrical and electronic equipment, communications, drugs) tend to experience faster growth rates. Firms within rapidly growing industries must respond quickly to the competitive environment to keep pace. These are the firms adopting the radical technologies and theoretically, those which must make major adjustments in manpower and organization as a result.

On the other hand, more technologically mature industries (e.g., steel, textiles, automobiles, agricultural products) may evidence a slower rate of responding to technological innovation (Burack and McNichols, 1973; Freeman et. al., 1982; Skinner, 1979). Slow or declining growth rates characterize these industries as do low rates of technological change. Firms in these industries stress the maintenance and refinement of current operations rather than research and development of new technologies. Unless some crisis situation develops, significantly different production processes are unlikely to be implemented. According to Hughes (1982), organizational resistance should occur to a lesser extent in these companies than in those incorporating the more radical technologies.

Organizational resistance may stem from a variety of sources (Katz and Kahn, 1978). First, technological innovation requiring changes in the organizational structure is likely to meet resistance from those who benefit from the current reward and power structures. Second, technology which alters particular jobs may be resisted by those with specialized skills and expertise as they face the possibility of obsolescence. Third, like individuals, organizations develop standard procedures for handling routine and problem situations in order to maintain smooth and efficient functioning. For individuals and organizations alike it is difficult to change established ways, especially when viewed as the best means of obtaining desired goals. Threats to established, standard procedures are therefore likely to be met with resistance by the organization as a whole.

What then can be done to reduce individual organization resistance to change necessitated by new developments in technology? Lav and Ruttenberg (1982) suggest three viable solutions: (1) establish good communication and cooperation between labor and management (e.g., notifying them early on of the innovation, the need for it, its effects on the workers, and compensation for these effects; (2) share the benefits gained by introducing the new innovations with the workers who implemented them; and (3) provide adequate compensation to displaced workers and programs to assist in their re-employment and adjustment. The input of the workers should be used in planning for and implementing the change. Supervisors and the union also should be notified early of the planned changes and attempts should be made to work closely with them during both the planning and implementation phases.

Impact on the Workplace and the Individual Job

Technological innovation may have both positive and negative effects on the workplace (Bass and Barrett, 1981; Skinner, 1979). Quality of worklife may be affected positively by the fact that new mechanization (e.g., office work) often reduces or eliminates tedious and repetitive tasks or actual physical work. The work environment in offices and factories is now cleaner, better lit, and more comfortable. Numerically-controlled machine tools, word-processing systems, containerized shipping, automatic component insertion are all examples of technological innovation. None of these technologies were developed to improve working conditions but indeed this has been a fringe benefit.

Quality of worklife may be affected negatively by mass production which may result in repetition and boredom. Assembly lines, conveyors, and automation may reduce job scope, dissolve work groups, and increase feelings of social isolation. Machinery may make jobs faster paced, increasing the pressure to keep producing. Fortunately, some of the negative consequences associated with technology may be avoided or counteracted by careful job redesign and a concern for the social as well as technical consequences of innovation (Huse, 1980).

An investigation by Rothwell and Zegveld (1979) clearly demonstrates the positive and negative effects of technology on the workplace. Examining the impact of technology on ten industries in six countries, they found a need for higher level management skills in eight. For six of the industries skill levels decreased or became more redundant. Specialist skills were displaced outside the factory in four of the ten industries. Three industries indicated job loss due to lack of technological change competitiveness, five industries reported a reduction in labor force, while eight of the industries reported jobless growth where output increased with the same or reduced labor force. Job satisfaction, however, was reduced in only one of the ten industries.

Rothwell and Zegveld also examined the impact of microelectronics on employment in numerous industries and offices. Overall they found a great deal of unemployment and de-skilling of jobs as a result and predict much of the same for the future. Thus, the need for retraining and updating of existing skills is critical in these industries. However, they also predict the creation of new jobs in progressive firms utilizing microelectronics technology in the development of new products. Failure to adopt such technology on the other hand may result in greater job loss due to lack of technical competitiveness.

In essence, technological innovation may affect the workplace
in several ways, some of which (1) render skills and knowledge
obsolete, (2) require the development of new knowledge and skills,
(3) change employee attitudes and satisfaction regarding the
workplace (positive and negative), (4) create new job openings and
potentiality for career mobility, and (5) create unemployment.

The issues of employment and unemployment are important when
discussing the older worker because older workers tend to have
greater difficulty finding new employment and may not have the same
retraining opportunities (Becket and Hollenshead, 1982; Gordus,
Jarley and Ferman, 1981; Jenkins and Montarquette, 1979; Sheppard,
Ferman and Faber, 1950). For example, older adults may lack the
educational background to compete with younger adults in certain
occupations even though they may have a wealth of on-the-job
knowledge and experience. Self-doubts and fear of failure may
prevent the aging adult from taking advantage of retraining
opportunities when their skills are not readily transferable to a
new situation. Some older adults may find their job-search skills
lacking if they have been employed in the same job for a number of
years. Some employers may perceive aging workers as less
productive, untrainable, as poor "long-term investments," and as
greater insurance risks, creating a reluctance to hire older
adults. Although research evidence demonstrates that older adults
are trainable, and that job performance is not related to age in the
majority of occupations, discrimination in hiring and retraining
still exists (Meier and Kerr, 1976; Davis, 1979).

Career Development--One Response to Technological Change

Although career development should be a lifelong process, the
individual whose job is altered by technological innovation may be
especially motivated to engage in new or renewed career planning.
Depending upon the degree to which the job is changed, the
individual may consider plans to simply update current job
knowledge, to engage in continuing education, or to undergo
retraining.

According to Walker (1978) career planning is "the personal
process of planning one's life work, which entails evaluating
abilities and interests, considering alternative career
opportunities, establishing career goals and planning practical
developmental activities" (p. 2). In the past, career planning
activities were left largely up to the individual. In more recent
years, however, organizations have begun to recognize that career
planning programs fill an integral space in the human resource
system (Wexley and Latham, 1981; Von Glinow, Driver, Brousseau and
Prince, 1983). For example, such programs may aid in reducing
turnover among entry level personnel by clarifying career paths
within the organization (Wexley and Latham, 1981).

The model we are about to propose suggests one way an organization might view the individual engaged in the career development and planning processes (see Figure 1). It assumes that a person's attitude toward a career development program is fundamental to his/her willingness to participate. Growth need strength, organization and superior support of the program, mobility attitudes and technological innovation are some of the factors proposed to influence attitudes toward career development programs.

The model incorporates Hall's model of career growth (1971). Hall's model suggests that career planning may be conceptualized within a goal setting framework. If the career goal chosen is challenging and relevant to the person's identity, and that person works independently to achieve that goal, then goal attainment should lead to enhanced growth of one's identity and self-esteem. Such enhanced self-esteem may then lead to greater commitment to future career goals.

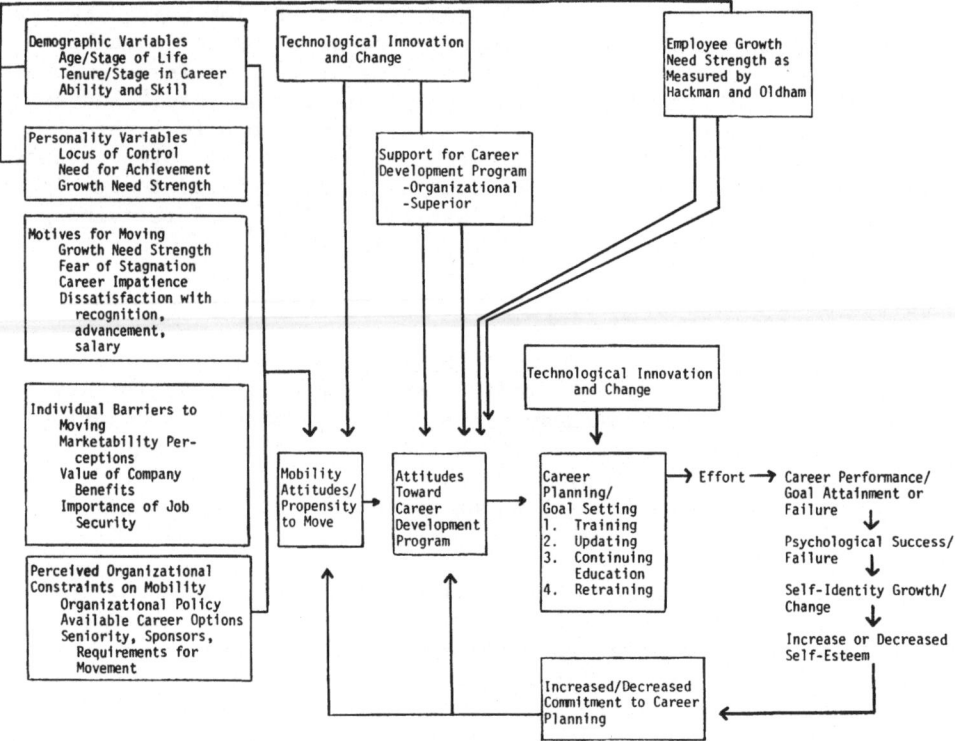

Figure 1. Career development model.

Consistent with Hall's model is the work done by Bandura on self-efficacy. Self-efficacy refers to people's judgments of their abilities to act in particular situations. Bandura (1982) suggests that self-efficacy judgments influence what activities people engage in, the effort they exert, as well as how long they will persist at such even in the face of adversity. Clearly the parallel is that just as goal attainment enhances self-esteem, so might it increase perceptions of self-efficacy and thus greater commitment to future career development. For older adults with doubts about their abilities to perform well in a retraining situation, successes early in the program may be crucial. Not only would their successes enhance their feelings of self-efficacy, but decisions to engage in future retraining may be enhanced as well. Failure could easily result in a decision to drop out of the retraining program as well as future programs, jeopardizing future career growth opportunities.

Attitudes Toward Career Mobility and Development

On logical grounds we may argue that a person seeking mobility within the organization or between organizations would show an interest in career development planning and related programs. For instance, the worker whose job is displaced by technological innovation may see upward or lateral movement within the organization as one possible solution to the problem. For this person, a career development program may be an important source of information regarding any extra training or education needed to move into these jobs and their career progressions. However, the decision to change jobs or careers and hence the decision to engage in career planning is influenced by many factors as we will see below.

Age or life-stage is one popularly discussed determinant of mobility attitudes and propensity to move. Numerous models suggest that the actual mobility rates of younger persons (up to around age 30) are much higher than those of older persons (Hall and Nougaim, 1968; Super, Crites, Hummel, Moser, Overstreet and Warnath, 1973), as the younger generations move about seeking their niche in life, and the older ones maintain theirs as they plan for retirement. Research on the actual mobility rates seem to bear out these hypotheses (Byrne, 1975; Saben, 1967; Sommers, 1977; Veiga, 1983). Veiga (1983) also found that age correlated significantly and negatively with propensity to move. With regard to career development, Gould (1979) predicted that career planning would be highest during Super's "stabilization period" (ages 31-44) when there is stable growth as the person attempts to secure his/her place within the occupation. Support, however, was not found for this hypothesis.

Tenure (seniority) or stage in one's career may also influence a person's decision to change jobs or careers. Schein (1971) describes three stages in a career: socialization; performance; and obsolescence versus the development of new skills. In the third stage, the obsolete person may be retained as "deadwood" with no options for mobility or may be retrained, transferred into a lateral position or forced into early retirement. Although career development could be helpful to the employee in all three stages, it seems likely that the person who would most strongly desire such help would be in the third stage of Schein's model. In terms of actual research, "mobility in the earliest stage of one's career bears an unequivocal relationship with one's later career" (Rosenbaum, 1979, p. 220; Veiga, 1981; 1983). From this we might conclude that those most likely to move or show an interest in career planning would be those whose skills have become outdated or obsolete and those who show a lifelong pattern of mobility.

The "individual barriers to moving" bear some relationship to these issues of age and seniority. As Veiga (1983) suggests, perceptions of one's own marketability may strongly influence the person's efforts to explore alternative career opportunities either within one's own organization or in an outside firm. Having held a particular position within the same organization for an extended period of time may only reinforce one's feelings of specialization and obsolescence rather than one's feelings of marketability potential. Similarly, the longer a person remains with a company and the older that person gets, the more likely he/she will think twice about risking any benefits accrued through the years in order to move to a new organization. The same may be true for the person who strongly values job security, in which case career development programs may be seen as a waste of time if there is little motivation to move.

Veiga (1983) identified five "motives for moving" which significantly influenced propensity to leave: fear of stagnation, career impatience, dissatisfaction with one's salary, recognition, and advancement. In the recent past, corporations have been fast-tracking young executives up their way in the organization. The saying went "if you haven't made it by 30 you're not going to make it" (Davis, 1979). More recently the trend seems to be shifting back to promoting the older, more experienced workers. Corporate expectancies for executive mobility are changing and in turn will influence individual worker's expectancies, satisfactions, and feelings of being on-time or off-time in their personal career timetables. Again we assume that people expressing a desire to move will react positively to the initiation of a career development program which conceivably could help them on their way.

Two personality variables are hypothesized to affect career mobility and development attitudes: need for achievement and locus

of control. The more widely researched of the two is locus of
control. According to Greenhouse and Sklarew (1981), exploration
"is a proactive attempt to understand and influence one's life" (p.
2). Thus, we might expect internals to engage in greater career
planning activity than externals. Research by Gould (1979) tends to
support this hypothesis. Neapolitan (1980) investigated
occupational changes in mid-career and found that people who change,
regardless of great obstacles, tended to reflect an internal locus
of control. A comparison group of people dissatisfied with their
careers but who did not change tended to perceive great risk beyond
their control which would doom any efforts to change. Beeher, Taber
and Walsh (1980) looked at locus of control and perceived channels
of mobility. They found that employees who had not had a recent
opportunity to change jobs were more likely to see mobility
opportunities as controlled by external forces such as luck or
favoritism, whereas employees with recent change opportunities
tended to perceive mobility as contingent upon performance (an
internally controlled factor). We may conclude that persons
believing that mobility opportunities are controlled by forces
beyond their control (an external locus of control) would be less
likely to view a career development program as beneficial to them,
and hence, less likely to participate in such.

The decision to change jobs or careers may also be influenced
by certain organizational constraints such as organization size,
structure, or technology (Vardi and Hammer, 1977; Vardi, 1980).
Vardi and Hammer describe three types of technologies: long-linked,
mediating, and intensive. Long-linked jobs resemble mass production
assembly line jobs where workers are semi-skilled and highly
interchangeable. Here it is hypothesized that job mobility is
predominantly lateral with seniority being the major determinant of
mobility. Jobs of a mediating technological nature typically serve
as links to two or more interdependent clients, such as banking or
insurance jobs. Mobility is hypothesized to be predominantly upward
in these occupations. Sponsorship or mentoring is seen as the key
to upward mobility here. Tasks within an intensive technology tend
to be complex and nonroutine whereby specialized skills and
progessional knowledge are required for successful performance.
Again mobility is primarily upward, but in these occupations
mobility should be perceived by employees to depend largely on
personal competence and skills. An empirical test of the theory
found support for the hypothesized directions of mobility, but only
partial support for the perceived determinants of mobility.
Employees in long-linked jobs also perceived education and training
as important to mobility opportunities. Employees in intensive jobs
perceived sponsorship as an additional determinant of mobility.
Relating this research to career development attitudes, we
hypothesize that employees who see mobility channels as closed to
them (e.g., where mobility depends more on seniority and sponsorship

rather than competence and skill per se) may be more reluctant to participate in education and training.

Technological Innovation and Career Development

The impact of technology on mobility attitudes and propensity to move has been alluded to above. If Vardi and Hammer's (1977) theory is correct, individuals in intensive (and perhaps long-linked) technology jobs will be more receptive to career development programs given that they perceive mobility channels as open to them.

We further hypothesize that in fast-paced industries with rapid technological growth, employees will be faced with a continuing need to update knowledge and skills, and will tend to hold greater interest in career development activities. For people in slow-paced industries, the relationship between technological innovation and career development attitudes should not be as clear. In sum, technological innovation may serve to stimulate career development interest in fast-paced industries but the converse may not necessarily hold true.

Growth Need Strength

Although no research has been done in this area, it is hypothesized that growth need strength (as measured by Hackman and Oldham's (1975) Job Diagnostic Survey) will influence an employee's attitudes toward a career development program. The Job Diagnostic Survey (JDS) measures the employee's desire to "obtain growth satisfactions from his or her work" (Hackman and Oldham, 1975, p. 163). On logical grounds we might argue that persons high in growth need strength (GNS) will response more favorably toward a career development program than those low in GNS. We advance this on the assumption that career development programs of any kind imply growth in one's job or career plans.

Type of Program and Organizational Support

Walker (1978) suggests that there are two types of career development programs. The first type focuses predominantly on promotability, potential, and career ladders without considering that not everyone is qualified for promotions, and without considering that not everyone is interested in moving up. Such programs tend to raise some employees' expectations unrealistically thus causing personal anxiety and disappointment when they fail to move up. It also may reduce their commitment to the organization resulting in diminished performance, organizational disruption, and possible turnover.

The second type of program Walker describes fosters more

realistic expectations on the part of the employees. It tends to focus on mobility within the organization, but only to the extent that it is realistic for that particular individual within that particular job. It focuses primarily on current job responsibilities, requirements, and performance results so that employees develop more realistic expectations about their capabilities and opportunities. Walker purports that such programs, by giving employees more realistic expectations, serve to enhance personal career planning capabilities and strengthen career commitments and future development planning. As a result, performance may be enhanced, turnover may be reduced, and utilization of talent within the organization may improve. According to Walker, the effective career development program is voluntary, realistic, recurring, and one which teaches the employee skills for evaluating and developing his/her own career plans.

Finally, employee attitudes are never formed in isolation. In order for employees to enter into a career development program with a positive attitude, they must perceive that the organization and their immediate superiors support such a program and will support their own efforts at career planning.

Training and Retraining the Adult and Older Worker

Training is defined as a planned effort by an organization to facilitate the learning of job-related behavior (Wexley, 1984). Cascio (1982) indicates that training and development activities lead to changes in skill, knowledge, attitude and social behavior related to the job or professional role.

Given the nature of technological change and its effects on the workplace as discussed above, lifetime retraining is necessary so that one can continue to build on previous knowledge and experience. Rapid technological change creates obsolescence of knowledge and skills among workers of all ages. Middle-aged and older workers may need updating, but today the need for retraining is shared by people in their 20s and beyond. Skills and knowledge gained from training may last for months, and up to a few years. Thus there is every reason to offer training opportunities to workers of all ages who will continue to work for extended periods of time.

A number of recent reviews emphasize the important role that training has in response to technological change (Dooling and Klemmer, 1982; Goldstein, 1982; Nickerson, 1982; O'Toole, 1977; Wexley, 1984). The ability of individuals to learn and benefit from training and education at all points in the life-span is well documented (Cross, 1981; Peterson, 1983; Willis, in press). Training in the work situation should be designed to optimize the opportunities for continuing development of the individual.

Training programs for workers are designed to: (a) improve self-awareness regarding their role and responsibility in an organization, (b) increase job-related knowledge and skill, and (c) enhance motivation to perform their job well. Training is an important way to achieve productivity, job satisfaction and organization revitalization (Bass and Barrett, 1981; Wexley and Latham, 1981). An individual may choose to enhance skills and knowledge outside the organization through continuing education and formal education programs. Such educational activities may facilitate current knowledge important to career development, or may facilitate a transfer to a new occupational setting.

A recurring concern expressed in the training literature is the need for organizations to conduct needs assessment for training as well as the evaluation of training efforts to determine the utility of various training activities (Cascio, 1982; Goldstein, 1982; Wexley, 1984). Such activities are critical to developing effective and successful training programs.

A major issue in the training of older workers is assuring equitable access to training opportunities. Training policies and practices may reflect informal age restrictions, standards and assumptions, which may exclude older employees. Employers remain reluctant to hire persons over 40 and offer training, or to train or retrain those already employed (Rosen and Jerdee, 1977; Sonnenfeld, 1978). Meier and Kerr (1976) emphasize that older workers have developed good work records, are healthy, are dependable, have good accident records and maintain productivity. Inappropriate beliefs about older employees lead to employers not taking advantage of good, available workers who happen to be older. Older workers themselves contribute to problems in training participation by being reluctant to volunteer for training or retraining.

Research in the industrial setting on the performance of older workers after training has found that their performance equals or surpasses the average performance of the younger group, as long as the older work had sufficient training time (Belbin, 1965, 1970; Belbin and Belbin, 1972; Meier and Kerr, 1976).

A good example of an attempt to integrate the industrial and developmental findings related to retraining older workers was the report of the Irish Airlines (Aer Lingus) when the company switched to new technology (Mullan and Gorman, 1972). A combination of techniques was used which emphasized: orientation, participation, discovery learning, an the communication process. The company found equal performance across all ages on the new equipment. This study shows the importance of a comprehensive, well-planned training program designed to consider the unique characteristics of the organization. It also highlights the importance of using the knowledge available on successful training methodology to enhance learner performance.

Various methods have been used to train older adults, such as: discovery learning (Sheppard, 1976); activity learning (Belbin and Downs, 1964, 1966); programmed instruction (Neale, Torye, and Belbin, 1968; Siemen, 1976); CRAMP technique (Downs and Roberts, 1977); and practice (Murrell, 1970). Discovery learning and activity learning have been popular because they were specifically designed to train older adults (Belbin, 1958; Belbin, Belbin, and Hill, 1957; Belbin and Shimmin, 1964; Belbin, 1970). Valasek and Sterns (1981) state that the method of training may not be the crucial aspect in the success or failure of a training program for older workers. Various methods appropriate for the type of task may work well for the older worker if the training program is <u>carefully designed</u>. This design requires the incorporation of evidence from all aspects of the training literature, with specific sensitivity to the special case of the older worker. This includes the discussion of whole versus part learning (Goldstein, 1974; Limon, 1980), the issue of training components of the task to criterion (Sterns and Sanders, 1980) and errorless discrimination learning (Terrance, 1964). The research evidence tends to support the conclusion that if the task is complicated, the individual will benefit from training which breaks the task into its component parts and then trains to criterion.

Conclusions

Well-designed training programs are important to the career development of trainees at any age, in responding to changes in the workplace due to technological change. Supervisors and managers need to be aware of potential barriers to participation by adult and older workers.

Older workers must not be excluded from short-term and long-term training opportunities because of negative attitudes about the training and growth potential of older workers. Supervisors need to recognize that some individuals may experience fear of the training situation. Older workers are concerned about being able to successfully complete the training program, or fear that the training situation might reveal inadequacies that can be observed by others.

Participation in career development programs is determined by many factors, which include: demographic variables, personality variables, motives and barriers to moving, and perceived organizational constraints. Organizations can help workers of all ages by supporting career development programs with well-designed training and retraining, sensitive to the needs of the aging worker.

REFERENCES

Bandura, A. (1982) Self-efficacy mechanism in human agency.
    American Psychologist, 37, 122-147.
Bass, B.M. and Barrett, G.V. (1981) People, work, and
    organizations. Boston: Allyn and Bacon, Inc.
Beckett, J.O. and Hollenshead, C. (1982) Older workers: Decision
    making in a plant closing situation. Gerontologist, 22, 261.
Beeher, T.A., Taber, T.D., and Walsh, J.T. (1980) Perceived mobility
    channels: Criteria for intraorganizational job mobility.
    Organizational Behavior and Human Performance, 26, 250-264.
Belbin, E. (1958) Methods of training older workers. Ergonomics, 1,
    208-221.
Belbin, E. and Belbin, R.I. (1972) Problems in adult retraining.
    London, England: Heineman Publishing Company.
Belbin, E. and Belbin, R.M., and Hill, F. (1957) A comparison
    between the results of three different methods of operator
    training. Ergonomics, 1, 39-50.
Belbin, E. and Downs, S.M. (1964) Activity learning and the older
    worker. Ergonomics, 7, 429-437.
Belbin, E. and Downs, S.M. (1966) Teaching paired associates: The
    problem of age. Occupational Psychology, 40, 67-74.
Belbin, E. and Shimmin, S. (1964) Training the middle aged for
    inspection work. Occupational Psychology, 38, 49-57.
Belbin, R.M. (1965) Training methods for older workers. Paris:
    Organization for Economic Co-operation and Development.
Belbin, R.M. (1970) The discovery method in training older workers.
    In H.L. Sheppard (Ed.), Toward an industrial gerontology.
    Cambridge, MA: Schenkman Publishing Company.
Burack, E.H. and McNichols, T.J. (1973) Human resource planning:
    Technology, policy, change. Kent, OH: Center for Business and
    Economic Research.
Byrne, J.J. (1975) Occupational mobility of workers. Washington,
    D.C.: U.S. Bureau of Labor Statistics, Special Labor Force
    Report No. 176.
Cascio, W.F. (1982) Applied psychology in personnel management,
    second edition. Reston, VA: Reston Publishing Company.
Cross, K.P. (1981) Adults as learners: Increasing participation and
    facilitating learning. San Francisco, CA: Jossey-Bass, 1981.
Davis, S.M. (1979) No connection between executive age and corporate
    performance. Harvard Business Review, 57, 6-18.
Dooling, D.J. and Klemmer, E.T. (1982) New technology for business
    telephone users: Some findings from human factor studies. In
    R.A. Kasschau, R. Lachman and K.R. Laughery (Eds.) Information
    technology and psychology: Prospects for the future. Praeger
    Publishers.
Downs, S. and Roberts, A. (1977) The training of underground train
    guards. A case study with a field experiment. Journal of
    Occupational Psychology, 50, 111-120.

Freeman, C., Clark, J., and Solte, L. (1982) Unemployment and technical innovation. Westport, CT: Greenwood Press.

Goldstein, I.I. (1974) Training: Program development and evaluation. Monterey, CA: Brooks/Cole Publishing Company.

Goldstein, I.L. (1982) Training in the 1970's: A view toward the 1980's. In R.A. Kasschau, R. Lachman and K.R. Laughery (Eds.) Information technology and psychology: Prospects for the future. Praeger Publishers.

Gordus, J.P., Jarley, P., and Ferman, L.A. (1981) Plant closing and economic dislocation. Kalamazoo, Michigan: W.E. Upjohn Institute for Employment Research.

Gould, S. (1979) Characteristics of career planners in upwardly mobile occupations. Academy of Management Journal, 22, 539-550.

Greenhouse, J.H. and Sklarew, N.D. (1981) Some sources and consequences of career exploration. Journal of Vocational Behavior, 18, 1-12.

Hackman, J.R. and Oldham, G.R. (1975) Development of the Job Diagnostic Survey. Journal of Applied Psychology, 60, 159-170.

Hall, D.T. (1971) Potential for career growth. Personnel Administration, 34, 18-30.

Hall, D.T. and Nougaim, K. (1968) An examination of Maslow's need hierarchy in an organizational setting. Organizational Behavior and Human Performance, 3, 12-35.

Hughes, T.P. (1982) Conservative and radical technologies. In Managing innovation. Lundstedt, S.B. and Golglazier, E.W. (Eds.) New York: Pergamon Press, 31-44.

Huse, E.F. (1980) Organization development and change. St. Paul, Minn.: West Publishing Company.

Jenkins, G.P. and Montarquette, C. (1979) Estimating the private and social opportunity costs of displaced workers. Review of Economics and Statistics, 61, 342-353.

Katz, D. and Kahn, R.L. (1978) The social psychology of organizations. New York: John Wiley and Sons.

Lav, I.J. and Ruttenberg, S.H. (1982) The human side of technological innovation: Labor's view. In Managing innovation. Lundstedt, S.B. and Golglazier, E.W. (Eds.) New York: Pergamon Press, 93-110.

Limon, L.P. (1980) Boost learning speed by presenting objectives in teachable part. Training, 13 (4).

Meier, E.L. and Kerr, E.A. (1976) Capabilities of middle-aged and older workers: A survey of the literature. Industrial Gerontology, 3, 147-156.

Mullan, C. and Gorman, L. (1972) Facilitating adaptation to change: A case study in retraining middle-aged and older workers at Aer Lingus. Industrial Psychology, 15, 23-29.

Murrell, F.H. (1970) The effect of extensive practice on age differences in reaction time. Journal of Gerontology, 25, 268-274.

Neale, J.G., Torye, M.N., and Belbin, E. (1968) Adult training: The use of programmed instruction. Occupational Psychology, 42, 23-31.

Neapolitan, J. (1980) Occupational change in mid-career: An exploratory investigation. Journal of Vocational Behavior, 16, 212-225.

Nickerson, R.S. (1982) Conclusion: Information technology and psychology--a retrospective look at some views of the future. In R.A. Kasschau, R. Lachman and K.R. Laughery (Eds.) Information technology and psychology: Prospects for the future. Praeger Publishers.

O'Toole, J. (1977) Work learning and the American future. San Francisco, CA: Jossey-Bass Publishers.

Peterson, D.A. (1983) Facilitating education for older learners. San Francisco, CA: Jossey-Bass Publishers.

Rosen, B. and Jerdee, T.H. (1977) Too Old or Not Too Old. Harvard Business Review, 55, 97-106.

Rosenbaum, J.E. (1979) Tournament mobility: Career patterns in a corporation. Administrative Science Quarterly, 24, 220-241.

Rothwell, R. and Zegveld, W. (1979) Technical change and employment. New York: St. Martin's Press.

Saben, S. (1967) Occupational mobility of employed workers. Washington, D.C.: U.S. Bureau of Labor Statistics, Special Labor Force Report No. 84.

Schein, E.H. (1971) The individual, the organization, and the career: A conceptual scheme. Journal of Applied Behavioral Science, 7, 401-426.

Shepphard, H.L. (1976) Work and retirement. In R.H. Binstock and E. Shanas (Eds.), Handbook of aging and the social sciences. New York: Van Nostrand Reinhold Co.

Sheppard, H.L., Ferman, L.A., and Faber, S. (1950) Too old to work--too young to retire. Washington, D.C: G.P.P. (for U.S. Senate Special Committee on Unemployment Problems)

Sieman, J.R. (1976) Programmed material as a training tool for older persons. Industrial Gerontology, 3, 183-190.

Skinner, W. (1979) The impact of changing technology on the working environment. Work in America: The decade ahead. Kerr, C. and Rosow, J.M. (Eds.) New York: Van Nostrand Reinhold Co., 204-230.

Sommers, D. and Eck, A. (1977) Occupational mobility in the American labor force. Monthly Labor Review, 100, 3-19.

Sonnenfeld, J. (1978) Dealing with the aging work force. Harvard Business Review, 56, 81-90.

Sterns, H.L. and Sanders, R.E. (1980) Training and education in the elderly. In R.R. Turner and H.W. Reese (Eds.), Life-span developmental psychology: Intervention. New York: Academic Press, pp. 307-330.

Super, D., Crites, J., Hummel, R., Moser, H., Overstreet, P., and Warnath, C. (1973) Vocational development: A framework for research. New York: Teachers College Press.

Terrance, H.S. (1964) Wave length generalization after discrimination learning with and without errors. Science, 144, (April), 78-80, 227-228.

Valasek, D.L. and Sterns, H.L. (1981) Task analysis and training: Applications from lab to the field. Paper presented as part of the Symposium on Industrial Gerontological Psychology: Why Survive? 34th Annual Scientific Meeting of The Gerontological Society of America, November, Toronto, Ontario, Canada.

Vardi, Y. (1980) Organizational career mobility: An integrative model. Academy of Management Review, 5, 341-355.

Vardi, Y. and Hammer, T.H. (1977) Intraorganizational mobility and career perceptions among rank and file employees in different technologies. Academy of Management Journal, 20, 622-634.

Veiga, J.F. (1981) Plateaued versus non-plateaued managers: Career patterns, attitudes, and path potential. Academy of Management Journal, 24, 566-578.

Veiga, J.F. (1983) Mobility influences during managerial career stages. Academy of Management Journal, 26, 64-85.

Von Glinow, M.A., Driver, M.J., Brousseau, K., and Prince, J.B. (1983) The designing of a career-oriented human resource system. Academy of Management Review, 8, 23-32.

Walker, J.W. (1978) Does career planning rock the boat? Human Resources Management, 17, 2-7.

Wexley, K.N. (1984) Personnel training. Annual Review of Psychology, 35, 519-551.

Wexley, K.N. and Latham, G.P. (1981) Developing and training human resources in organizations. Glenview, IL: Scott, Foresman, and Co.

Willis, S.L. (in press) Toward an educational psychology of the older adult learner: Intellectual and cognitive bases. In J.E. Birren and K.W. Schaie (Eds.) Handbook of the Psychology of Aging. New York: Van Nostrand Reinhold Company.

TECHNOLOGICAL ADVANCES FROM A HUMAN FACTORS POINT OF VIEW

H. McIlvaine Parsons

Essex Corporation/Lehigh University
Alexandria, Virginia/Bethlehem, Pennsylvania

Since "human factors" can be defined as dealing with the interactions between technology and individuals, it follows that this field should play a major role in examining the relationships of aging and technology. This interest has indeed been demonstrated in a special issue of Human Factors on aging in 1981, in the existence of a Technical Group on Aging in the Human Factors Society, and in various presentations at the Society's Annual Meetings.

From a human factors perspective, technology affects individuals, including the aging, in several ways. It may substitute some device or machine, beneficially or detrimentally, to do what the individual has been doing; that is automation. It may augment (or diminish) what the individual has been doing, help the individual recover some capability (or hinder this), or change what was done, positively or negatively. Technology may also, for better or worse, do something that is entirely new. Change agents of "hard" technology are mechanical, electronic, or chemical; "soft" technology techniques include education, training, behavior modification, and functional assessment.

Human factors application, itself a technology, has goals parallel to these aspects of technology. It tries to determine what human activities should or should not be automated and what human activities should or should not be supplemented by some technological process or innovation. When a process or innovation is introduced, human factors tries to make sure it is designed so humans can cope with it effectively and safely, fitting it to the human, or tries to shape the behavior of users, fitting humans to the process.

The ways technology affects aging individuals can be examined according to technology's introduction in various contexts: transportation (e.g., automobile driving, mass transit, selective public conveyances); work (e.g., factories, offices, services); housing (e.g., special residences and residential components, including consumer products); health care (e.g., hospital environments, prosthetic devices); communications (e.g., electronic networks, incapacitation signalling); entertainment (e.g., cable television, computer games); and safety/security (e.g., anti-crime protection, accident prevention). All of these have seen human factors interventions, or should have, and developments without the human factors label are described in various catalogs and handbooks (Baker & Krauser, 1982; Sargent, 1981).

Still another human factors approach is to look at those characteristics of aging individuals that set them apart from others. For example, what particular limitations do they have that can be related in one way or another to technology? These have been reviewed by Fozard (1981) and Fozard & Popkin (1978), and human factors investigators have conducted research in sensory, cognitive, anthropometric, biodynamic, and other capabilities and limitations in the aging population.

To provide an overview from a different viewpoint, this paper outlines the kinds of impacts technology has on aging individuals within a framework of nine categories I have used before (Parsons, 1981): (1) daily activities; (2) locomotion; (3) social interaction; (4) feelings; (5) perception; (6) motivational variables; (7) health and sefety; (8) learning; and (9) manipulation. Some (for example, social interaction, feelings, motivational variables) may seem outside the customary human factors purview--though Welford (1981, pp. 106-108) discusssed motivational variables and aging--yet they should be included if the human factors domain does indeed consist of the relationships between individuals and technology.

Daily activities are whatever is not included in other categories that an aging individual does at work or at leisure, at home, in the workplace, in recreational or entertainment settings. Activities, especially in the workplace, have often been called motor, psychomotor, or cognitive "performance," traditionally a major human factors focus. Locomotion may be either micro--as within some environment, or macro, as between locations at a distance. Social interaction includes conversations and other communications between individuals, family and connubial associations, and organizational interactions at work or elsewhere. Under feelings come emotions, moods, satisfactions and dissatisfactions, comforts and discomforts. By perception I mean the way past experience influences sensory inputs, overlapping with cognition in the processing of language and imagery, especially inside our heads. Motivational variables include incentives and

deterrents (or if you will, reinforcers and punishers) and
conditions that affect their strength. Presumably, the meaning of
health and safety is self-evident. Learning may be purposeful or
not, involving knowledge or skills. Finally, by manipulation I mean
that an individual may react reciprocally to some technological
process by participating or not, or to some product by buying or
selling or disregarding it, or to some process or product by
changing it in some fashion. I submit that to view technology and
aging comprehensively from a human factors point of view, we must
consider all of these impacts, and each intensively.

DAILY ACTIVITIES

    How does or might automation affect aging individuals' daily
activities? In the factory, automation of the transfer of heavy
objects--through forklifts, moving belts, and robots--means there is
less need for muscular strength among many workers. An older person
whose strength has diminished can become a robot operator at a
control panel or a robot maintainer. As a variety of augmentation
technology, human factors engineering must be applied to these tasks
(Parsons & Kearsley, 1982), as well as to those in factory
operations that have not been automated. Some technological
augmentations can be especially helpful to aging workers: improved
lighting, less noise, better workspace layouts, and more effective
tool designs--as Bromley (1966) advocated.

    I do not have at hand instances of how office automation or
augmentation may benefit older clerical workers, but it seems likely
that one aspect of introducing video display terminals (VDTs) may
have been detrimental to the activities, as well as the feelings and
health, of older workers. Poor design of CRT displays and
inappropriate placement related to illumination sources have created
serious glare effects, to which aging individuals are especially
susceptible. Reduced performance and muscle strain resulting from
bad chair and chair-table design probably have been greater among
older workers than their younger associates.

    Some of the recent automation in the service sector, enabling
people to carry out transactions at home, might seem to be welcome
to some older individuals but has been unfavorably regarded by many
older women. According to a survey by the American Association of
Retired Persons (Kerschner & Chelsvig, in press), "Our respondents
have stressed that they want to maintain social ties through grocery
shopping and banking. They do not want to have technology replace
their needs for socializing."

    One of the most important daily activities of anyone, young or
old, is sleeping. Research on sleep has advanced through new
technology. Polygraphic techniques, especially

electroencephalography, have revealed various differences between
younger and older individuals asleep (Miles & Dement, 1980; Spiegel,
1981), such as older persons' "inability to sustain sleep across the
night" (Webb, 1982, p. 585). Sleep itself can benefit from
technology, e.g., hypnotic drugs and mattress/foundation design.
Yet the latter has "remained virtually unmentioned in the literature
about beds (for the aging) or about the aged (Parsons, 1981, p.49),
"though aging people have special requirements in the way their
bodies are supported. Discomfort may be more likely due to
reduction in pelvic padding and greater skin sensitivity." Koncelik
(1976) wrote that research and development studies should be
undertaken to determine the many functional aspects of the beds in
nursing homes, such as deflection depth, spring rate, suspension
members, and materials. (Sleep laboratories and sleep disorder
centers have disregarded this technology for sleepers in general,
not just the aging.)

From one viewpoint in gerontology (disengagement theory), it is
inevitable that as people age their activity levels tend to drop.
From another (activity theory), such diminutions should be
resisted. (The construct "activity" has variously been construed as
social interactions, as physical exercise, as involvement in
community or other organized activities, and so forth.) When
environments become relatively impoverished, as in nursing homes,
special efforts may be needed to raise activity levels; for example,
many homes have activities directors and programs. Some residents
no longer feed or bathe themselves or otherwise take care of
hygienic needs, and behavior modification technology has
successfully revived such residential activities--as it could raise
or maintian activity levels in general for the aging.

New electronic technology has also been brought to bear.
Weisman (1983) recently reported a project at the Hebrew Home of
Greater Washington, where staff from the Home and the Georgetown
University School of Community Medicine introduced four computer
games to 50 residents whose average age was 85. According to
Weisman (p. 362), the games became very popular. In one game,
Little Brick Out, a player turned a knob to manipulate a bat on the
screen to hit a ball which could knock "bricks" out of a wall, the
object being to demolish the whole wall. The computer sent a stream
of messages such as "Sam, your score is not so good. Keep trying."
"Everyone was able to achieve some success, and most residents
improved with repeated trials." In another game, Ribbet, residents
pressed a button to make a frog on the screen leap up and catch a
passing butterfly with its tongue. "The visual and auditory
reinforcers were immediate after each successful catch of the
butterfly." This case study indicated that intrinsic and extrinsic
consequences contingent on some specific behavior can increase the
frequency of that behavior--a fundamental notion of operant
psychology/behavior analysis/behavior modification that human

factors scientists and practitioners should take to heart.

In gerontological studies, operant technology has associated reinforcers (what I call "consequators") with some activity to raise its level (frequency), on the presumption that more of that activity will in various ways benefit the individual, and often an institution's staff as well. For example, a nursing home resident may regain physical strength, self-esteem, or overall enjoyment of life. What needs more emphasis is that the very consequators that increase activity are also its products, and they provide additional stimulation.

Such feedback stimulation has several important characteristics in addition to being motivational (as incentive or deterrent). It is "discriminative," or in cognitive terms, informational. It can be emotion-evoking. It follows doing something and is contingent on what was done (it does not otherwise occur), though one is rarely aware of the contingency. It should be distinguished from non-contingent stimulation an individual receives before doing anything; that non-contingent stimulation may also be discriminative and emotion-eliciting (and is more likely to produce awareness).

In our daily activities we receive both kinds off stimulation, minute by minute, hour by hour. But the aging may receive less--less of the antecedent, non-contingent variety because their physical and social environments provide less, and less of the subsequent, contingent variety because the aging are less active. Indeed, they are less active because of both deficits.

A deficit in non-contingent stimulation (sensory deprivation, loneliness) has at times been considered responsible for accelerations in the aging process (aside from effects on activity). But apart from motivational effects on activity, little thought has apparently been given to the deficits in contingent stimulation and how such deficits affect aging.

According to Oster (1979,pp. 365-366), bodily homeostasis depends on total sensory input and responses to it. "If the sensory stimulus (the trigger creating the mechanism of homeostasis) is removed," he said, "there is no stressor effect and the cycle of response is eliminated. The result is a decrease in the hypothalamus-pituitary response and therefore a decrease in the production of adrenocorticotropin (ACTH) and corticoids." He added that his research on sensory deprivation in elderly patients "demonstrated that decreased sensory input caused changes in the central nervous system and musculoskeletal system which were compatible with absence of the hypothalamus-pituitary response."

These conjectures about sensory inputs suggest investigating the relative importance of antecedent and consequent stimulation and

the latter's contingency relationship, its sensory modalities
(including proprioceptive stimulation from activity itself), and its
parameters such as intensity and variety. It seems advisable to
study in aging individuals activity as a generator of stimulation as
well as stimulation as a generator of activity.

LOCOMOTION

Technology has originated much macro-locomotion--for example,
air travel and space flight, as well as migration to the sun belt as
a result of air conditioning, and it has automated much
micro-locomotion, for example the automobile, a substitute for
walking. Should the aging drive automobiles or use buses as much as
they do, for short trips, instead of walking? According to Leon
(1981), vigorous walking benefits the cardiovascular system by
getting more blood to the heart, making the heart better able to
pump blood and blood vessels better able to carry it; it also firms
and strengthens body muscles and lowers body weight, helps
metabolize sugar, and lowers tension. I suggest it also increases
the sensory inputs I discussed earlier, both antecedent and
consequent, in comparison with those received as a passenger in a
vehicle. For the aging individual, the decision whether or not to
automate walking is not a matter of creating hard technology--cars
and buses are here--but one of applying soft technology--inducing
the individual to walk instead of ride. The issue is dramatized on
the golf course, frequented by many affluent as well as modest-means
Americans. Does the golfer walk and carry his/her own clubs? Walk
and have them carrried by a caddie? Walk but pull/push a wheeled
container for the clubs? Or ride with the clubs in an electric
cart, walking only from prescribed paths to where the ball was hit?
Though their choices may be limited for various reasons, golfers can
decide how much to walk, influenced by the effort involved as a
deterrent in competition with the beneficial exercise achieved as an
incentive--see Welford (1981) for an analogous analysis.

When the golfer returns to his/her suburban home, a radio
device may open the garage door, obviating the need to get out of
the car, locomote to the door, and open it manually. Though for
some older homeowners such manual operation is too strenuous, for
others the choice lies simply between automation and a mixture of
effort and inconvenience, two of the motivational variables that
have accounted for many labor-saving devices in modern technology.

In the total scheme of things such automation for the aging
golfer and homeowner may not mean much, but consider the residents
of nursing homes, or those who may become such. Walking may benefit
them physiologically, sensorally, and socially (it increases contact
with others). But many aging individuals prefer locomotion by
wheelchair, an important technological device, even when medical

constraints on walking are absent. Wheelchairing requires less
effort. Several years ago I undertook a project at the Hebrew Home
of Greater Washington to find out whether soft technology in the
form of behavior modification (operant conditioning) could induce an
elderly resident to change from wheelchair locomotion to walking.
The reinforcer was systematic, contingent admiration and approval,
but some hard technology was also needed, a wheeled walker with a
hinged seat. The resident persisted in using it instead of the
wheelchair because whenever he felt tired he could drop the seat and
rest on it. (He could flip the hinged seat from the upright to the
horizontal position but could not reach down to reposition it when
he wanted to resume a trip until I rigged a string and pulley for
him--a simple human factors engineering fix.) At first he took only
a few steps, but the intervals between sitting gradually became
longer. Eventually, entirely on his own, he made trips around the
Home he had never attempted by wheelchair, and finally he abandoned
the walker for a cane.

This episode was not a profound breakthrough of great
technological complexity but does suggest the need to combine soft
and hard technology in helping aging people.

SOCIAL INTERACTION

Technology affects social interaction in aging individuals in
too many ways to describe here, but let me mention some less obvious
manifestations.

One form of social interaction is social control. Domination
by the old was a marked characteristic of primitive societies. Old
men told the young what to eat, when to marry, whom to marry; they
specified rules and administered punishments. Their authority was
strengthened by religious beliefs, including ancestor worship.
Older individuals were the repositories and perpetuators of tribal
history and tradition, of knowledge and skills needed for survival.
Though the aged are still revered in China (Green, 1983) and some
other cultures today (note Hoos's reference to the Eskimo), in
Western countries gerontacracy--rule by the old--persists only in
the judiciary, clergy, and academia. What happened? Parsons
(1915), who was to become an eminent anthropologist, suggested that
it succumbed--gradually, to be sure--to technology, that is, the
invention of printing--moveable type--by Gutenberg or others about
1454, brought by Caxton to England in 1476. "The keeping of written
records was a blow to gerontocracy" (p.63) because the elders no
longer were the only institutional memories, though "long after
writing and even after printing was invented the old held their
own. Their accumulated experience was still a big asset; for their
environment was comparatively unchanging." But when society became
less stable, as in modern times, their influence waned.

If printing technology did indeed enfeeble gerontocracy, what can one predict about the effects of new, electronic means of language-related information creation, storage, and dissemination? Will expert-based knowledge systems, data banks, and electronic networks weaken gerontocracy's last vestiges, especially in academia? Will automation (especially artificial intelligence) supplant professors in courses in well-established factual and relational knowledge?

Another aspect of social interaction that modern technology affects and may affect further is lovemaking. This is a touchy subject, yet society has progressed considerably on the road to candor since 1899, when a Chicago gynocologist, Denslow Lewis, tried to publish his clinical findings about sexual intercourse--decades ahead of Kinsey--and was denounced by university professors and rejected by the Journal of the American Medical Association. Though some may be shocked that hard technology can or should be brought to bear on sexual love between aging husband and wife, prosthetic devices and hormonal injections have been developed for males, and counseling and behavior modification as soft technology have been used for couples.

It has been well established (Butler & Lewis, 1976; Ludeman, 1981; Verwoerdt, Pfeiffer, & Wang, 1969; Pfeiffer, Verwoerdt, & Wang, 1969; Weg, 1983), that many aging individuals of both genders retain their sexual capabilities and inclinations well into old age. A decade or so ago it was calculated that more than 50 per cent of women over sixty were widows, about 15 per cent of men widowers. This imbalance, presenting problems for the widowed who still desire sexual expression, may be what instigated a project rumored at the First Annual International Robot Conference in June of 1983 in Long Beach, to develop a male robot who can be programmed to function sexually in response to vocal commands and tactile stimulation, with a humanoid body and face that could be tailored to resemble whatever male a customer desired, including the appearance of the deceased husband; the robot might also produce synthetic speech based on recordings. According to Modern Maturity (1983, August-September, p.16), two women artists in Georgia have been creating "life-sized human soft-sculptures, made-to-measure men-about-town," perhaps anticipating the robots.

FEELINGS

Though technology can elicit many kinds of emotions and moods in individuals using its processes or products, historically human factors scientists and practitioners have restricted their interests to discomfort, stress, and job satisfaction, and even these have been virtually ignored in their studies of the aging (Chapanis, 1974; Fozard, 1981). Gerontologists, however, including

investigators of environmental design for the aging (Lawton, 1983),
have included feelings in a construct called "psychological
well-being"--negative affect (anxiety, depression, agitation, worry,
pessimism), happiness, and positive affect (e.g., pleasure). Schulz
(1982) has published an overall look at emotions in the aging.

Technology can help alleviate negative affect in the aging,
such as fear of crime--which is greater than the actual criminal
incidence--through protective devices; discomfort, through better
design of seating, bedding, illumination, and other features of the
environment; and depression, through anti-depressant drugs--though
these may be overexploited in institutions. It can produce positive
affect, such as pleasure in watching television or playing computer
games. One must hope that technology will increasingly try to make
the elderly <u>feel</u> better.

PERCEPTION

Sensory deficits that develop with aging can be considered
under the heading of "perception" (Fozard, 1981) but they can also
be placed in "daily activities." What interests me stems from the
definition I gave of perception at the start: the influence of past
experience on sensory inputs. With aging comes more and more
experience, some varied, some repetitious. How in the aging does or
might technology affect the unison of the past and present?

Much of the research on human cognition seems to have been
based on reaction times, but there have been two uncontrolled
variables in the reaction time research on older people, namely,
instructions to subjects and information or motivational feedback.
Past research on simple reaction times has shown that feedback can
have major effects, even extending reaction time to several
seconds--what I have called "the precautionary pause"--a performance
feature that merits more human factors consideration (Parsons,
1976). Might age-differential effects due to instructions to
subjects or feedback have contributed to age differences in reaction
time and thereby confounded experimental results?

MOTIVATIONAL VARIABLES

My definition of motivational variables at the start as
incentives and deterrents--Welford (1981) calls them benefits and
costs--should have indicated my behavioral bias. As I noted, a
considerable body of research, mostly in nursing homes and
hospitals, has shown how operant consequation techniques applied to
the aging can make desirable behavior, such as self-care, more
frequent and undesirable behavior, such as incontinence, less
frequent (see for example, Hussian, 1981; Ince, 1976; Bellucci &

Hoyer, 1975; Hoyer, Mishara, & Riedel, 1975; Hoyer et al., 1974;
McClannahan & Risley, 1974; Baltes & Zerbe, 1976; Lopez et al.,
1980; papers at the Nova Behavioral Conference on Aging, 1978; and
articles in the International Journal of Behavioral Geriatrics,
established in 1982.)

An older person's behavior can be changed by reinforcing some
alternative behavior; this is the customary operant approach. I am
more interested in learning how in the everyday world an older
person's behavior is modified by environmental deterrents (aversive
consequences) and how behavior can be changed by purposefully
removing such deterrents--a matter emphasized by Skinner (1983).
For example, if an older person falls (or has some other accident),
he or she may incur damage that deters the individual from repeating
the performance that led to it--for example, walking. A vehicular
accident may deter driving, or at least result in more cautious
driving, possibly too cautious. Perhaps accumulated experience of
this sort, including vicarious experience, especially affects the
aging--in marked contrast, of course, to teen-age drivers. Might
there be some generalization of avoidance or caution from one
example of technology, the automobile, to others? Failure, with its
attendant confusion, is another deterrent. Might failure in
operating a word processor deter an older person, more than a
younger one, from trying again or from coping with some other
electronic device?

If an aversive consequence deters some older person from doing
something he or she might enjoy or benefit from, the removal or
diminution of that extrinsic or intrinsic deterrent may result in
desirable performance, provided it has some favorable consequence
(reinforcer). We should look for "hidden inhibitors" arising from
technology; we should also be concerned about intrinsic deterrents,
such as effort or inconvenience, that might be lessened through some
technological innovation. As muscular strength and stamina decline
with age, the greater is the exertion required for many tasks.
Again, consider walking. Suppose on the walls of corridors of
nursing homes there were, at intervals, hinged seats and vertical
grab bars next to them. A resident who was deterred by the effort
of walking the entire length of the corridor might undertake the
trip if it was possible to sit and rest once or twice en route.
Environmental design, a technology practiced mostly by architects,
has sadly neglected the effort variable.

Other important hidden inhibitors to older people are risks;
technology could minimize these too. Risks are not just hazards.
Quoting an unpublished paper by L.E. Gelwicks, Golden (1973, p.
137), wrote that due to the notoriously poor guidance and
communication systems in subway and other transportation terminals,
"is it any wonder that the older person prefers to stay at home than
risk making a fool of herself, or even worse, getting lost?" Were

terminal designers ever to exploit human factors knowledge and
pretest their system with prospective passengers, especially the
aging, such transportation technology would be improved.

HEALTH AND SAFETY

Just as it can for other age groups, technology can minimize
not only hazards but also the severity of accidents suffered by the
elderly, whose greater fragility makes some accidents more
damaging. Human factors technology can also help maintain health or
restore it. Let me illustrate.

It has been estimated there are about 500,000 active bedsore
cases in the U.S. at any one time, mostly in hospitals and nursing
homes, many involving elderly patients. Bedsores, or decubitus
ulcers, develop rapidly when body weight is localized for an
extended period, compressing blood vessels and thereby causing
deterioration of skin, subcutaneous tissue, and underlying muscle.
Pressure sores are most likely to form where the skin lies over a
bony prominence but can occur wherever long-standing pressure occurs
(Lindan, 1961). Though many victims are hospital patients such as
paraplegics, kyphosis of the skeletal structure and weaker
musculature may keep aging individuals who are lying in bed from
making the many postural changes, awake and sleeping, that otherwise
safeguard us from ulcer formation (Parsons, 1981).

There have been two technological countermeasures against
pressure sores, to replace manual turning of the body by nurses at
least every two hours. (1) Air-fluidized beds, waterbeds, and mud
beds distribute body weight evenly, but each body area experiences
some pressure. (2) Alternating-surface beds keep changing the bed
areas that support the body; in a sense, they automate manual
turning. One type rocks the body gently from side to side. In
another, a small electric motor inflates and deflates alternate
tubes in an air mattress so pressure is never sustained for more
than a few minutes at a time on any one skin area. The latter type
seems to have some advantage in cost and maintenance. These could
be called human factors innovations.

LEARNING

Some of my illustrations of technology for aging have stressed
soft rather than hard technology and low tech rather than high
tech. Computers are definitely high tech. They not only support
the teaching of various skills and types of knowledge in
computer-aided instruction, but also themselves constitute subject
matter to be learned and skills to acquire.

Kearsley and Furlong (in press) recently completed a program of teaching computer basics and word processing to 17 groups of 10-15 men and women each, between 60 and 90 years old, in the Arlington/Alexandria, Virginia area. Each group met four times for two or three hours each in churches, clubs, and senior centers, at no cost to the participants--of whom my wife was one. Kearsley (personal communication) told me that though his students may have taken a little longer than younger people, they seemed to have no cognitive difficulty in understanding instruction and learning to operate a Commodore microcomputer. One reason for taking longer, he said, was that so many were unfamiliar with keyboards and cassettes --parts of the larger technology. He reported that participants especially liked to program, apparently because of the feedback they got, and found off-the-shelf programs less challenging.

In an interesting instance of peer instruction, Edward C. Varnum, who received his B.A. in 1933, has been teaching several courses in "Home Computers" at Mesa Community College, in Mesa, Arizona. About one-half of the students are over 62. These need to be paced more slowly than the younger ones, many of whom are already familiar with microprocessors (Varnum, personal communication); mixing age groups seems not to be a favorable arrangement--apparently the younger students dislike it.

MANIPULATION

Varnum's project is one modest illustration of how the aging themselves can manipulate technology, rather than technology controlling the aging. To some extent, at least, the relations between the aging and technology can and should be reciprocal. The aging should at least participate in determining how technology should be applied in their lives and have some measure of control. An example of such participation and control where the technology involved was housing, in Simi Valley, California, was reported by Lang (1978, p. 545):

> Consistent with current thinking a building developer
> with his architect presented plans for a senior housing
> development to a group of senior citizens to obtain
> their approval. The group was so critical of their
> plans, their recommendations made such sense that the
> original plans were scrapped. A Housing Conference
> sought their recommendations. The housing needs
> requested by senior citizens were found to differ
> significantly from local building codes, HEW and HUD
> regulations.

Neither Lang's paper nor the builder's report (Griffin Development Co., 1977) referenced any guidelines from HEW, HUD or other governmental sources that could have helped the senior citizens and the builder and architect analyze and establish requirements, if only because satisfactory guidelines did not exist. Nor do they yet (Parsons, 1981), though several contributions have come from private sources (Koncelik, 1976; Lawton, 1975; Raschko, 1982). How can the aging control the technologies that affect them unless they are supplied the knowledge with which to act?

Possibly this vacuum has existed because the aging themselves have not been in positions to fill it. Examine the age distributions among professional and managerial personnel in state and federal agencies (and in the projects they fund) dealing with technological and research concerns of older people, in particular Housing and Urban Development and Health and Human Services (formerly HEW), including the Administration on Aging, Institute on Aging, and National Institute of Mental Health. Without suggesting that only the aging should be responsible for investigating and helping the aging (or that only women should be responsible for investigating and helping women or blacks for investigating and helping blacks), older people should at least be well represented. Are they?

REFERENCES

Baker, G. T. & Krauser, C. K. A catalog of products and services to enhance the independence of the elderly. Philadelphia: Institute on Aging, Drexel University, 1982.

Baltes, M. M. & Zerbe, M. B. Independence training in nursing-home residents. Gerontologist, 1976, 16, 428-432.

Bellucci, G. & Hoyer, W. J. Feedback effects on the performance and self-reinforcing behavior of elderly and young adult women. Journal of Gerontology, 1975, 30, 456-460.

Bromley, D. B. The psychology of human aging. Baltimore: Penguin, 1966.

Butler, R. N. & Lewis, M. I. Sex after sixty. A guide for men and women in their later years. New York: Harper & Row, 1976.

Chapanis, A. Human engineering environments or the aged. Gerontologist, 1974, 14, 228-235.

Fozard, J. L. Person-environment relationships in adulthood: Implications for human factors engineering. Human Factors, 1981, 23, 7-27.

Fozard, J. L. & Popkin, S. J. Optimizing human development. End and means of an applied psychology of aging. American Psychologist, 1978, 33, 975-989.

Golden, H. M. The dysfunctional effects of modern technology on the adaptability of the aging. Gerontologist, 1973, 136-143.

Green, P. S. Growing old in China. Modern Maturity, 1983 (August-September), 58-59.

Griffin Development Co. Report of the Senior Citizens Conference on
    Housing. Tarzana, CA: Griffin Development Co., 1977.
Hoyer, W. J., Kafer, R. A., Simpson, S. C., & Hoyer, F. W.
    Reinstatement of Verbal behavior in elderly mental patients
    using operant procedures. Gerontologist, 1974 (April),
    149-152.
Hoyer, W. J., Mishara, B. L. & Riedel, R. G. Problem behaviors as
    operants: Applications with elderly individuals.
    Gerontologist, 1975, 15, 452-456.
Hussian, R. A. Geriatric psychology. A behavioral perspective.
    New York: Van Nostrand Reinhold, 1981.
Ince, L. P. Behavior modification in rehabilitation medicine.
    Springfield, IL: Charles C. Thomas, 1976.
Kearsley, G. P. and Furlong, M. Computers for kids over sixty.
    Reading, MA: Addison-Wesley, in press.
Kerschner, P. A. & Chelsvig, K. A. The aged user and technology.
    In The physical and mental health of aged women. Springer
    Publishing Co., in press.
Koncelik, J. A. Designing the open nursing home. Stroudsberg, PA:
    Dowden, Hutchinson, & Ross, 1976.
Lang, C. L. Seniors plan senior housing. In Proceedings of the
    Human Factors Society-22nd Annual Meeting, Santa Monica, CA:
    Human Factors Society, 1978.
Lawton, M. P. Planning and managing housing for the elderly. New
    York: Wiley, 1975.
Lawton, M. P. Environment and other determinants of well-being in
    older people. Gerontologist, 1983, 23, 349-357.
Leon, A. S. In K.W. Sehnert (Ed.), The family doctor's health
    tips. Meadowbrook Press, 1981.
Lindan, O. Etiology of decubitus ulcers: An experimental study.
    Archives of Phys. Med. Rehabilitation, 1961, 42, 774.
Lopez, M. A., Hoyer, W. J., Goldstein, A. P., Gershaw, N. J., &
    Sprafkin, R. P. Effects of overlearning and incentive on the
    acquisition and transfer of interpersonal skills with
    institutionalized elderly. Journal of Gerontology, 1980, 35,
    403-408.
Ludeman, K. The sexuality of the older person: Review of the
    literature. Gerontologist, 1981, 21, 203-208.
McClannahan, L. E. & Risley, T. R. Design of living environments
    for nursing home residents. Gerontologist, 1974, 14, 236-240.
Miles, L. E. & Dement, W. C. Sleep and aging. New York: Raven
    Press, 1980.
Oster, C. Sensory deprivation and homeostasis. Journal of the
    American Geriatric Society, 1979, 27, 364-367.
Parsons, E. C. A warning to the middle-aged. New Review, 1915
    (June 1), 3(7), 62-64.
Parsons, H. M. Caution behavior and its conditioning in drivers.
    Human Factors, 1976, 18, 397-408.
Parsons, H.M. Residential design for the aging (for example, the
    bedroom). Human Factors, 1981, 23, 39-58.

Parsons, H. M. & Kearsley, G. P. Robotics and human factors:
    Current status and future prospects. Human Factors, 1982, 24,
    535-552.
Pfeiffer, E., Verwoerdt, A. and Wang, H-S. The natural history of
    sexual behavior in a biologically advantaged group of aged
    individuals. Journal of Gerontology, 1969, 24, 193-198.
Raschko, B. B. Housing interiors for the disabled and elderly. New
    York: Van Nostrand Reinhold, 1982.
Sargent, J. V. An easier way. Handbook for the elderly and
    handicapped. Ames, Iowa: Iowa State University Press, 1981.
Schulz, R. Emotionality and aging: A theoretical and empirical
    analysis. Journal of Gerontology, 1982, 37, 42-51.
Skinner, B. F. Intellectual self-management in old age. American
    Psychologist, 1983, 239-244.
Spiegel, R. Sleep and sleeplessness in advanced age. (Advances in
    Sleep Research, Vol. 5). Jamaica, NY: SP Medical and
    Scientific Books, 1981.
Verwoerdt, A., Pfeiffer, E. and Wang, H-S. Sexual behavior in
    senescence: II. Patterns of sexual activity and interest.
    Geriatrics, 1969, 24, 137-154.
Webb, W. B. Sleep in older persons: Sleep structures of 50- to
    60-year-old men and women. Journal of Gerontology, 1982, 37,
    581-586.
Weg, R. B. (Ed.). Sexuality in the later years. New York: Academic
    Press, 1983.
Weisman, S. Computer games for the frail elderly. Gerontologist,
    1983, 23, 361-363.
Welford, A. T. Signal, noise, performance, and age. Human Factors,
    1981, 23, 97-109.

HUMAN FACTORS AND TECHNOLOGY:

THE USER SETS THE PACE

Robert A. Vecchiotti
Organizational Consulting Services, Inc.
St. Louis, Missouri

Arnold M. Small
Human Factors Department
University of Southern California
Los Angeles, California

It is the thesis of this paper that the applications of advanced technology in association with the information revolution must be based on much more than just availability of advanced equipment. The latter, a monolithic approach, abrogates contributions of such relevant disciplines as systems science and human factors to achieve balanced and assured overall payoffs.

In a society that frequently looks to technology as a remedy for problems such as in health care, education, manufacturing and office operations it is increasingly important to pay attention effectively to user requirements and characteristics, including those of older adults. However today many of the criteria used to justify the purchase of technology, especially in the office environment, focus only on increasing speed, keeping pace with the competition, encouraging excessive growth and in general responding to technical features made attractive by eager salesmen. This hardly assures the success predicted.

Those demographic changes already underway which exhibit a future increase in the number of workers 35 to 54 in the next decade and a reduction in those under 25 provide a stage setting that demands urgent attention to the middle aged and older worker in all respects. Nowhere is this more true than in the office environment. As office technology planning and the older work force converge, then is the time to determine the changes in work patterns and effective person-machine-environment interfaces. If the right questions are asked, the answers obtained, digested and compared,

the requirements identified and translated into design parameters
and features, realistic applications of high technology can bring
out the full potential of both the old and young in the future work
force.

The tools and techniques are available for obtaining answers.
They include analyses of performance objectives, functions, tasks,
workloads, and linkages; data generating experiments; and the delphi
method in addition to the well-developed quantitative research
design techniques. The data are needed by decision makers in
considering trade-off features within the person-machine-environment
system. In this connection the "power down" trend which puts more
decision making and resources down in the office, plant etc., should
provide opportunity for semi-autonomous work groups to emerge within
the larger companies. This trend may be advantageous in affording
older persons the opportunity to work at that location, whether
home, office or camper, which best meets their needs and
capacities. In fact, based upon demographic forecasts of the work
force over the next three or four decades, older executives,
managers and employees may be asked to or volunteer to remain
actively employed to maintain viable organizations.

The general shift toward knowledge-based operations is well
documented as is the exponential rise in total information and
data. This all signals widespread necessity for and planned control
and implementation of those changes which are inherent in current
trends, many of which affect the older members of the work force.

No one discipline has all the puzzle pieces in this dramatic
change. The user/equipment interface is a product of
inter-disciplinary teams. Such teams can develop tailored systems
to meet specific needs. It is rare that off-the-shelf hardware and
software can meet each user requirement. The tailoring process has
been made more cost effective by modularization, i.e., building a
system in "chunks." Not every user needs a full system capability 24
hours a day.

Today there is another key focus in building a user/equipment
system. When using task analysis, it is important to include what
the user and the computer hardware do with the information being
exchanged. In other words, one must pay more attention to how
people think and how people acquire and use information. Cognitive
processes need to be made more visible. This is especially true in
view of the fact that managers today manage the informed, not the
uninformed.

In terms of user requirements, it is becoming increasingly
important to focus attention on the middle manager of whatever age.
Only recently have attempts been made to address the impact of
office technology on management and organization design. The

impetus for these efforts has come from data processing managers who
are confronted by a situation where expanding capacity has led to
expanding management concern, not expanded use. The concern is
partially assuaged by recent attempts of data processing managers to
use more "natural" language rather than the highly specialized
jargon of data processing in the user/hardware interface. There has
also been increasing interest in what managers do and what
information they need to do it.

There are now several empirically-derived descriptions of what
managers do. Henry Mitzberg, in his book ("The Nature of Managerial
Work," 1973), identified three roles played by management. The
major roles are interpersonal, informational, and decisional. A
manager's interpersonal role requires that he or she be a
figurehead, liaison to other groups, and a leader. The
informational role includes monitoring the environment,
disseminating information, and being a spokesperson for the group
and its work. The manager is also a nerve center for information in
that he or she identifies what is relevant, stores it, and retrieves
it on demand for use in carrying out day-to-day activities. The
manager serves as a human processor for all incoming and outgoing
information. Lastly, as a decision maker, a manager is expected to
be a resource allocator, negotiator, disturbance handler, and
entrepreneur.

The roles Minzberg outlines are interlocking. A management
position as a nerve center gathers information through a network of
sources, and sifts and sorts the information into categories
relevant to the objectives of the group. It is then disseminated to
others for action or serves as the basis for a series of decisions
about the information and its importance.

The manager's role is also influenced by the level in the
organization and the environment in which it occurs. The higher the
position, the less technical and more strategic are the information
needs. The more strategic the needs, the more external factors
enter into decision making. Dr Eliot Jaques of the Bruner Institute
in England has developed a system of organization development which
predicts the time span appropriate to various layers of management
from first level management to the chief executive officer whose
time span could reach out 15 to 20 years into the future (Jacques
and Stamp, 1983). Indeed, the CEO represents a key position in that
its focus is to assure development of an organization that can
capture market share and remain viable while the environment
changes.

Top managers tend to be more informal and less structured than
the first echelon, which tends to spend more time on the current and
specific issues in more formal and structured ways. Matching
technology to these different roles, levels, and conditions must

take into consideration the differing requirements at each level of management.

   If the user/manager were to set the pace in the addition of office technology, what requirements can be identified for computer hardware and the man/machine interface? Here is a partial list of general requirements:

1. Executive positions require less technical data, and more summaries and trends.

2. Any system should be tailored to the unique indicators (non-routine) and information used by a manager to measure progress and spot potential problems.

3. The system should be reliable, efficient, and very user friendly,--even congenial.

4. Management should have software flexible enough to permit restructuring information in several alternative formats.

5. Software should be able to generate consequences of present actions to assist in thinking about options before a decision is made.

6. Management considers intuition very important; it should be integrated with any technical support system by override provisions or direct input of information.

7. The decision support system should be able to handle both a stable environment and change with minimal disruption of operations.

8. Feedback on results should be included, especially when building and evaluating alternative scenarios.

9. The system should support the development of subordinates through record-keeping and prescriptions for growth based on individual assessment (computer-based career pathing).

10. The system must support the development of subordinates by on-line interaction with follow-up schedules tailored to the work/activity calendars of both manager and subordinate.

11. Provision for personal development with enrichment opportunities embedded in the system should be included.

12. Opportunity for peer dialogue and exchange of information should be provided.

13. Contact with outside sources may be required through a combination of telephone and video terminals.

14. Menus and catalogs for additional sources of information should be available in the software.

15. Opportunity to build networks within the organizational system should be provided.

16. A built-in consensus system may prove to be extremely valuable, especially when seeking advice from experts at remote sites.

17. Provisions for portability would be useful and would make the system accessible from home or remote location as required.

18. Provision for continuous survey and data gathering on what managers are doing may be useful in identifying emerging requirements for system upgrading.

19. On-line dialogue of system users and system maintainers would be valuable in assuring that operating problems are resolved efficiently.

The list could obviously be added to and continuosly updated. Please note that the list contains few references to hardware design. The focus is clearly on what managers and older workers need as a starting point to help them carry out their responsibilities.

To meet these requirements a multidisciplinary approach is necessary. Specialists from data processing, human factors, personnel, and the user should be responsible for needs assessment, requirements analysis and definition, system development, and implementation and evaluation.

Finally, as the population continues to age, and the number of entry level young people decline, greater involvement of retired people may be needed back at work. Yet they may never have to leave their homes or travel for training sessions. Much of the tasks necessary to come On-line will be embedded in the system itself.

Satisfying the complex and changing operational needs of the future workforce requires that we examine user requirements in detail. The tools and techniques are available. The human factors perspective provides a framework to allow the user to set the technology development pace. Why be concerned about anticipating future user requirements? Because each of us may spend the rest of our lives working with systems we can't use, and our growth and development could be slowed dramatically in the coming decades by technological illiteracy.

REFERENCES

Jacques, E. and Stamp, G.   1983.   Level and type of capability in
    relation to executive organization.   Final technical report.
    U.S. Army Procurement Agency, Europe.   Contract No. DAJA
    37-80-C-0007.
Mitzberg, H.   1973. The Nature of Managerial Work.   New York: Harper
    and Row.

PEOPLE IN FUTURE FACTORIES AND OFFICES: WITH AN INTRODUCTION TO SOME SPECIAL OPPORTUNITIES AND PROBLEMS FOR AN AGING WORKFORCE[1]

Gordon H. Robinson
Dept. of Industrial Engineering
The University of Wisconsin-Madison

Gerald Nadler
Dept. of Industrial and Systems Engineering
The University of Southern California

James G. Peterson
Dept. of' Industrial Engineering
The University of Wisconsin-Madison

A major difficulty with the topic of future factories or offices is that the mind usually grasps it visually, as a static picture at a specific time--the year 2000 has been popular. This conceptual limitation is significant in debates and planning regarding new technological possibilities and new opportunities and problems for people at work.

This snapshot view of future organization is, at best, incomplete. It ignores continuing developments in technology, such as those that produced the microprocessor, and encourages debate about the desirability of specific renderings of technological possibilities, forms unlikely to appear in any event, and far less likely to be influenced by the debate.

We focus here on the configurations and purposes of the social systems that will relate to these future organizations.

Three issues must be considered:

1. A new factory or office does not appear suddenly in full operational maturity, but rather is continually designed and redesigned, implemented, constructed, and rebuilt.

2. It is not automatically programmed to improve. Continual
   energy and direction must be employed if it is to adapt
   successfully to changing needs and potential.

3. It will never be without people--for the reasons above,
   and to complement the latest innovations in the technical
   system.

The factory and office are a human phenomenon. Every step from
conception to eventual destruction is for, by, and because of
people.

THE FACTORY OR OFFICE AS A TECHNICAL SYSTEM

The technical production or service delivery system lies at the
heart of the organization. It is the breadwinner. No organization
can survive if this system fails to produce something of adequate
value. Adequate value is not a simple issue, however, as
organizations operate increasingly in complex, interconnected
environments that are changing rapidly and unpredictably. Many
technical systems, envisioned just prior to the energy crises of the
1970s, now look quite obsolete. Long-range survival now demands
technical decisions and plans much more complex than those of the
time when most of the currently operating organizations were
designed.

There are four sets of decisions to be made: 1) the need or
purpose to be achieved by the organization; 2) the product or
service to be offered; 3) the technology (science or "know-how")
available for the production of this product or service; and 4) the
specific technical system chosen to make the product or deliver the
service.

The first three are highly interconnected. Products may have
to await technological possibilities; new developments and products
depend on the needs the organization seeks to fulfill; and new
technologies may suggest new products. The fourth, how we will
render the technology in our specific organization, presents many
more choices than commonly realized. The choice here depends on the
characteristics of the social system required to operate the
technical system, and on a wide range of economic factors.
Engineers and managers are only now learning that specific technical
systems are not uniquely specified by the product or technology.
There is no one best way--at least not until the needs of the social
systems are jointly considered with the needs of the technical
systems.

The Volvo, Kalmar factory (Gyllenhammar, 1977) illustrates some
of these choices. At Kalmar, two major technical decisions were

effected by considerations of the present and future social systems
that would operate the plant. A unique architectural design,
encouraging the perception of smaller, independent "shops,"
represented a specific design choice from among the technological
possibilities. The usual method of conveying automobiles during
assembly (the "line") was deemed too constraining for the needs of
the social system and a new technology of computer controlled carts
was developed. It is important to note that neither of these
decisions was related directly to the product, nor for that matter
to the usual economic measures. Both were in fact rather costly and
involved some risk. The appropriateness of these specific decisions
is not the issue here, of course. The questions asked and issues
raised concerning the social system's needs and the reflection of
these needs onto the technical system is our interest.

Gyllenhammar (1977) feels strongly that large, multinational
organizations have two areas of responsibility if they are concerned
with long-range growth, or even survival. One is responsibility for
the people--their social system. The quality of life for these
people, now and in their future, affects Volvo, and is affected by
Volvo, through what is now commonly known as quality of working
life. (The other responsibility is toward the material resources
that society chooses to make available; in this case, for the
production of automobiles.)

It has thus become more clear that what we used to think of as
decisions based purely on our proprietary choice of product and the
correct, optimal technology are not those kinds of decisions at
all.

THE FACTORY OR OFFICE AS A SOCIAL SYSTEM

The future factory and office will depend more, rather than
less, on the people within it. Higher and more sophisticated levels
of mechanization and automation have critical needs for people to
assist with breakdowns, predictable maintenance, and redesign. Thus
while there will be fewer people engaged in direct production there
will probably be many in allied, supportive functions. That is, the
production function will be more richly interconnected with other
functions and organization units, and more dependent on them. The
more interesting issue is what these people will be doing, and
consequently what will be their tasks, jobs, roles and necessary
organization.

Twenty years ago, in a well-reasoned article, Jordan (1963)
spoke of the need for people in automated systems to provide manual
backup in the event of machine failure. The importance of this
function became dramatically apparent within a few years as several
U. S. space missions survived only because of interventions by their

on-board humans, who up to then had been viewed by many designers as simply cargo. As we have moved farther along in the sophistication of our technical systems we have also moved farther from any significant possibility of manual backup operation. This is particularly true in advanced manufacturing, where the nature of the product, the form of the transformations, and the necessary rates of production prevent almost all possibility of human intervention in the production flow. It will probably become true in "automated" offices as well, as we already see in the difficulties bank tellers and airline reservation agencies have when their computers are "down."

What then are people doing on the shop floor as we move into highly automated factories? We can begin to examine this question by looking at our most advanced and technologically sophisticated factories. One of these is a chemical processing plant discussed by Davis and Sullivan (1980). In this plant 150 employees are involved with $300 million in capital investment, $2 million per worker. This ratio in itself begins to suggest some issues for the possible design of the social system. What responsibility does this worker have for his or her $2 million machine? Or in terms of costs, what can or can't you afford to pay for in a social system that is so relatively inexpensive?

The design of this plant, and a number of others with similarly high degrees of automation--such as food processing and paper making--was greatly influenced by an emerging body of conceptual knowledge and technique known as sociotechnical systems (Trist, 1981). Starting with studies of the impact of new, highly mechanized machines in the British coal mines in the late 1940s; extending these insights to include new roles for supervisors in experiments with automated weaving looms in India in the early 1950s; and continuing through important new ways of characterizing the interactions between the technical and social systems with studies in paper making in Norway in the mid-1960s (Engelstad, 1979); this theory has turned aside much conventional wisdom in regard to both individuals and the organizations in which they are embedded. Of greater importance to the issues before us here, however, is that the sociotechnical view gives us increasing understanding of what the social part of the system is in fact there for. Prior to this new understanding most designers simply viewed the people as doing--temporarily--that which they could not yet automate. Forty years of experience with automation have highlighted people's curious tendency to stick around. We are now beginning to understand why.

Cherns and Wacker (1978) discuss the potential extension of Parson's macrosocial model to the industrial organization. They present four functions as necessary for the continued survival of the organization: (1) production (they use the term "goal attainment" but we have found it unnecessarily confused with "goal setting" in discussions with industrial managers); (2) maintenance and development; (3) adaptation; and (4) integration.

Production is the function that adds value to some resource that in turn can be sold at a profit. The new roles for the social system in highly automated production are now fairly well understood. These roles center on variance control--setting right that which has gone wrong--and learning to do it better. This need for people will be discussed in more detail below.

Adaptation actually encompasses two functions: protection of the production system from outside influences that would reduce its effectiveness and thus the profits; and the introduction of new technology and continued technical system improvements that will enhance the production potential. Adaptation thus forces us continually to balance the need for change with the need to let production alone so that it can produce with high efficiency. Some, such as Hayes (1981), have noted that control of this balance is possibly one of the things the Japanese do differently, if not better, than we. Decision making within this function obviously incorporates many nonquantitative variables and values. Automation will probably never be particularly prevalent here, although computers will help greatly with structuring the decisions and projecting the possible consequences.

Maintenance and development provides the organization with an appropriate social system. Major responsibilities here involve the selection, training, and career management of the people. The motivations of people at work, their expectations of other people at work and of the organization, and their individual differences would appear to make any systematic automation of this function rather difficult, if not actually incongruent with its goals. These functions become more complex and consume more resources in high technology organizations in which we are likely to see such organizational changes as semi-autonomous work groups and the need for greatly increased training in both skills and knowledge.

Integration is of course much of what traditional management does. It is important to point out, however, that newer organization designs, particularly in highly automated plants, have an increasing amount of integration and coordination at lower levels, levels previously thought not properly to contain these functions. This movement of decision making and authority downward, and the often accompanying flatter organizations, results less from considerations of quality of working life, or employee satisfaction, than from the requirements of the technical system, although it usually improves both. Automation of managerial functions is unlikely, and again has a degree of incongruency for the people to be managed. Goal setting, of course, must remain a human function.

Returning to the production system itself, the notion of the
social system controlling variances has already been introduced.
This concept has been very helpful in understanding the form and
needs of the social system in highly automated plants. It is
important to point out that we cannot, in any but the most trivial
of production systems, reach that point of eliminating or
automatically controlling all of the variances. Continued evolution
of product and technology alone preclude this possiblity. In the
chemical processing plant previously mentioned (Davis and Sullivan,
1980) an examination of this issue led to the design of a learning
organization, such that control of things gone wrong or adjustment
to new demands would be as effective as possible. It is worth
noting at this point that the highly automated technical system
originally planned for this plant was deliberately degraded in order
to facilitate this learning. This singular act may foretell a very
different future for automation than technological possibilities
alone would indicate.

Organizations are increasingly forced to function in what Trist
(1980) has described as a "turbulent" environment. He suggests that
the appropriate design for success, or simple survival, in this
environment is one increasingly looking toward rapid adaptation or
continual redesign. The social system, augmented by a rather thin
technology, must perform these functions (Robinson and Peterson,
1983).

DEMANDS ON THE PEOPLE IN THE FACTORY

What the people are to do has been outlined. The next question
is who these people should be. It is already clear that scientific
management's "replaceable-parts" person will not do in today's high
technology settings. It is perhaps fortunate then that such people,
in what supply they ever existed, are in short supply today (Davis,
1980).

The needs here become clear as we see that the role is
replacing the job (Davis and Taylor, 1979). Role includes the
social system's rich communication and coordination needs that are
increasingly important with advanced technology. With this role we
also see an increasing need for responsibility. No longer can the
operator wait for supervisory request. The discretionary parts of
the job increase, while the prescribed parts are reduced. Even
today we systematically underevaluate the importance of worker
discretion, a fact painfully recognized by the British Rail System
as their operators brought it to a halt with "work to rule."

Within the general concepts of role and responsibility, the
functional needs for the social system point to the more specific
characteristics that the people must possess. The control of the

variances in the technical system requires three things in addition
to responsibility: skill, knowledge, and the willingness to exert
authority that has been delegated (Davis and Wacker, 1982).
Knowledge, as opposed to a collection of skills, seems increasingly
necessary with advanced technology--knowledge of the product or
service being produced, of the process by which the product is
shaped and of the quantities, qualities, and costs necessary for the
organization's continuation or growth.

    The social system's "adaptation" function can be conceptualized
as the people controlling the variances in production, solving
problems in "quality circle"-like units and certain staff
functions. The new needs for the staff engineers will be discussed
below. The social system's "integration" function now extends from
the bottom to the top of the organization. Cooperation through the
understanding of common goals and assumptions becomes critical as
these flatter organizations move to a mode of "us against the
product" rather than "us against them." The social system's
"development" function must provide and maintain all of the
preceding, and therefore must have a keen understanding of all of
these new social system needs, as well as the skill and knowledge to
produce the necessary people.

IMPLICATIONS FOR THOSE INVOLVED IN DESIGN

    This article has argued that as technology has become more
complex and automated, the social system required to operate and
support it also becomes more complex. Roles replace jobs;
coordination, cooperation, and knowledge demands affect all of the
people. We therefore have more complex technical and social
systems, and they are more richly interconnected. Not only is the
future factory or office a more complex system, in the systems
theoretic sense, but it is increasingly viewed as an organic rather
than a deterministic structure--more like a forest than a piece of
clockwork. The environment is all pervasive, and the factory itself
is continually evolving. For example, the Fanlac Fuji plant, the
factory in Japan claimed by many to be the most advanced
illustration of robotic manufacturing, has signs all over the plant
exhorting the 60 to 70 employees to cut total costs by 30 percent.

    The design and the management of this system require skills,
tools, and strategies that are not in great supply--and for that
matter not even well-defined. The responsibility thus rests on our
educational and training systems to uncover these needs and produce
the needed designers and managers. The fact that both design and
management are diffused down to lower levels in advanced
organizations changes the nature of the possible solutions but not
the problem.

We hesitate simply to say that we need systems designers, for according to job descriptions or degree program credentials, these people abound. Few of these people today have the credentials we actually find necessary. The numerous failures of orderly and timely implementation of new technological ideas or new organizational forms is the clue. To most current engineers, for example, the term "system" means a fairly complex machine, a mathematical algorithm, or a computer.

Four issues are relevant.

1. The system is more than its parts, and cannot be understood, much less designed well, if this is not explicitly taken into account.

2. The system is sociotechnical, with interacting technical and social systems, and is virtually guaranteed to be suboptimal if either part is considered in isolation.

3. The design is never finished, partly because of the continuing advances in technology such as the microprocessor, and partly because change takes time, energy, and resources.

4. Implementation of redesign often requires the design of a temporary, additional organization.

These are the issues the designer must understand, and must have mastered the necessary tools and techniques to effect. Such a designer is thus a generalist--coordinating, facilitating, organizing, and planning.

This designer also coordinates the needs for the total range of stakeholders, a set much larger than in the past. Government agencies, community groups, consumer agencies, unions, and so forth will have a stake in the design, and as such have the ability to offer assistance, or bring our plans to a complete halt. (The latter has often been the result in recent years in situations involving freeways, dams, power plants and potential abusers of natural resources.)

PEOPLE IN WORK GROUPS AND REDESIGN: POSSIBILITIES AND OPPORTUNITIES

The emergence of organizations designed as sociotechnical systems results in two phenomena of interest to the issue of an aging workforce. First, these organizations prepare and facilitate a broader spectrum of people to actively solve problems, plan for new outcomes and participate in the implementation of these changes. Second, semi-autonomous work groups emerge as a common and

useful configuration of the social system. These groups will often
contain varying levels of skill, knowledge and experience among
their members.

Employee age becomes a design parameter affecting both the
dynamic/participatory environment and the composition and function
of work groups. "Experience" is, of course, a positive attribute of
age. It is especially positive when it is combined with
purpose-oriented design approaches, rather than with critical
analysis approaches (Nadler, 1981). With both of these design
features this experience could express itself in several useful
ways, as: 1) offering a greater variety of methods or solutions, 2)
offering specific knowledge as to mechanisms of some solutions, and
3) offering experiential knowledge on the problems of
implementation. Whether managers and designers will chose to, and
be able to offer an environment to utilize these attributes remains
to be seen. An incorrect reward structure or poorly developed
career patterns or inappropriate approaches and concepts for design
(Reich, 1983), could easily lead to an experience producing the
"dynamic conservatism" common among middle managers--the experienced
people in organizations today.

The Semi-Autonomous Work Group

Age as a design variable has potentially important implications
for the composition and function of the semi-autonomous work group.
Both what the group does, and how it manages itself, are affected.
A fairly wide range of skills, knowledge of their relationship to
the overall product or service being produced and the ability and
willingness to take responsibility and independent action are the
necessary qualifications for a properly formed work group.
Experience (age) can produce all of these. At best, the younger
worker will have some of the skills, and, of course, possibly some
newer ones to add to the repertoire. This broader range of
abilities could allow a newly formed group to take on its requisite
response assignments more quickly and thus function at its designed
"autonomous" level at an earlier point in time.

Training new workers and increasing the skills of existing
workers falls within the range of activities of many work groups.
These are tasks obviously suited to more experienced
workers--especially as it is recognized that "knowledge" as well as
specific skills is useful.

The range of tasks, and the group's "autonomous" abilities to
allocate itself to them, could easily allow compensation for those
disadvantages of age (reduced sensory abilities, physical strength,
etc.) in exactly the same way that the lack of skill or knowledge is
accomodated in the new, young workers. Even in a shipping and
receiving group, with functions including a fair amount of heavy and

tiring work, new designs include the group itself performing testing, inventory control and record keeping--which are tasks easily performed by older workers (Davis, 1980). And, of course, much of the heavy work is mechanised. Designs usually include some minimum number of skills that must be mastered, and often pay is tied to these skills.

Participative Redesign

An aging work force could both participate in--and promote participation in--the process of continual redesign. Three factors indicate the potential importance of experienced workers in these processes. First, due to their accumulated experience through involvement in a number of earlier redesign projects, older workers have implicitly learned the "technology" of redesign (Robinson and Peterson, 1983), including skills in planning, group facilitation and management, design strategies and implementation issues. They could be the "systems" people whose need was suggested earlier. As such, they complement the new, highly specialized skills that may be brought by younger technical people.(2)

Second, this knowledge of how to redesign is a skill, and hence could be taught to younger workers in the organization. The final point is that older employees could be "change agents" (Nadler, 1981), and hence act as key people in promoting the continual redesign necessary in the face of the rapidly changing environment most organizations find themselves in today.

In Summary

It appears distinctly possible that the learning from accumulated experience (age) could be singularly useful in newer organization designs and with advanced technology products or techniques. It also appears that the organization designs now being seen in advanced plants offer several mechanisms to explicitly utilize this ability and to minimize the problems of performance loss due to age. It may well be that those organizations learning to best use this resource will be the most productive. The specifics of the organization design will be important. Obviously the mix of ages is an important variable. The career and reward structure mentioned earlier is probably the most significant issues. Unless the aging employee sees the future roles, is encouraged to prepare to take them on and has some measure of security during this process, we will see a mixed set of results, at best.

SOME PROBLEMS TO BE RESOLVED

Uncertainties and problems abound in trying to understand and bring about the potentials for using the aging workforce outlined above. Two clues to our current state of ignorance are: 1) our inability to utilize the variable "age" in our present organizations, and 2) the new and experimental nature of the sociotechnical organizations that represent the future.

The wide array of organizational needs for people, designs that may better utilize the people, and the possible roles for aged people presented here produce several testable/measureable hypotheses for research. Evidence with significant organizational change suggests, however, that this research will largely be of the "action" variety (Clark, 1972), rather than in the laboratory. There are, however, literally hundreds of "experiments" with new organizational configurations going on in the world today. "Age," in almost all of these, is a random, uncontrolled variable and its effects are seldom, if ever, measured. It would be difficult, but certainly within reason and our abilities, to look carefully at what is now going on with the age issue as an independent variable. The current anxieties concerning the new roles for "middle managers" is a problem that would also gain from this research.

This research could be exciting in that it would necessarily cut across a number of disciplines. Gerontology, sociology, planning and design, engineering, ergonomics, and psychology would have to be integrated to provide the research agenda that could adequately assess the effect of the older worker in work groups and continual organizational redesign--and, of course, in other contexts in future organizations.

The large increases in the proportions of older workers will begin within 15 years. This next 15 years will also see dramatic changes in the forms of work organizations and the structure of job. Whether we are able to take advantage of this historic coincidence is the question.

NOTES

1. A more detailed view of the future automated factory, including discussion of the continual planning and design processes, can be found in G. Nadler and G.H. Robinson, Design of the automated factory: More than robots, in Robotics: Future Factories, Future Workers, R. Miller (Editor), Annals of American Academy of Political and Social Science, 1983.

2. The technology of planning and design does not change rapidly, as it does for some "high technology" products or manufacturing methods. This is partly because people are an important issue and partly because this technology is still, and for

some time will remain, largely an art--evolving rather than being replaced.

REFERENCES

Cherns, A. B. and Wacker, G. J. 1978, Analyzing Social Systems, Human Relations, 31.

Clark, P. A. 1972, Action Research and Organizational Change, London: Harper and Row.

Davis, L. E. 1980, Individuals and The Organization, California Management Review, 22.

Davis, L. E. and Taylor, J. C. 1979, Design of Jobs, 2nd Edition, Santa Monica, California: Goodyear.

Davis, L. E. and Wacker, G. J. 1982, Job Design, Chapter 2-5 in Handbook of Industrial Engineering, G. Salvendy (Editor) McGraw-Hill.

Davis, L. E. and Sullivan, C. S. 1980, A Labour-Management Contract and Quality of Working Life, Journal of Occupational Behavior, pp. 29-41.

Englestad, P. H. 1979, "Sociotechnical approach to problems of process control." Chapter 19 in L. E. Davis and J. C. Taylor (Eds.), Design of Jobs, Santa Monica, California: Goodyear.

Gyllenhammar, P. G. 1977, People at Work, Addison-Wesley.

Hayes, R. H. 1981, Why Japanese Factories Work, Harvard Business Review, July/August.

Jordan, N. 1963, Allocation of Functions Between Man and Machines in Automated Systems, Journal of Applied Psychology, 47, pp. 161-165.

Nadler, G. 1981, The Planning and Design Approach, Wiley.

Reich, R. 1983, "The Next American Frontier," The Atlantic Monthly, March and April.

Robinson, G. H. and Peterson, J. G. 1983, Groups at Work: A Sociotechnical View, Proceedings of the Human Factors Society, 27th Annual Meeting, Human Factors Society.

Trist, E. L. 1980, The Environment and System-Response Capability, Futures, April, pp. 113-127.

Trist, E. L. 1981, The Evolution of Socio-Technical Systems, Ontario: Ministry of Labour.

A CAPABILITY-DEMAND APPROACH TO THE AGED IN TECHNOLOGICAL

ENVIRONMENTS: A CASE FOR IMPROVED TASK ANALYSIS

Martin V. Faletti and M. Cherie Clark

Stein Gerontological Institute
Miami Jewish Home and Hospital for the Aged
Miami, Florida

INTRODUCTION

The Symposium on Aging and Technological Advances considered a
broad range of issues involving technological advances and the aging
human, an apt choice considering that industrialized societies are
being impacted upon by the conjunction of human aging and
technological advance on an unprecedented scale. It appears clear
that technology will continue to change how we live and work on a
day to day basis. Whether these changes operate to our detriment,
or advantage, as we age will depend on the extent to which the
demands of technological environments can be reconciled with the
capabilities which we as aged operators bring to our transactions
with these environments. This discussion is focused primarily on
the application of human factors research to the study of the aged
user's capabilities and its implications for design and application
of technology to activities of daily living (ADL).

FUNCTIONAL CAPABILITY IN THE OLDER USER

While the prospect of the increasing numbers of older aged
(Taeuber, 1983) suggests that adapting technology to the older user
will involve designing for users with levels of physical and mental
capabilities which are reduced relative to younger populations, the
ability of older users to function in any environment is more likely
to be affected by the level of their capabilities relative to
environment demands. The issue is less one of how users change with
age but rather how the capability levels resulting from these
changes relate to what the environment and its component

313

technologies demand for successful performance of daily tasks and
activities.

Human factors routinely views task performance as being a
function of the congruence between operator capabilities and
environment demands relative to the task to be performed. As a
result, human factors research offers a range of methods and
techniques designed to accomplish capability-demand analyses of
performance in task settings as a means of specifying configurations
for technological systems involving human operators (e.g., Oborne,
1982). The application of such an approach to performance of daily
tasks and activities is likely to provide a more balanced view of
the factors affecting aged functioning in technological environments
and can indicate directions for design engineering which can enhance
the performance of the older user.

TASK ANALYSIS: DEFINING CAPABILITIES AND DEMANDS

This paper stresses the need to refine and apply existing tools
and techniques for task analysis to daily activities. Specification
of tasks is a critical first step in capability-demand analyses
because task specifies the functional relationship between the
person and the existing or simulated environment. In effect,
neither person capabilities nor environment demands exist in the
absence of task transactions. That is, person characteristics
(e.g., anthropometric stature, biomechanical force, array processing
speed) can be operationalized as capabilities only in the context of
their relevance to some function or use of the environment for a
particular purpose. Similarly, environment characteristics (e.g.,
object weight, handle size, control loading can be operationalized
as demands only when they are directly involved in something the
operator is trying to accomplish.

Because individuals are highly variable in the ways in which
they accomplish daily activities, the objective of task analysis
should be the systematic decomposition of the activity. Further,
task descriptors should specify the components of the person and the
environment involved in the task. this approach can be illustrated
by considering the development of task taxonomies for describing and
contrasting shopping for food in the grocery store (i.e., the
on-site environment) versus shopping by computer (i.e., a
microprocessor mediated environment).

First level task descriptors, most closely related in content
to the activity, might include finding the right aisle or product
subfile, reading labels, prices, nutrition information, and
selecting the item and putting it in the cart or placing the subfile
entry into a user select file.

Second level task descriptors would be more specific to the on-site or microprocessor environment. On-site descriptors would reflect person components of the task such as scanning shelves and objects, locating desired items and physical movements such as reaching for and grasping these items in order to transport them to a cart. Person descriptors for the microprocessor mediated environment would include the person scanning textual information and making keystrokes to enter commands involved in selecting desired items. On-site environment components relevant to the capabilities might include label colors and object shapes as cues for item identification and selection as well as shelf height and item weight as demands for physical activity. Microprocessor environment components may reflect little demand for gross physical activity, but might include demands for subtle discriminations required by screen text displays as well as fine touch and motor control involved in using a keyboard. Assessing the relative advantages of both environments for older users essentially involves trade offs in types and levels of demands presented by each environment versus the capabilities of the users.

Third stage task decomposition focuses on refined (sub)tasks, described in terms of person and environment components which are involved in operating specific technologies associated with each environment. On-site subtasks might include biomechanical motions and force exertions associated with accessing items from shelves or moving a cart of increasing inertia. Microprocessor subtasks might include visual identification of text and keys required to accomplish command entry and physical dexterity and force exertions involved in accomplishing sequences of deystrokes. The person components involved in making keystrokes might describe the operator's position relative to the device, movement of the hand and positioning the fingers, and force exertions required to actuate the key. Environment components might include the position of the device (height relative to the operator), the dimensions and layout of the device to be used, and the force loading/feedback cues of the keys.

Given specific and comparable person and environment components involved in the task, relevant person characteristics such as fluctuation in finger positioning and force exertions during hand positioning can then be measured in samples of older users on whom this technology may be focused. These operator characteristics can be directly related to environment design parameters (e.g., size and layout of keyboards) and thus provide more precise estimations of congruence between operator capabilities and demand levels associated with alternative configurations for technological environments.

The research discussed below attempted to apply this type of task analytic approach to the study of one prototypical ADL, meal

preparation, as a basis for developing empirical data on relevant
person capabilities and environment demands. The major objective of
this pilot effort was not to evaluate performance associated with
different technologies or environments. Rather, the study was
designed to examine (1) differences in task relevant person
capabilities in samples of older women who differed with respect to
independence in performing meal preparation tasks and (2) the role
of environment demands in reduced activity performance.

CAPABILITY - DEMAND ANALYSIS OF A DAILY ACTIVITY

Design

     The decomposition of an activity specifies k tasks involved in
the activity and, for each task k, i person components and j
environment components which provide respective links to measurable
person and environment person and environment characteristics
(Faletti, 1984). Development of task taxonomies associated with
successful performance of the activity was accomplished by selecting
and observing a representative sample of independent older women who
accomplished meal preparation without assistance.

Sampling Activity Performance

     Intuitive construction of a taxonomy of tasks involved in a
daily activity is likely to reflect variations dictated by a range
of factors. For meal preparation, these factors might include
taste, diet restriction, and/or kitchen configurations. Unlike work
settings where task taxonomies might already exist or be known in
enough detail to develop simulations to accomplish capability-demand
analyses, daily activities in community settings are less well
defined. This research approached the problem by first sampling
from person-environment transactions reflective of success at the
activity. Specifically, a sample of some 125 older women preparing
meals independently in the community completed meal logs for 10 days
and questionnaires regarding use of technologies, frequency of meal
preparations and other meal related information. A sub-sample of 40
cases from this pool was selected and, for each case, a
representative meal preparation was videotaped to provide a
permanent visual data record as a basis for the task analysis.

Task Analysis of a Daily Activity

     The research on meal preparation utilized a field portable,
multiplexing videotape system (c.f., Wellens, Revert, and Faletti,
1982) to capture the entire meal preparation sequence accomplished
by subjects in the independent sample. The multiplexing allowed two
images from different cameras (i.e., different views) to be
synchronously recorded on the same tape thus giving two views of

subject activity. These tapes were rated by multiple judges to
assess the reliability and replicability of task analysis schemas.
Meal preparations by five pilot cases were used to develop the task
analysis procedure, assess inter-judge reliabilities and performance
of the task protocol, and refine the procedure used for all subjects
in the independent, videotape sample (40 cases).

Many discussions of kitchen tasks emphasize the location of
tasks and, in particular, height of counters and cabinets. Thus,
task categorizations for meal preparation found in some handbooks
(e.g., Woodson, 1981) often simply describe the tasks in terms of
where it took place (e.g., the sink, the counter), but do not
specify what types of actions are performed or, more importantly,
what types of objects are involved in these actions. In the present
research, actual transactions reflected in the videotapes suggested
that these non-location compnents, such as object handling, were a
frequent feature of task activity; thus focusing attention on
demands associated with these and other non-architectural features
of the environment.

At its present stage of development, the refined procedure
consists of three phases: (1) segmenting the stream of behavior
into discrete and identifiable tasks (task definition), (2)
specifying person and environment components involved in each task
(task characterization), and (3) specifying parameters of person and
environment components in comparable terms (component
characterization). It should be noted that, in this pilot effort,
the schema focused on instrumental activities involved in meal
preparation. Clearly, an analogous system for decomposition of
information processing tasks is needed because many emerging
technologies place significant information processing demands on the
operator.

Utilizing Bennett's (1971) conceptualization of language as the
vehicle for defining and characterizing tsks, the operator (elder)
is, in grammatical terms, the "subject" and the first phase of
analysis involves the development of a system of verbs and objects
which describe respectively what the subject does (transaction) and
where or with what the subject performs the transaction
(environment). Thus the system used to define distinct tasks viewed
an activity in terms of subject (person), verb (transaction), and
object (environment). Descriptors for the definition phase of
analysis are meal preparation specific and thus admittedly violate
Companion and Corso's (1977) emphasis on generalizability of task
descriptors (e.g., reposition items, cut/chop, mix/stir).

The second phase of task analysis aimed at a task
characterization which was (1) expressed in descriptors
generalizable beyond meal preparation activity and (2) specified
relevant person and environment components in the transaction. The

task characterization protocol sought to define actions and objects
in ways which better specified the person actions and environment
components involved in the task transaction. A number of components
(e.g., grip type, object, location) are used with schemas for both
transport and manipulative tasks while other components are
selectively relevant (e.g., barriers to access in transports).
Action descriptors characterize a generic type of movement
accomplished as part of the task; usually to impart some motion to
an object. Grip type descriptors reflect three major grip types
defined by Drury and Coury (1982): (1) the power grip (fully closed
hand), (2) the hook grip (use of palm and fingers without gripping
by the thumb), and (3) the precision grip (fingers and thumb
pinching the object). The cradle grip was added to address those
cases where the object is supported from the bottom and not
"grasped," but rather is loosely contained within the hand. Like
the action descriptors, object descriptors reflect a first level of
characterization which distinguishes the objects as classes. This
class designation, together with the specific frame (on the
videotape), were used to complete a more detailed characterization
of the object as part of component characterization.

     Third stage component characterization schemas currently being
applied to videotape data include use of an XYZ coordinate system
which can be used to describe more generically the motion(s)
involved in task actions; that is, motion associated with operating
a particular appliance or utensil or motion capabilities of an
operator's hand. Such refinements utilize descriptors which can be
directly applied to describing either what a person can do
(capability) or what the environment requires to be done (demand);
thus providing the comparability required for empirical approaches
to capability-demand analysis.

Estimating Reliability of Task Descriptors

     This approach sought to develop task analysis as both a
judgmental as well as an empirically based activity. While the
selection of descriptors and systems was admittedly a creative
endeavor involving judgments on the part of the investigator, the
permanent videotape records supported the use of empirical
assessments of reliability in developing and refining the system.
Reliability of the task definition and characterization system
described above was assessed using both the percentage of agreement
among raters and kappa, a measure which adjusts proportions of
agreement for chance (Light, 1971; Conger, 1980). The descriptors
used in the task definition phase were, as noted above, specific to
the activity to facilitate their use by raters in segmenting the
behavior stream into specific tasks. Task definition thus produces
a sequential listing of tasks (i.e., task type and target object
descriptors) indexed by frame. Four raters, blind to each other's
rating, accomplished task definition for each of four videotapes as

a basis for assessing reliabilities. The overall pairwise kappa for four raters over four cases was .41, ranging over cases from .14 to .60. These reliabilities did not support use of a single rater task definition system. Rather, the use of a two judge, consensus rating procedure was employed for all task definition. Pairs of raters, again blind to each other's ratings, characterized each defined task in a list corrected to reflect rater consensus as a basis for assessing the reliability of characterization descriptors. While these reliabilities were generally higher, ranging from .51 (action type descriptors) to .97 (object type), characterization procedures also used a two judge, consensus rating system.

TASK TAXONOMIES IN CAPABILITY-DEMAND ANALYSES

The utility of task analysis data for accomplishing capability-demand analysis of daily activity performance can be illustrated by considering the manipulative tasks from the taxonomy and some selected data on grip strength. The task data reflect significant proportions of tasks which involve a standing work position (e.g., at the main counter or sink) and the accomplishment of manipulative tasks (e.g., repositioning items, cutting items). The variation in objects and manipulative actions appear to be a more significant source of demand than variation in task locations. A second feature of the data is the frequency of precision grips and, to a lesser extent, power grips. While not shown here, data also suggest that the right hand "acts" and the left "positions." Specifically, most holding and suspending actions involve the left hand in conjunction with more dynamic actions by the right hand. Also, power grips are more associated with the right hand. These results focus attention on the extent to which current food packaging and preparation technology require manipulative capabilities including joint flexion and grip strength to grasp items of various sizes and configurations as well as to move objects in particular ways.

CAPABILITY-DEMAND CONTRASTS: HAND GRIP AND OBJECT SIZE

One aspect of manipulative capability, basic power grip force (fully closed hand around a handle), was measured using a hand dynamometer with a five position, adjustable grip (Jamar Adjustable Hand Dynomometer, Asimov Engineering Co., Los Angeles, California). While these and other capability assessments (e.g., stature, joint flexion, pulling force) were accomplished in a field setting (i.e., the residence of the individual) in sessions paced to avoid excessive fatigue, data are clearly subject to more sources of possible performance variation relative to more controlled laboratory sessions with replicated and averaged trials. Further, because the handle of this five position dynamometer remains fixed

at the set position, the grip force assessed is more isometric than
dynamic.

The five positions provide a rudimentary, but illustrative,
approach to examining the relative effect of object size on
effective grip force for samples of older women. Specifically, the
position sizes relate generally to measured grip surface size ranges
of classes of objects handled in meal preparation. While data were
collected on both hands, Figure 1 presents mean power grip force for
the right hand for each of four samples of older women at each of
the five handle sizes. The independent group (n=50) are fully
independent older women (mean age = 73). The community care/mobile
group (n=70, mean age = 79) reflect women in day care or meal
programs. While they are not able to prepare meals, they are able
to travel to acquire this support. The community care/residence
group (n=74, mean age = 80) include congregate and home delivered
meal cases. Mobility is more impaired in these sample
participants. The nursing home group includes mentally intact,
ambulatory residents (n=47, mean age = 85). The dashed lines
indicate point estimates of grip force (mean Kg) for "elderly" U.S.
males and for general populations of industrial and military
females.

There are two significant aspects of these data. First,
analysis of variance and a posteriori (Tukey A) tests of means
indicated that, for means associated with each of the five
positions, the independent sample means were significantly different
from the cluster formed by means for the other three nonindependent
samples. Second, while the sample of independent older women could,
on the average, exert greater forces, the highest mean grip force
attained (regardless of specific handle position) was substantially
below levels associated with the two contrast populations (i.e.,
elderly males and general population females). These results
suggest that age associated reductions in grip force do not
necessarily associate with inability to perform tasks where this
capability is relevant. While there does appear to be a significant
difference in this capability which is associated with independence
in performing relevant tasks, the absolute difference between the
independent sample and the non-independent samples is narro; thus
offering the possibility of supporting significant performance
changes with moderate reductions in handling demands where grip
force is a significant component.

The issue of the extent to which environment features affect
performance is indirectly suggested by the generally curvilinear
relationship between grip force an handle size. Specifically, the
data in the left hand portion of Figure 1 indicate, for all groups,
greatest average grip force levels for mid-size positions and
reduced levels for the smallest and larger handle sizes. The right
hand portion of Figure 1 presents sample means for a hand grip

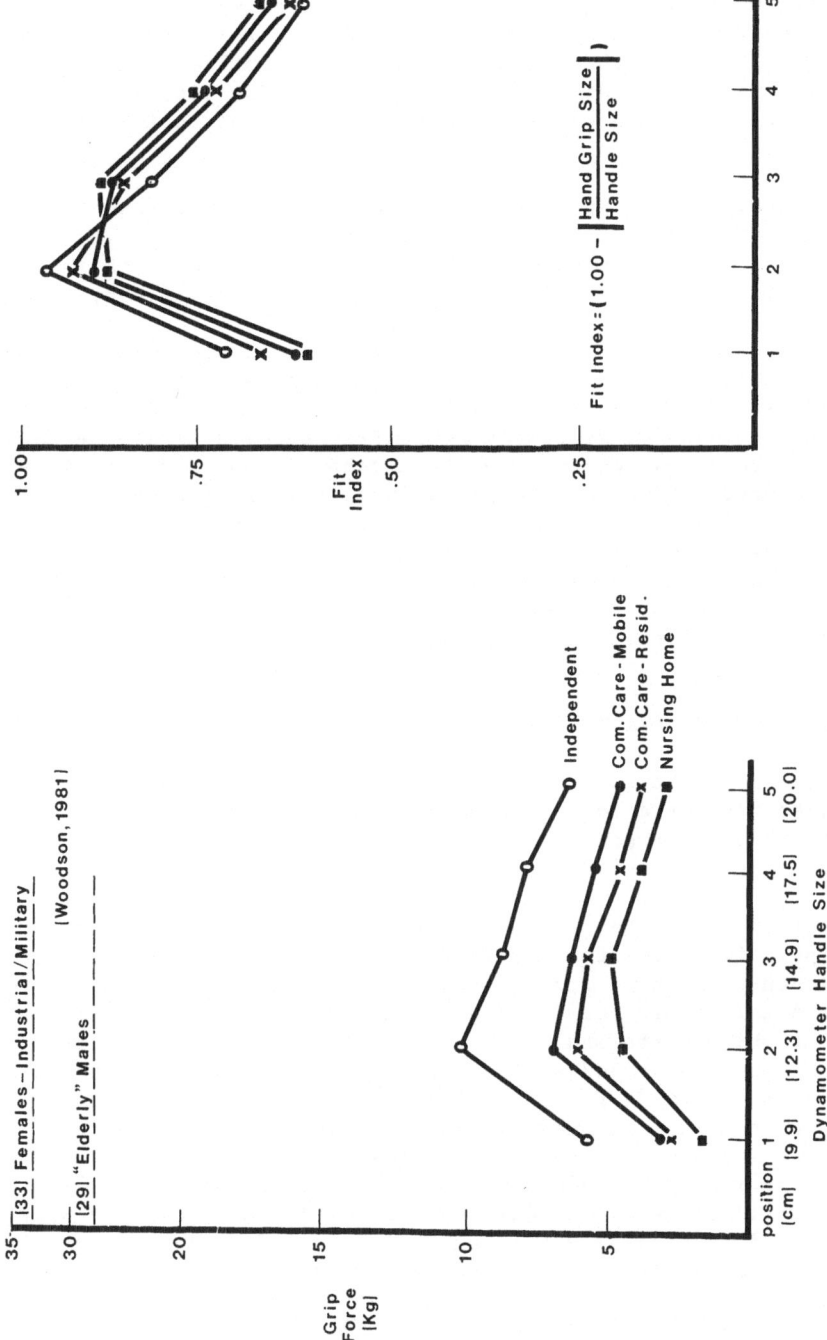

Figure 1. Grip force (left) and person-environment fit index (right) for hand grip task, for four samples.

size/handle size "fit" function calculated from hand measurements of
participants. The hand size estimates the inside circumference of
the closed hand (power grip) and this is placed in ratio to external
circumference of the dynamometer handle at a particular position.
The most striking feature of the data is the similarity among ratios
for groups and hands; with the second handle size position, and to a
lesser extent the third, being associated with the best "fit"
between hand and handle.

Measurement of grip surface size on a preliminary sample of 100
objects handled in meal preparation indicates that utensils,
frequently occurring objects, have grip surface sizes less than the
smallest handle size position of the dynamometer while many food
packages and containers have grip surface sizes in the range of the
three largest handle sizes. Only about eight percent of the objects
assessed presented grip surface sizes comparable to handle sizes
where subjects achieved the maximum force exertions. To the extent
that this trend holds as remaining samples of objects are measured,
it appears that most objects in the kitchen do not reflect sizes
which are most conducive to applications of maximal grip force in
handling. While these observations based on simple grip force are
clearly limited in view of the more dynamic force exertions involved
in object handling, corollary evidence by Rohles (1983) revealed
differences in dynamic force exertions (i.e., twisting jar lids)
associated with variations in object (i.e., lid) size.

CAPABILITY-DEMAND ANALYSIS IN TECHNOLOGY APPLICATIONS

There appear to be several ways in which task analytic
approaches can serve to sharpen and enhance the contributions of
human factors and ergonomic research to the problem of the aged in
the technological environment. First, a task based approach serves
to focus attention on what tasks are actually required in work or
daily activities. The task data reflecting a significant proportion
of manipulative tasks focus attention on products and tools used in
the kitchen; something rarely treated in design of new housing for
older adults. While seemingly mundane, these daily activities are
realms in which performance reductions have major implications for
the older individual and society in genral. Second, task analysis
can provide the basis for constructing empirical assessments of
person and environment characteristics which are relevant to
performance of one or more types of tasks. To the extent that task
refinements reflect more generic person or environment
characteristics, resulting data can have relevance to tasks which
are involved in several different activities of daily living.
Third, task analysis can effectively define what role technologies
can, and ought, to play. The above example suggests potential roles
for manipulative aids and package/utensil design in enhancing the
task effectiveness of reduced grip capabilities in older operators.

Fourth, knowledge of tasks and relevant capabilities and demands can also serve to enhance the effectiveness of our technological design and engineering. The task profiles form the meal preparation work, if nothing else, define quite specifically areas in which assistive devices, from a better can opener to a domestic robot, might improve the kitchen as it currently exists in many residences. The extension of this approach to the range of daily activities can support design engineering and technological development which can make significant strides in adapting technological environments to older users--our future selves.

NOTE

This report is based on research supported by the National Institute on Aging Grant #R01-AG-2727.

REFERENCES

Bennett, C. (1971). Toward empirical, practicable, comprehensive task taxonomy. Human Factors, 13, 229-236.

Companion, M. and Corso, C. (1977). Task taxonomy: Two ignored issues. Proceedings, 21st Annual Meeting, Human Factors Society, 358-361.

Conger, A. J. (1980). Integration and generalization of Kappas for multiple raters. Psychological Bulletin, 88, 322-328.

Drury, C. and Coury, B. (1982). Container and handle design for manual handling. In R. Easterby, K. Kroemer, and D. Chaffin (Eds.), Anthropometry and biomechanics: Theory and application. New York: Plenum Press, 259-268.

Faletti, M.V. (1984). Human Factors research and functional environments for the aged. In I. Altman, J. Wohlwill, and M. P. Lawton (Eds.). Human behavior and the environment: Vol. 7. The elderly and the environment. New York, NY: Plenum Press, 191-234.

Light, R. J. (1971). Measures of response agreement for qualitative data: Some generalizations and alternatives. Psychological Bulletin, 76, 365-377.

Oborne, D. J. (1982). Ergonomics at work. New York, NY: John Wiley and Sons, Ltd.

Rohles, F. (1983). Opening jars: An anthropometric study of the wrist-twisting strength of the elderly. Proceedings of the 27th Annual Meeting of the Human Factors Society, 1, October, 112-116.

Taeuber, C. M. (1983). America in transition: An aging society. (U. S. Bureau of the Census, Current Populations Reports, Series P-23, No. 128). Washington, D.C.: U. S. Government Printing Office.

Wellens, A. R., Revert, R. and Faletti, M. V. (1982).   A videotape
    field observation system for behavioral research.   Unpublished
    Manuscript, University of Miami.
Woodson, W. E. (1981). Human factors design handbook.   New York, N.
    Y.: McGraw-Hill.

IMPACT OF MICROPROCESSORS ON THE QUALITY OF LIFE: COMPARATIVE

NEEDS OF OLDER VS. YOUNGER GENERATIONS

John Lyman

Psychology Department and
Engineering Systems Department
University of California
Los Angeles, California

INTRODUCTION

The impact of the microprocessor is that its ubiquitous applications in industrial production, agriculture, health delivery systems, the office, the home and in personal functions has a rational
potential for producing a true Utopia for human-kind. Used with two
counterparts not yet as far along in applications development, viz.,
nuclear fusion and genetic engineering, spectacular productivity
increases can be projected. These increases should make it possible
to provide all the material resources needed by the population of
any country in the world. The highest imaginable material standard
of living without reducing or damaging the resources of our planet
is realistically a possibility. Steps toward this goal in terms of
eliminating dirty, boring, repetitive and dangerous jobs and reducing
working hours and years are already a part of first world cultures.
A society to come later can provide the average individual with the
time, resources, and opportunities to achieve fulfillment of personal needs, both creative and routine. Abolition of poverty, and with
it, the tyranny of daily work for sustenance is almost inevitable in
the framework of the knowledge revolution that is upon us. Providing, of course, that humans individually and in social aggregates
can keep the urge to destroy each other under control.

In this early and exciting stage of the microprocessor and information society scenario it is not possible to project accurately
what specific turns our various and diverse human cultures may take
on a grand scale. But it is possible to look at some background and
some of the potential areas where the microprocessor and related
microelectronics are currently taking us.

QUALITY OF LIFE

Whether for youth or for the aged and despite the appalling negative statistics in some local areas, the quality of material life is improving all over the world. The mutual export and import of materials and finished goods includes knowledge related products that are slowly but effectively changing the ways in which massive numbers of persons view their personal world. Local standards for what makes up "the good life" are being impacted and eroded through communication. The only universal standards for "the good life" appear to be a state of feeling healthy, being satisfied·with your status and being able to do what you want to do within your capabilities. There is no one ideal environment in which these standards may be obtained for everyone. There are many possible states in which the quality of life may be considered high. This makes it especially difficult to know about how new technologies may affect a particular situation. It seems to be a reasonable assumption, however, that if the basics, viz., adequate food, opportunity for free social intercourse and freedom from disease exist reliably in the infrastructure of a given society, the quality of life for that society may be high somewhat independently of an advanced state of material wealth. It thus may be true that technologies that enhance and provide these basics are truly the focal point for universal applications. This, I realize, is somewhat a banal statement, but it is the starting point for moving from the local conditions of a south sea island paradise to a scope that includes the entire world. It is my thesis that the outgrowth of applications based on the microprocessor are sufficiently versatile to accomplish such a goal.

MICROPROCESSOR DEVELOPMENT

The microprocessor is very much a product of modern times. In one tiny package this family of devices ties together virtually all the major developments in information theory computers and microelectronics since the mid 1930's with roots that go back more than a century. A special and crucial part of this history can be dated rather directly to 1948 when John Bardeen, Walter Brattin and William Shockley developed the transistor at Bell Laboratories. With its small size and low power requirements it prepared the way for integrated circuits. As the densities of components could be made greater on these early devices it became a challenge to see how far the process could be carried. In 1971, the U.S. company, Intel Electronics, developed the first microprocessr, a device made on a silicon chip less than 0.5 cm per side, that could perform all the logic and arithmetic functions necessary to produce an all electronic, small sized calculator. This chip, incidently, was produced for Japan, where the first hand held electronic calculators were manufactured. By 1976 production of an eight bit computer that

had 20,000 components on a single silicon chip was under way. Today
it is commonplace to manufacture chips with 100,000 components and
there is every reason to believe that chips with multi-million
components can be achieved by the end of this decade.

As the number of microelectronic components per chip has
increased the price has gone down. The power of computers has
increased about 10,000 times over the past 15 years while the price
of each component has decreased by a factor of about 100,000. A
single transistor cost about $10.00 in 1960; in a chip today a
transistor costs only a tiny fraction of one cent. One of the real
impacts of the microprocessor is that it is almost absurdly
inexpensive for the complexity and range of applications of which it
is capable. If there is a secret to the success and future promise
of the microprocessor it is that it can, indeed, become a close part
of the personal environment of everyone on earth. The rooms full of
equipment and the enormous power requirement of the early mainframe
computers held no promise for such things to come.

GROWTH OF APPLICATIONS

The insidiously extensive penetration into our daily lives of
microelectronics is occurring in a background where we can find
robots in the workplace, automation of clerical functions,
electronic banking and mail and perhaps most importantly, the
beginnings of a revolution in traditional educational processes.
Table 1, based on data to January, 1983, shows the rate of growth of
microelectronic equipment being delivered to various fields of
application.

The relatively low percentages for consumer audio, home
appliances and television illustrate the point that there is little
remaining room for growth in those areas--penetration is approaching
100% already. Smaller, lighter and more complex variations will be
the theme for the future in many consumer products. Well over a
billion calculators in all the shapes and sizes we see everyday have
been produced in the past half decade. Electronic watches, both
digital and analog are now the standard products of the industry,
almost entirely displacing mechanical movements. The current
popular craze with video games is evolving into new uses for home
television as an information center, with the power and flexibility
of sophisticated video displays pointing toward selective two way
communications utilities as commonplace for services of all kinds
and descriptions. With this, of course, goes the explosive increase
in the popularity of the personal computer, both for home and office
use. With current prices starting at under $100.00 and still
trending downwards, it doesn't take much imagination to realize that
the desk console of the computer-at-home, linked to a television
screen (probably with color and 3-D), will in the future become

Table 1.   Growth of Microelectronic Equipment Deliveries
           (Source: Curnow and Curran, 1983)

Equipment	Per Cent per Annum
Data processing systems	22
storage	20
peripherals	24
Office copying	30
Word processing	32
Office, other	20
Communications: telecommunications	29
radio/television	10
data communications	12
Industrial controls	24
Test and measure	16
Automobile	30
Medical equipment	20
Other controls	18
Consumer audio	8
Home appliances	8
Personal devices	10
Video recorders	4
Games, etc.	16
Television	2
Defense	34

an essential basic central utility for most households. At issue is
not so much what the technological trends themselves will be but how
we are going to be able to cope with them through the changes in
life style that will inevitably result. For a moment's thought,
some of the affected areas that have emerged, even at this early
stage, are listed in Table 2.

HUMAN-COMPUTER INTERACTIONS

From the human standpoint, at its best, microprocessor use is
transparent to its user. Highly reliable and virtually unaffected
by the exigencies of normal use it will, like the many electric
motors that work for you day in and day out in your home, your car,
and your workplace, require little or no personal knowledge about
the way it works or how it is configured into a particular system.
In a total living or working area management system, optimally
efficient environmental control for lighting, air quality,
temperature, water supply and waste distribution will evolve

routinely for new construction and as retrofitting for old construction. Silent in operation, for most applications it will remain as unthought-of as the sand beneath your feet whence it came. The natural role for microprocessor function is in the areas of instrumentation, communication and automation. With its capability for programmed storage and transmission of information in the forms of data, voice, telex, text or image it is a device for gluing together energy and materials in forms yet unimagined. Indeed, no matter how enthusiastic one becomes in speculative review, the substantive backup becomes more impressive.

But for all the speculative possibilities, there is a here and now with respect to the future. The now is already the future for persons who have lived through the widespread impact of the introduction of radio and television along with the myriad of changes and innovations in home appliances, automobiles, highways, supermarkets and such. The now for the youth of our time is focussed on different goals and different life styles than was the case for the youth of the teens and twenties of this century. Yet the basics have much in common even if the trappings are different. Human physiology and basic human cultural needs are rather independent of the times and the attendant technologies, though the behavioral patterns may differ substantially in content as seen by other generations. With less than four percent of the U.S. population currently engaged in farming and the expectation that only two percent will be so engaged by 2000 A.D. we can see a model for the future, with a similar reduction taking place in the industrial workforce as automation and robotics replace the need for "factory hands." Coupled with a trend toward a healthier and longer lived population, and a population that doesn't want to retire into limbo, a real question arises as to what changes in the traditional work ethic will come about.

For the time being, the trend is toward increased participation in the service industries by the workforce. A good quality of life that is heavily based on an orientation towards "things," as is the case in the first world countries, requires many supporting services, from plumbers to hair dressers. And though automation factories may evolve to take over much of the production for mass consumption, someone has to maintain the machinery that runs the factory as well as perform planning and management functions. Even the most optimistic projections do not yet include robots that maintain robots in any but highly limited situations. No matter how the picture is described, however, the concept of total employment in the tradition of an eight hour day, 40 hour week is not for the future. More leisure time, fewer duties that are job related, and more opportunity for personal development is built in, largely through the impact of microprocessors in configurations that will become increasingly smarter and more economical.

Table 2. Representative Microprocessor Applications
(Source: King, 1983)

The electronic watch and calculator;
The personal microcomputer;
Improved functioning of the internal combustion engine;
Increased fuel efficiency;
Domestic appliances of many kinds, such as programmed
  washing machines and dishwashers, sewing machines and
  eventually, the domestic robot;
Information selection and retrieval;
Automatic translation and interpretation;
Novel traffic control systems;
New systems of public transportation;
Computer aided design;
Multi-purpose computer controlled machine tools;
Central control of large industrial systems
  (oil refineries, chemical plants and steel works);
The automated factory;
The automated office;
New systems of banking, transfer of money, insurance, etc;
Environmental monitoring;
Optimization of agricultural yields resulting from
  computerized analysis of factors influencing growth;
Electronic mail;
New information and communication systems;
The teleconference;
Medical diagnosis and prosthesis;
Computer aided educational systems.

An emerging, probable scenario is that as experience with the
information age develops many personal human services will be better
supplied by machines. For some of these, like washing your car,
human participation may not be missed. But other services, such as
accurate medical tests and diagnosis, may have traditions that are
difficult to change.

A fear, perhaps unfounded, is that as more information and more
individual interaction with machines becomes a major occupier of
each person's waking hours, strong forces toward alienation will
develop. Fears concerning negative effects of television, video
games, and now, computer aided learning, have yet to be conclusively
resolved. Like the fears concerning the effects on livestock when
the railroad was introduced, they may just fade away as we get used
to the changes in our mental and physical environment.

One trend that has uniquely emerged is that as computers come into more use, the capacity of both the machine and operator can grow. Thus a person may not only learn new ways of doing something more efficiently, but he or she may also start demanding new functions from the machine. The flexibility inherent in such shared human-computer interactions adds new, not previously programmed, information enhancing capabilities. For the aged and the handicapped these new, personalized, capabilities, may often provide the critical functional difference between effective and ineffective performance.

MARKET FACTORS

Special applications of microprocessors are emerging in everything from toys to communications satellites. These special applications require that a market be sufficiently large to justify the initial expense of designing and producing custom chip configurations. For young and old alike, many products have equivalent appeal and thus expand the market potential, but for highly specialized items identification of markets is a critical constraint. This has been a problem for many potential applications for the handicapped as the markets have not provided sufficient economic justification for the transition from research to manufacturing. Hearing aids, heart pacemakers, portable kidney dialyzers and soon, automatic insulin delivery systems (artificial pancreas) have very large markets and hence have been developed and manufactured. All are based on specialized microelectronic devices. Yet there are many other prosthetic and function aiding devices that have not left the research stage or even begun it because of lack of market potential. One factor that has kept markets small is that customs duties and special tariffs have discouraged development of worldwide marketing strategies. Political as well as technical factors must thus be taken into account for dissemination of microprocessor products.

Having the Library of Congress as a resource at your beck and call by means of a pocket sized portable data handling device is an exciting and technically feasible idea but the bottom line for us all is to have good health. For the young and aged alike in this highly body conscious society it is in the health sector that the microprocessor may have one of its strongest impacts. Portable personal devices for measuring blood pressure, heart rate, energy expenditure, etc. are rapidly entering the armamentarium of helpful objects for everyday living. Biofeedback techniques for aiding with the handling of physical and mental stress are readily available and are safer and often more effective than drugs. Video instruction in health techniques and inexpensive diet control calculators encourage good habits of nutrition.

In both subtle and blatant ways microprocessors are shaping the quality of life for persons of all ages. The foundations have been laid for far reaching future changes. Health, well being and opportunity for self development appears to be in the process of reaching new heights that will positively affect most of human kind for generations to come.

REFERENCES

Curnow, R., and Curran, S., 1983, The technology applied, in: G. Friedrichs and A. Schaff, eds., Microelectronics and society: A report to the Club of Rome. New York: New American Library.

King, A., 1983, Introduction: A new industrial revolution or just another technology?, in: G. Friedrichs and A. Schaff, eds., Microelectronics and society: A report to the Club of Rome. New York: New American Library.

TECHNOLOGICAL INTERVENTIONS FOR CHANGES IN HEARING AND VISION

INCURRED THROUGH AGING

John F. Corso

State University of New York (Cortland)

INTRODUCTION

The scientific literature on aging clearly establishes that psychophysical processes and human performance undergo gradual but continuous modification throughout life (Corso, 1971). This finding has two major implications for human engineering: (1) that predictable behavioral changes which occur as a function of age should be taken into account whenever an elderly person or group is involved in a particular working or living function; and (2) that the application of behavioral data derived from young adults may be inappropriate in the development of design characteristics for machines, systems, tasks, or environments intended for use by older people (Corso, 1981a). The age-related changes which occur in all human sensory/perceptual systems, and their functional significance, have been described in detail by Corso (1981b); and Fozard (1981) has presented some implications of these changes for person-environment relationships.

The present paper, therefore, has two primary purposes: (1) to focus upon aging auditory and visual functions and their relationship to problems within the field of human factors, and (2) to suggest the manner in which technological advances may be utilized in human factors to offset these aging deficits.

A CONCEPTUAL MODEL FOR HUMAN FACTORS

Human factors as a discipline is concerned with the design of machines, man-machine systems, operational tasks, and physical environments in accordance with the characteristics, behavioral

333

capacities, and overall limitations of individuals and organized
groups (Corso, 1981a).

Auditory Tasks

Within this definition, there are three broad functional
categories which involve the auditory modality: (1) auditory
signaling systems; (2) auditory tracking; and (3) speech
communication. The data on age-related auditory changes suggest
that no significant modifications are required in the design of
auditory signaling systems, unless the precise directional location
of the sound source is demanded, or the specific identification of a
particular alarm from an ensemble of potential alarms is required
(Corso, 1977b, 1977c).

Auditory tracking is a viable tracking mode under certain
operating circumstances for normal hearing operators, but the data
on presbycusis suggest extreme caution in the use of this technique
with older persons. Since auditory tracking ordinarily involves the
variables of frequency and intensity, singly or in combination,
older persons would be expected to perform relatively poorly due to
their decreased ability in frequency discrimination (Konig, 1957)
and intensity discrimination (Pestalozza and Shore, 1955).

Speech communication is the area of human factors in which the
data on age-related changes in hearing are most relevant. The
information in speech is conveyed physically in acoustic waves which
are characterized by frequency, amplitude, and time patterns. Thus,
any factor which significantly affects these variables will alter
speech intelligibility, whether the factors reside with the message
source, the transmission system, or the receiver. The present
paper, however, will consider primarily the psychoacoustic findings
in presbycusis.

Visual Tasks

In human factors there are four broad functional categories
which involve the visual modality: (1) visual detection, identification, and estimation of objects or patterns, or their representations, and their associated static or dynamic characteristics;
(2) interpretation of visual displays, including those related to
system status monitoring, continuous system control, briefing or
debriefing procedures, search operations of reconnaissance and sur-
veillance, and decision-making; (3) monitoring visually-coded
message panels and alerting systems; and (4) reading conventional
printed material, including labels, books, operating instructions,
maintenance manuals, and check lists.

Collectively, these functional categories involve the fundamental psychophysical properties of vision. These include: visual

thresholds; static and dynamic visual acuity; visual discrimination
of size, shape, distance, and speed; color perception; contrast discrimination; light and dark adaptation; critical flicker frequency;
and glare. Age-related changes occur in all these processes and the
present paper will illustrate the manner in which technological
changes may be utilized to offset some of the observed deficits.

The Model and Its Application

     The functional processing of auditory and visual information
and other forms of input by the human operator is presented in
Figure 1. The significance of Figure 1 for human factors is that it
identifies specifically the major loci at which technological aids
may be introduced into man-machine systems with the potential for
improving the operator's performance. These include: the display
format for encoding and presenting input information; devices for
enhancing the properties of significant functional variables and
their relationships during the course of information processing; and
methods for the efficient and effective transmission of the operator's decision or action output to other segments of the given
system.

AGE-RELATED CHANGES IN AUDITION

     The most common auditory disorder in the population is
presbycusis, i.e., the alteration of hearing functions associated

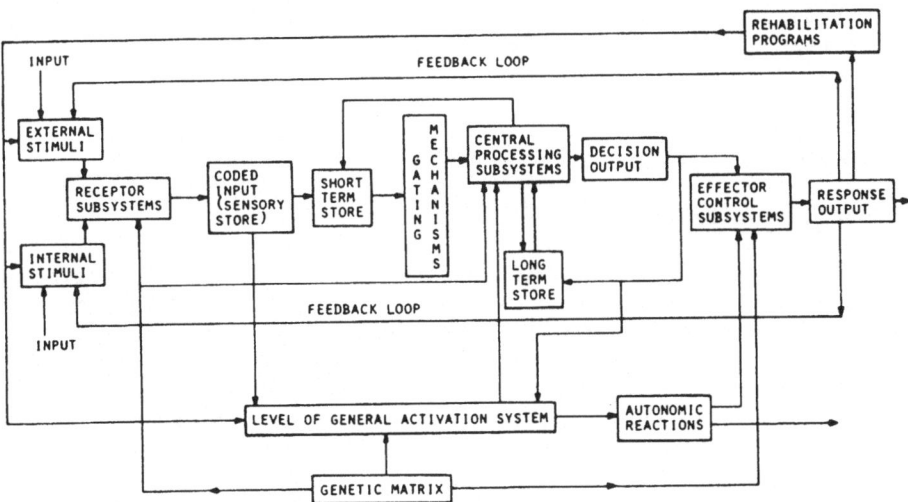

Figure 1. Hypothetical diagram of functional stages involved in
          human information processing. From Corso, (1981b).

with physiological aging of the auditory system (Corso, 1977a). It
ranks second only to arthritis among the chronic health conditions
which affect the elderly (Harris, 1978).

The major psychoacoustic findings in presbycusis, with
relevance for human factors, have been reported previously by Corso
(1981b). These include changes in the basic psychophysical
functions: pure tone thresholds, frequency discrimination,
intensity discrimination, and absolute judgments for pure and
complex tones; and changes in speech intelligibility: speech
reception threshold, speech discrimination in quiet, and speech
intelligibility under adverse listening conditions. Speech
intelligibility has also been found to be related to temporal gap
detection (Corso, J.F., Corso, G.M. and Corso, C.A., 1981).

AGE-RELATED CHANGES IN VISION

Visual deficiency ranges from an acuity level at which it is
possible to read with a special optical aid to total blindness.
Approximately one of every 20 persons suffers from a significant
uncorrectable visual impairment which precludes reading ordinary
newsprint with either eye, even with glasses (Kupfer, 1975). The
major causes of blindness involve cataracts, degenerative processes
in the retina and optic nerve, glaucoma, diabetes, and other complex
etiologies associated with aging.

Corso (1981b) has presented a detailed review of the primary
age-related psychophysical changes which occur in vision and are
related to problems within human factors. These include visual
thresholds, static and dynamic visual acuity, depth perception,
brightness discrimination, color perception, dark adaptation, visual
masking, critical flicker frequency, and glare.

TECHNOLOGICAL AIDS FOR ALTERED HEARING

Communication Systems

In existing communication systems improved auditory performance
for elderly personnel can be achieved by providing optimal design
characteristics for the message source, the transmission system, and
the message receiver, given a particular operational environment.
Without introducing additional components, speech communication can
be improved by adequate attention to significant language factors at
the message source including: word familiarity, word length,
vocabulary size, phonetic composition, and sentence structure.
Within the transmission system, peak clipping and low-pass filtering
may be used effectively in high intensity noise fields; specially
designed amplifiers, automatic gain control units, peak limiting

devices, noise shields, noise-cancelling microphones, and contact microphones have also been found to improve performance. At the destination, individual headphones, loudspeakers, protective ear devices, volume controls, squelch controls, side tone circuitry, and foot-operated talk-listen units may be incorporated in the basic speech communication system (Kryter, 1972).

Hearing Aids and Auditory Recoding

The most widely-known device for improving the auditory effectiveness of older persons is the electronic hearing aid, which serves as a sound amplifier. However, modern aids can now provide a miniaturized assembly with amplitude compression (limiting), reduced frequency distortion, and directional microphones. Furthermore, speech discrimination can be improved by designing the amplification system to meet individual needs as determined from specific audiometric patterns of hearing loss. Corso (1977a) has concluded that binaural (stereophonic) aids are psychoacoustically superior to monaural devices since they improve intelligibility and directional hearing, but not all hearing losses can be assisted by current technology.

Since individuals with presbycusis and certain other forms of hearing impairment tend to have better hearing for low frequencies than for high, numerous attempts have been made to improve speech intelligibility by lowering speech frequencies (Braida et al., 1979). The systems for accomplishing this include: (1) linear frequency lowering, (2) nonlinear frequency lowering, and (3) selective frequency lowering. Theoretically, it is reasonable to expect that some improvement in speech perception should be obtained through these techniques, but the results from evaluation experiments have been disappointing. Nevertheless, frequency lowering has produced some positive results (Block and Boerger, 1980).

Speech-Visual Interactive Systems

Since the intelligibility of words is enhanced by the presence of visual cues from the talker (Woodward and Barber, 1960), a special eyeglass has been designed to aid in speechreading (Upton, 1968). In this system a wearable speech analyzer is used to extract phoneme information which is then displayed visually in binary code by means of miniature incandescent lamps bonded directly on the wearer's eyeglass lens. The wearer sees the phoneme symbols in the form of a dynamic light pattern superimposed around the speaker's lips. Use of the device has improved the identification of key words in sentences by 10 to 20% (Gengel, 1976).

A major advance in visual-speech interactive technology is the teletype. A teletype device is now available that permits a

hearing-impaired person to communicate over the telephone via a
tele-typewriter, with the coded message response being converted to
a printed output (Texas Instruments, Inc., 1979). Since a person
with a severe hearing impairment often possesses poor voice
intonation, the ability to communicate through a keyboard is a
significant technological achievement. Keyboard entry permits the
activation of voice synthesizers which can generate speech of higher
quality than that produced by the mute or deaf person (Phonic Mirror
Handivoice, 1977).

Auditory to Tactile Conversion Devices

Another approach to assist the hearing impaired individual is
to employ the tactile representation of auditory information. The
speech coding strategies for tactile aids fall into three
categories: (1) direct transmission, in which the acoustic signal is
transmitted directly to the "listener" by means of a tactile
vibrator; (2) spectral coding, in which the time-varying
distribution of speech energy is obtained from a set of filtered
frequency channels and transmitted to the "listener" by a suitable
code in a set of corresponding tactile vibrators; and (3) feature
coding, in which a selection is made from the important acoustic
features of speech perception, e.g., fundamental frequency
variations, format frequencies, time relations between different
acoustic elements, etc., and the selected features are transmitted
to the "listener" via a suitable code in tactile vibrators.
Regardless of the coding strategy implemented, tactile speech
transmission is hampered by poor temporal resolution (Keidel, 1973)
and poor frequency discrimination (Gault, 1926).

Auditory-tactile conversion devices fall into two groups:
(1) systems that stimulate a finger or fingers, and (2) systems that
stimulate other parts of the body. Within these two groups, various
numbers of vibrators have been applied, i.e., from one or two to a
matrix of 8 x 36 electrodes. Spens (1980), as the only subject,
compared several existing tactile systems using the swedish numerals
one to nine for the speech material. The best results were obtained
with a 6 X 16 stimulator matrix on the forefinger and a
time-frequency-amplitude code. Speech intelligibility for the
numerals approached 80%.

The ability to identify vowels in vibrotactile transformations
of consonant-vowel syllables has been measured recently for two
types of displays: (1) a spectral display (frequency by intensity)
and (2) a function display of vocal tract area (vocal tract location
by cross-sectional area) (Green et al., 1983). Each display was
presented to the fingertip via tactile transducers within a matrix
of 24, 12, or 8 frequency channels, with six intensity levels in
each channel. Two forms of the spectral display were used: a
spectral contour and a histogram. Vowels in a /b/V/ context can be

identified when as many as 24 or as few as 8 spectral channels are
used. However, if the area of the 12- or 8- channel displays is
reduced to occupy 1/2 or 1/3 of the 24-row tactile matrix,
performance is deteriorated. For 24 channels, vowel identification
reaches 90.5% for the spectral contour and 85.6% for the histogram.
The area function display was found to be inferior to both forms of
spectral display.

Artificial Auditory Organs

In certain kinds of nearly total deafness, the hair cells of
the cochlea are non-functional, but the VIIIth nerve remains
essentially intact. Therefore, attempts are being made to stimulate
the nerve endings directly by means of electrodes implanted in the
cochlea. The cochlear implant is not a hearing aid in the
conventional sense, nor is the intent to replicate the original hair
cell/nerve ending interconnections.

At least five schemes for cochlear implantation have been
attempted and are under investigation. The five schemes encompass
two basic kinds of cochlear prostheses: singlewire electrodes and
multiple electrode systems. The results of House et al. (1979) and
of Bilger (1977) supported the early view that single channel
systems could only transmit information about speech rhythm and some
information about environmental sounds, while speech perception
would require the use of a multichannel approach. However, recent
studies with a single channel stimulator have been extremely
encouraging, with subjects able to extract information on
voiced-unvoiced, fricative-nonfricative, and transient-nontransient
speech sounds (Hockmair-Desoyer et al., 1981).

In multichannel systems, some investigators have used spectrum
coding (e.g., Chouard, 1980), while others have tried feature coding
(e.g., Tong et al., 1979). For a single subject, the latter
technique has yielded the correct identification of 65% of spondee
words in a known set of 16, and 15% of unknown simple sentences.

The primary problem with cochlear implants is the restricted
dynamic range between threshold and the maximum tolerable level
(approximately 12-15 dB) and the restricted frequency range over
which pitch changes can be heard, i.e., up to 400 Hz but sometimes
up to 1000 Hz (Bilger, 1977). The data currently available indicate
that electrical stimulation of the cochlea should be a more
effective auditory recoding technique than tactile stimulation. A
single channel cochlear implant yields better frequency
discrimination and better time resolution than a simple tactile aid
(Risberg, 1982). However, the very limited dynamic range of
electrical stimulation will require the development and
implementation of a more effective amplitude coding technique
(Pfingst, Burnett, and Sutton, 1983). Patients with cochlear

implants can at least detect the presence of acoustic surroundings and perform telephone communication of a limited type if a simple code is established between the implanted patient and the telephone partner (Berliner, 1978).

TECHNOLOGICAL AIDS FOR ALTERED VISION

Braille Print

The most common design aid for the visually handicapped is Braille print. This consists of a 2 X 3 array of raised dots which are coded to represent each letter and numeral. As the fingers are moved across the print, letters and words are discriminated tactually by means of the position of dots, distances between dots, and size of dots.

In Grade II Braille, higher reading speeds are obtained by using single symbols to represent common function words, prefixes, suffixes, etc. To avoid the problem of excessive bulk, Braille tests are now being stored on punched paper tape or magnetic tape. The tape is read by a device which then displays the text one line at a time by means of an array of electromechanically energized pins (Dupress et al., 1968). To reduce the cost of manually produced Braille texts, various computer output devices have been developed to emboss Braille symbols directly onto page format. Skilled readers can read up to 200 words per minute with Grade II Braille (Allen, 1978).

Optical to Tactile Conversion Devices

The tactile representation of printed material can also be accomplished by direct translation with an optical to tactile converter (e.g., Optacon). This is a portable device which contains a hand-held probe for scanning a page while a finger of the other hand is stimulated by a vibrating array of rods. The probe contains an integrated matrix of 144 phototransistors which produce a coded spatial representation in a small 24 X 6 assembly of tactile stimulators (Lindvill, 1973). After an average instructional period of 58 hours, 112 subjects attained a mean reading speed of 12.3 words per minute (Weisgerber et al., 1974), which is much slower than Braille.

Optical to Auditory Conversion Devices

Another approach which carries out direct one-to-one transformations of input to output is, for example, the Stereotoner. This instrument uses a small optical probe for scanning along a line of print and generates a coded output of ten musical tones presented to the reader through stereophonic

earphones. Depending upon experience and other personal factors,
reading speeds from 7 to 90 words per minute can be attained
(Weisgerber et al., 1975).

Auditory outputs have also been obtained from other classes of
input sources consisting of typesetters' tapes, magnetic disks, or
microgroove recordings. The output, e.g., the Spelltalk system,
consists of a series of speechlike sounds representing the 26
letters of the alphabet, but the sounds are not the spoken names of
the letters represented. Spelltalk, therefore, is a code which
differs from spelled speech. With 50 word vocabularies, reading
speeds of 120 words per minute or higher have been reported
(Detwiler et al., 1973).

Another class of reading aids provides an auditory output, but
these aids are too complex to be considered as direct translation.
The aids are called reading machines, e.g., the cognodictor (Mauch,
1976). This machine recognizes print via an array of 64 photocells
in conjunction with a special optical process and generates an
output in the form of a five-bit code transcribed from 31
pre-recorded spelled-speech letters. Buffer storage up to eight
letters is provided between input and output units, i.e., between
optical character recognition circuitry and the spoken output, with
speed of approximately 90 words per minute. Other developments in
this field have produced a reading machine reported to speak at a
rate of 200 words per minute (Kurzweil, 1975).

Text-to-Speech Reading Systems

Although text-to-speech reading systems provide an auditory output, they employ complex high performance conversion procedures which
require the use of at least a medium-sized computer. The systems are
outside the range of personal reading machines and are intended for
large libraries in which blind subscribers can be provided with
requested texts in the form of intelligible synthetic speech,
comparable to talking books. For one system of this type inexperienced listeners are reported to understand nontechnical narratives
at speaking rates of 130 words per minute (Nye et al., 1975).

Aids for Driving and Personal Mobility

Personal mobility of the blind is typically dependent upon
auditory cues reflected from environmental objects and surfaces.
Several electronic sensory aids are now available which convert
ultrasonic energy or light signals into auditory or tactual cues as
replacements for the traditional long, white cane. One device is
the laser cane which detects objects without actual contact and
thereby extends the distance of sensitivity. Another device is the
Sonicguide which presents binary auditory signals to the visually

handicapped pedestrian from nearby objects (Shingledecker and
Foulke, 1978).

For driving, some individuals with severe deficits in static
visual acuity have been licensed by wearing telescopic spectacles
(Booher, 1978). Two miniature Galilean telescopes are mounted in
the upper section of each carrier lens which contains the person's
normal refractive correction. The lens is used for general visual
functions and, by a lowering of the head, the telescopic lens may be
used for reading signs and for examining the details of
environmental objects.

Artificial Visual Organs

A major consequence of aging in the visual system is the
formation of cataracts (Corso, 1981b). The incidence increases
progressively after age 50 and approaches 95% of the population 85
years or older. The cataract operation is the single most common
surgical procedure performed in the United States. Treatment of the
cataract involves removal of the lens, which leaves the patient
aphakic. Three types of substitutes are available to restore visual
acuity: spectacles, contact lenses, and intraocular lenses.
Optically, the intraocular lenses are most effective, but the
personal needs and characteristics of the patient should determine
which correction will be used. There are four classes of
intraocular lenses, as determined by the location of the lens within
the eye. Approximately 120,000 lenses are implanted annually
(Amchin and Leflar, 1979) from an estimated 400,000 cataract
extractions (Jaffee, 1977).

REFERENCES

Allen, J., 1978, An approach to reading machine design. Human
    Factors, 20,(3): 287-293.
Amchin, J. and Leflar, R., 1979, Hazards of intraocular lenses.
    Washington, D.C.: Health Research Group.
Berliner, K., 1978, Cochlear implants. Paper presented before the
    Texan Speech and Hearing Association. San Antonio, Texas.
Bilger, R. C., 1977, Psychoacoustic evaluation of present
    prostheses. Annals of Otology, Thinology, and Laryngology,
    Suppl. 38, 86: 92-140.
Block, R. and Boerger, G., 1980, Horverbessernde verfahren mit
    bandbreitenkompression. Acustica, 45: 294-303.
Booher, H. R., 1978, Effects of visual and auditory impairment in
    driving performance. Human Factors, 20 (3): 307-320.
Braida, L. D., Durlack, N. L., Lippman, R. P., Micks, B. L.,
    Rabinowitz, C. M., and Reed, C. M., 1979, Hearing aids: A
    review of past research on linear amplification, amplitude
    compression, and frequency lowering. American Speech and
    Hearing Association, Monograph No. 19.

Chouard, C., 1980, The surgical rehabilitation of total deafness
    with the multi-channel cochlear implant: indications and
    results. Audiology, 19: 137-145.
Corso, J. F., 1971, Sensory processes and age effects in normal
    adults. Journal of Gerontology, 26: 90-105.
Corso, J. F., 1977a, Presbycusis, hearing aids, and aging.
    Audiology, 16: 146-163.
Corso, J. F., 1977b, Information processing for pure tones in older
    adults. Journal of the Acoustical Society of America, 61: S
    87, (Abstract)).
Corso, J. F., 1977c, Information processing for complex tones in
    older adults. Journal of the Acoustical Society of America,
    62: S 3, (Abstract).
Corso, J. F., 1981a, Human engineering for the elderly. Academic
    Psychology Bulletin, 3: 197-201.
Corso, J. F., 1981b, Aging sensory systems and perception. New
    York: Praeger Publishers.
Corso, J. F., Corso, G. M., and Corso, C. A., 1981, Prediction of
    speech discrimination from temporal discrimination measures.
    Paper presented at the Fifty-second Annual Meeting of the
    Eastern Psychological Association, New York.
Detwiler, J. S., Longini, R. L., and Sullivan, K. R., 1973, The
    function of Spelltalk in reading for the blind. IEEE
    transactions. Systems Man and Cybernetics, SMC-3, 405-410.
Dupress, J. K., Baumann, D. M. B., and Mann, R. W., 1968, Towards
    making Braille as accessible as print. MIT Engineering
    Projects Laboratories, Report No. DSR-70249-1. Cambridge,
    Mass.
Fozard, J. L. 1981, Person-environment relationships in adult-hood:
    Implication for human factors engineering. Human Factors, 23:
    7-27.
Gault, R. H., 1926, Touch as a substitute for hearing in the
    interpretation and control of speech. Archives of
    Otolaryngology, 3: 123-135.
Gengel, R. W., 1976, Upton's wearable eyeglass speedreading aid:
    history and current developments. In S. I. Hirsh, D. H.
    Eldredge, & S. R. Silverman (Eds.), Hearing and Davis: Essays
    Honoring Hallowell Davis. St. Louis, MO.: Washington
    University Press, 221-299.
Green, B. G., Craig, J. C., Wilson, A. M., Pisoni, D. B., and
    Rhodes, R. P., 1983, Vibrotactile identification of vowels.
    Journal of the Acoustical Society of America, 73: 1766-1778.
Harris, C. S., 1978, Fact book on aging: A profile of America's
    olde population. Washington, D.C.: The National Council on
    Aging, p. 11.
Hockmair-Desoyer, I. J., Hockmair, E. S., Burian, K., and Fisher, R.
    E., 1981, Four years of experience with cochlear prostheses.
    Medical Progress through Technology, 8: 107-119.
House, W. F., Berliner, K. I., And Eisenberg, L. S., 1979, Present
    status and future directions of the Ear Research Institute

cochlear implant program: Acta Otolaryngologica, 87: 176-184.

Jaffee, N., 1977, Current concepts in ophthalmology. New England Journal of Medicine, 299: 235-237.

Keidel, W. D., 1973, The cochlear model of skin stimulation. In Devices, The Psychonomic Society, 27-32.

Konig, E., 1957, Pitch discrimination and age. Acta Oto-Laryngologica, 48: 475-489.

Kryter, K. D., 1972, Speech communication. In H. P. Van Cott and R. G. Kinkade (Eds.). Human Engineering Guide to Equipment Design. Washington, D.C.: U.S. Government Printing Office, Chap. 5.

Kupfer, C., 1975, Report of the National Advisory Eye Council, Vision Research Program Planning Committee. U.S. Department of HEW-PMS-NIH-NET, DHEW Publication No.(NIH) 75-664 (195).

Kurzweil, R. D., 1975, Company prospectus. Cambridge, Mass.: Kurzweil Computer Products, Inc.

Lindvill, J. G., 1973, Research and development of tactile facsimile reading aids for the blind (the Optacon). Stanford Electronics Laboratories, Stanford University, Stanford, California.

Mauch, R. R., 1976, Quarterly Progress Report for the period January 1, 1976-March 31, 1976. Dayton, Ohio: Mauch Laboratories.

Nye, P. W., Ingemann, F., and Donald, S. L., 1975, Synthetic speech comprehension: A comparison of listener performances with and preferences among differsis on the basis of different tests of auditory function. Laryngoscope, 65: 1136-1163.

Pfingst, B. E., Burnett, P.A., and Sutton, D., 1983, Intensity discrimination with cochlear implants. Journal of the Acoustical Society of America, 73: 1283-1292.

Phonic Mirror Handivoice, 1977, Electronic voice system generates messages for vocallly model a comparable adult age segment ten or even five years hence. This makes it important to be aware of the demographic, health, education and social changy handicapped. Electronics, 50(23):32-35.

Risberg, A., 1982, Speech coding in aids for the deaf: an overview of research from 1924 to 1982. Speech Transmission Laboratory, Royal Institute of Technology, Stockholm, Sweden, Quarterly Progress and Status Report 4, 65-98.

Shingledecker, C. A., and Foulke, E., 1978, A human factors approach to the assessment of the mobility of blind pedestrians. Human Factors, 20 (3): 273-286.

Spens, K. E., 1980, Tactile speech communication aids for the deaf: A comparison. Speech Transmission Laboratory, Royal Institue of Technology, Stockholm, Sweden, Quarterly Progress and Status Report 4, 23-29.

Texas Instruments (TI) News Bulletin No. 769, 1979, New Communication Service Available to Hearing-Impaired TIers.

Tong, Y. C., Black, R. C., Clark, G. M., Forster, I. C., Millar, J. B., O'Loughlin, B. F., And Patrick, J. F., 1979, A preliminary report on a multiple-channel cochlear implant operation.

Journal of Laryngology and Otology, 93: 679-695.

Upton, H., 1968, Wearable eyeglass speech reading aid. American Annals of the Deaf, 1: 222-229.

Weisgerber, R. A., Everett, B. E., Rodabaugh, B. J., Shanner, W. M. and Crawford, J. J., 1974, Educational evaluation of the Optacon as a reading aid to blind elementary and secondary students. Final Report Contract No. OEC-0-72-5180. Palo Alto, California: American Institute for Research.

Weisgerber, R. A., Everett, B. E., and Smith, C. A., 1975, Evaluation of an ink print reading aid for the blind: The Stereotoner. Final Report contract No. V101 (134) P-163. Palo Alto, California: American Institute for Research.

Woodward, M. F. and Barber, C. G., 1960, Phoneme perception in lip-reading. Journal of Speech and Hearing Research, 3: 212-222.

THE OLDER ADULT AS COMPUTER USER[1]

Alan A. Hartley
Scripps College
Claremont, California

Joellen T. Hartley and
Shirley A. Johnson
California State University
Long Beach, California

Computerization of the contemporary workplace has been rapid and extensive. It is likely that this change will have a major impact on the older worker. In the laboratory, older adults are less able than younger adults to master new material (Hartley, Harker, and Walsh, 1980). In field studies, they have been characterized as less likely to adopt innovations (Phillips and Sternthal, 1977). The exposure of older adults to computer use raises questions for both applied and basic research. Can older adults master common computer applications? Can techniques be found that will improve or speed mastery? Does familiarity with one application transfer to others of the same class? How does knowledge develop as the older adult learns? How is the knowledge base accessed to solve specific problems, to carry out specific tasks?

The application chosen for initial investigation of these questions was the use of word processors--the class of computer programs that allow entry of textual material, storage, updating and editing. A word processing program (called EDITOR) was written that surreptitiously recorded the user's actions and the time taken to select and execute them. Groups of healthy older and younger adults (aged 65-75 and 18-30 years, respectively) without prior computer experience completed twice-weekly training sessions over several weeks in the use of EDITOR. Instruction was computer-assisted. It included presentation of the concepts and functions, examples, simple practice exercises, and more complex editing problems. At regular intervals, learners were asked to provide a written summary describing their knowledge of the computer and EDITOR. In two studies, experimental manipulations included (1) exposing

individuals who had learned to use EDITOR to training in the use of
a popular, commercially-available word processing program and (2)
allowing immediate practice with each variant of a function rather
than presenting the whole function before permitting practice.

The results of the initial study show that older adults are
able to acquire expertise in the use of a computer word processor
and to apply that expertise to solve problems. After 12 hours of
instruction (6 sessions) there were no differences between older and
younger learners in recall of information about EDITOR or in the
correctness and efficiency with which computer operations were
carried out. Older adults did, however, require more time to select
and carry out the appropriate procedures, as would be predicted from
previous research (Birren, Woods, and Williams, 1980). Older adults
also required more assistance while carrying out editing tasks.
Transfer of training was positive and equivalent for older and
younger adults when compared with naive, age-matched control
groups. The second study confirmed that recall of information about
EDITOR was similar in older and younger adults. In this study,
assistance during problem-solving sessions was severely restricted,
and as a result older adults were less effective in carrying out the
editing tasks than they were in the first study. Taken together,
these two investigations show that the older adult can master the
use of computer technology. Information is acquired at the same
rate as younger adults. The older adult's utilization of this
information is, however, somewhat slower and somewhat less effective
than the younger adult's. Ready access to assistance during the
learning process facilitates performance in older adults, especially
during the early phases of learning. The findings suggest that with
appropriate training the technologizing of the workplace need not be
a threat to the older worker.

[1] This research was supported by a grant from the Spencer
Foundation, Chicago, Illinois.

REFERENCES

Birren, J. E., Woods, A. M., and Williams, M. V., 1980, Behavioral
     slowing with age: Causes, organization, and consequences, in
     "Aging in the 1980s: Psychological Issues," L. W. Poon, ed.,
     American Psychological Assoc., Washington, D. C.
Hartley, J. T., Harker, J. O., and Walsh, D. A., 1980, Contemporary
     issues and new directions in adult development of learning and
     memory, in "Aging in the 1980s: Psychological Issues," L. W.
     Poon, ed., American Psychological Assoc., Washington, D. C.
Phillips, L. W. and Sternthal, B., 1977, Age differences in
     information processing: A perspective on the aged consumer,
     Journal of Marketing Research, 14:444.

TECHNOLOGY AND THE OLDER PERSON: AGE, SEX AND EXPERIENCE

AS MODERATORS OF ATTITUDES TOWARDS COMPUTERS

Iseli K. Krauss and William J. Hoyer

Syracuse University
Syracuse, New York

Although computer-based technology has become increasingly pervasive in our personal lives as well as in the workplace, we know very little about individual differences in attitudes and reactions to this technological onslaught. We were particularly interested in learning about older people's degree of willingness to incorporate computer-based technology into their lives. The main purpose of the present study was to examine the individual difference factors of age, sex, and amount of computer experience as moderators of attitudes toward computer-based technology.

Twenty-five older women (mean age = 61.3 years) and 20 older men (mean age = 63.8 years) responded to our questionnaire. They were employees and the spouses of employees who were taking part in a pre-retirement program of a large corporation in Syracuse, New York, and individuals attending a senior center program on computers. The young adult comparison sample (mean age = 21.9 years) consisted of 20 women and 16 men undergraduate students from Syracuse University. The questionnaire consisted of three parts designed to assess 1) current contact with computer-based technology in everyday life (outside of the workplace), 2) potential uses of a personal computer, and 3) attitudes towards computers. To measure current contact and potential uses of computers, subjects checked items from a list of 10 responses. To measure attitudes towards computers, participants answered 12 questions on a Likert-type scale; sample items were: "Computers make me nervous," "I avoid computers whenever possible," and "Computers are fun to use."

In response to the questions regarding contact with computer-based technology in everyday life, statistically significant sex differences and age differences were found in the number of

responses given. As can be seen from Figure 1 , men reported more
contact than women; and, as expected, young adults had more contact
than older adults with computer-based technology. For some of the
"contact" items (e.g., "Do you use an automatic bank teller?") and
for the total of contact items, young men reported significantly
more contact as compared to older men and to both age groups of
women; and young and older women were not significantly different
from each other.

With regard to the question, "If I owned a personal computer I
would use it to ...," there was a statistically significant sex
difference as well as a significant age by sex interaction (see
Figure 2). Young men reported more "uses" than all other groups on
the following items: To play games, to create computer graphics, to
file addresses and recipes, and to keep financial records. Men
differed from women at both age levels on the following items: To
educate self and family, and to make writing easier. There were no
differences among the groups for the "help me with my checkbook" and
"help me with my taxes" items. A statistically significant sex
difference and no age differences were obtained for attitudinal
responses (see Figure 3). The correlation between positive
attitudes and amount of contact was significant ($r = .35$, $p < .01$) as
were the correlations between age and contact ($r = -.48$, $p < .01$) uses
and contact ($r = .26$, $p < .02$) and attitudes and uses ($r = .49$,
$p < .01$).

It can be concluded that the individual difference factors of
age, sex, and amount of experience with computers all serve to
moderate attitudes toward computer-based technological change in
adults. The most important and surprising findings were that men
and women differed at both age levels with regard to attitudes
toward computers and potential uses for computers.

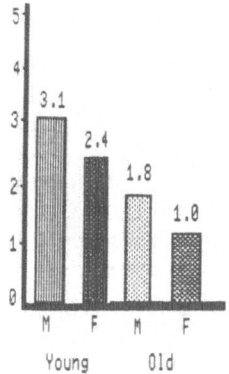

Fig. 1. Mean con-
tact for young and
old men and women.

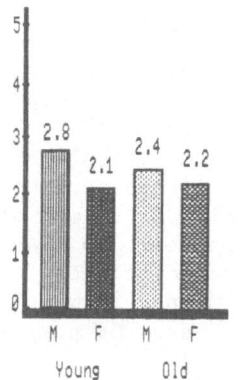

Fig. 2. Mean atti-
tudes for young and
old men and women.

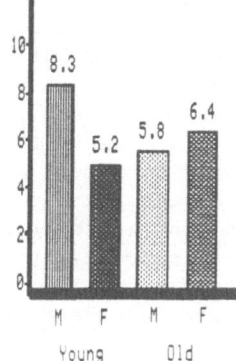

Fig. 3. Mean number of
uses for young and old
men and women.

RESEARCH ON HUMAN FACTORS: GERMAN SOLDIERS

AND FUNCTIONAL AGE LIMITS

Friedrich W. Steege

Federal Ministry of Defense
Bonn, Germany

The main responsibility of the Psychological Service of the German
Federal Armed Forces is the development, application, and use in
instruction of scientifically proven psychological knowledge, with
special emphasis on the individual soldier's needs. One of the
general research topics at present, not peculiar to the German
Forces, is the enhancement of the quantity and quality of military
personnel. Manpower planners of the Western nations have to face a
more or less dramatic decrease in numbers of recruitable personnel
in young age groups. This situation compels every official in
charge of research coordination and/or personnel management to
exploit every possibility of personnel acquisition, and to make
optimum use of the personnel being acquired.

Here is the place where personnel research and development in
general and human factors research, or research in the field of
ergonomics, meet. The special problem a task force on manpower
planning in our Ministry is presently facing is the question of
whether the future personnel (beginning in the late 1980's) can to a
considerable extent be made up of enlisted or professional soldiers
serving longer than they do today. The problem is to broaden the
traditional perspective of functional age limits and apply it to
areas other than aviation psychology or some other specialties
requiring high perceptual-motor speed and coordination. One of the
special questions in this context is: how old can a military leader
be and still meet the needs of the forces?

The following cautious estimates have been derived from
psychological, psychophysiological, and behavioral and social
research findings which are in part contradictory: (i) military
leaders on unit level should not be older than 30 to 35 years of

age; (ii) military leaders on higher levels (battalion and above) should not be older than 42 to 45 years of age; (iii) officers in highly complex human-machine-systems (characterized by quick information processing and high decision density) should not be older than 40 years of age.

As we know, it is a most difficult undertaking to find functional age (limit) indicators in the age frame specified (40 and more). One cannot run the risk of having proficiency declining below a certain threshold in many specialties. We therefore try to differentiate carefully the dimensions of analysis. The most important in our opinion (besides the sensory-motor dimension) are the cognitive/information processing dimension and the motivational/emotional dimension. The latter includes stress resistance factors.

Another area of concern in this field of research in our Ministry is earlier legislation of special age limits for adult professional training. One job accession regulation in our organization prohibits such training after age 40 because it is judged not to be cost-effective. We have started to question such setting of limits.

In this context, a review by Lehr and Olbrich on "A Framework for the Psychological Analysis of Age Limits in the German Federal Armed Forces" is presently being written. It is expected to provide more solid ground for policy decisions. Furthermore, a NATO Defense Research Group program concerning the definition and measurement of what we have called psychological fitness, and contract research on the "bandwidth of attention" have to be mentioned here.

What we have not dealt with yet is the influence of technological advances on future recruiting of soldiers, and applying age limits. Only recently, manpower planners have asked how future technological progress will perhaps facilitate the allocation of personnel to military specialties. In this connection one thinks of robots taking over some special mission.

Generally, there still seems to be reason to contemplate obvious gaps between advanced weapons technology and application of new technologies to life sustaining purposes. Also, much effort will be necessary, especially in military settings, to bridge the even more serious gaps between progressing technology and human characteristics, attitudes, understanding, and fears.

APPRAISING THE PERFORMANCE OF OLDER WORKERS

Chandra M.N. Mehrotra

College of St. Scholastica
Duluth, Minnesota

Performance appraisal should receive critical and effective
priority in personnel management. This recommendation is based on a
number of reasons: the aging of the workforce, the desire of many
individuals to continue working past normal retirement age, the
disclosure rights granted by the Freedom of Information Act, the
1978 amendments to the Age Discrimination in Employment Act, the
large magnitude of individual differences among older workers,
demands for equal employment opportunity, the growing emphasis on
communication and participation in employee relations, and the
motivation of organizations to make the best use of the available
work force. Performance evaluation provides an accurate measure of
how well a person is performing on the job. This information allows
the management to (a) make personnel decisions regarding salary,
productions, layoffs, etc. and (b) help employees improve
performance, plan work, develop skills for career growth, and
strengthen the quality of their relationship as managers and
employees. This dual purpose of performance appraisal creates
problems for managers as they find it difficult to serve both as
judge and as counselor simultaneously. Many organizations have,
therefore, developed separate appraisal programs for performance
evaluation and performance planning and review (Meyer, Kay, and
French, 1965).

The Freedom of Information Act and the 1978 amendments to the
Age Discrimination in Employment Act have created a legal mandate
for performance evaluation and employers are looking anew at their
appraisal programs in the light of these laws. In order to meet the
legal requirements the appraisal program should be reasonable,
relevant and reliable. Behaviorally anchored rating scales score
high on all three of these criteria. They are reasonable because

353

employees are generally involved in developing the scales; and
reliable because the anchored scales aid consistent interpretation
of performance (Cheedle, Luthans, and Otteman, 1976; Campbell, et
al., 1973).

Walker and Lupton (1978) have provided specific guidelines
which would be helpful in meeting the basic requirements of
reasonableness, relevance, and reliability. These guidelines
include (a) apply specific performance standards; (b) assure that
the program is rationally designed; (c) document performance
evaluation; (d) administer the program systematically, and (e) train
the supervisor.

Records play a critical role in the courts (Schuster and
Miller, 1981). Written evaluations, specific criteria, and frequent
evaluations at regular intervals lead to success for employers in
court cases. However, frequent evaluations conducted by multiple
raters result in accumulation of large amounts of data on individual
employees during their tenure of employment in the organization.
Computer technology can be usefully employed for maintaining the
records of performance appraisal, monitoring the comparative
efficiency of various procedures, and scanning the records in order
to identify employees for new assignments and responsibilities. The
use of computers can increase the availability and utilization of
data relevant to important decisions for older workers. This
data-based approach to making decisions will play an important role
in the world of work where rapid technological advances are creating
new roles for workers by altering the distribution of work between
the man and the machines.

It is necessary that we continue to do research that will
enable us to achieve the realistic goals of equity and age
neutrality. Some suggestions for research studies follow:

1. Since ratings are the most common tool used in performance
   appraisal, there is a need to study the individual difference
   correlates of performance ratings. Examples: How do the raters
   of different ages rate the older workers? What training programs
   can be designed for the raters so that their ratings will be
   based on performance rather that on ratee's age?
2. There is a need for providing an age analysis of the national
   workforce by occupation and skills so that a company can
   ascertain how its workforce stacks up against national norms in
   different classifications (Jacobson, 1980).
3. We should experiment with different types of training programs
   for those involved in conducting performance appraisal. This
   research should aim at developing training programs with
   long-term effectiveness.
4. It will be useful to study the type of errors made by the
   evaluators in appraising the performance of older workers. This

information can be utilized in providing feedback to the
appraisers and in developing better training programs.

If we indeed believe that the value of an individual worker to
the employer can only be judged on the worker's merit and not by his
or her chronological age, we should continue to conduct research
that will lead to the development of efficient systems of
performance appraisal.

REFERENCES

1. Campbell, J. P., Dunnette, M. D., Avery, R. D., & Hellevik, L.
      V., "The Development and Evaluation of Behaviorally Based
      Rating Scales". Journal of Applied Psychology, 1973,
      15-22.
2. Cheedle, W. M. Luthans, F. and Otteman, R. L., "A New
      Breakthrough for Performance Appraisal", Business Horizons,
      August 1976, 66-73.
3. Jacobson, B., Young Programs For Older Workers: Case Studies in
      Progressive Personnel Policies, New York: Van Nostrand,
      1980.
4. Meyer, H. H., Kay, E. and French, J. R. P., Jr., Split Roles in
      Performance Appraisal, Harvard Business Review,
      January-February 1965, 12-17.
5. Schuster, M. H. and Miller, C. S., Evaluating the Older Worker:
      Use of Employer Appraisal Systems in Age Discrimination
      Litigation, Aging and Work, Fall 1981, 229-243.
6. Walker, J. W. and Lupton, D. E., Performance Appraisal Programs
      and Age Discrimination Law, Aging and Work, Spring 1978,
      73-83.

HUMAN FACTORS IN AGING: ISSUES FOR ADULT EDUCATION

James E. Thornton

The University of British Columbia
Vancouver, B.C., Canada

As adult educators our education and training efforts in the later years should be based on images of the older person as person. If these images are guided largely by problem or deficiency assumptions, and they often are in our study of aging in the workplace, we negatively stereotype both the person and their needs, and also the later years as a period of time. Images of the older person should allow for aging to imply some purpose. As adult educators we should assume that the person is designed to learn and to adapt through the life-span: the dynamics and structures for learning are embedded in the neurophysiological mechanisms of the organism and influenced by psychosocial and sociocultural factors of the environment.

The last stages of the human life cycle often seem out of step with the mood of expansion, growth, and dynamic change characteristic of the technological forces shaping the present period of human history. Perhaps human development has always lagged behind human technology. Perhaps the major forces presently shaping our futures are taking us into one type of time and space dynamic of constant change, and perhaps these same forces are colliding with the finite boundaries of another time and space called old age. The juxtaposition of aging and technology in the human experience creates a philosophical dilemma and an educational conundrum: our culture lacks an image of the second half of life, but our culture has great belief that technology will produce its own solutions. We have no map or strategy of the cultural chaos being created by technology, certainly none to guide educational thought for the later adult years. What is it that education needs to accomplish in the later half of life? If we could answer this question, if we could achieve an informed consensus, life-long

357

learning and life-span education might become significant forces in
the context of aging and technology. Adult education would become a
force of social and cultural change and of individual adaptation,
not "retroactive altruism." Adult education would become an
instrument for anticipatory adaptation in all its form.

Adult education interventions (processes, programs and
technology) are value-laden about what is functional and normative,
and are loaded with ethical considerations and consequences when put
in the service of means or ends of social policy. If conceived as
means adult education is instrumental in adaptation and change
strategies. Examples are: 1) career preparation and advancement, 2)
social role acquisition, 3) enhancement of human productivity in the
workplace, 4) enhancement of informed consumerism, 5) facilitation
of community development, and 6) enhancement of mental and physical
health among designated populations. If conceived of as ends adult
education is an instrumentality for expressive needs in the
development of human potential and cultural transmission with
relatively few explicit purposes to measure its success. Examples
are: 1) transmission of the collective culture, 2) enhancement of
physical and mental well-being within a general concept of leisure,
and 3) enhancement of lifelong learning as an end in itself.

As means or ends adult education is not secured by public
policy in our many educational agencies or workplace. This is likely
for two reasons: lifelong education has not been mandated as
essential to social stability and growth, and there is not an
explicit image of education for adulthood except in response to
problems individuals have and to increased time at leisure. The
point is that adult education, whether conceived of as lifelong
learning, life-span education, recurrent education, or whatever,
must be seen as an essential strategy in the continuing development
of the individual and in the orderly transformation of society.
This perspective on adult education is crucial as the expansion of
knowledge, the complexity of social change and the enormity of
technological developments change our focus of adult education and
learning from what to know to how to know.

For adult education to become an effective intervention in
social and technological change it should adopt strategies which are
proactive and anticipatory and which help individuals and groups
become architects of futures offered them by change. For a coherent
social policy to emerge respecting the functions and allocations of
resources to adult education, a paradigm is needed relating it to
life-span development. Present literature does not provide that
paradigm of aging and adult education. Such a paradigm requires
social images of older people as people which can guide our
decisions and inform our choices about education for the later half
of life. We know enough about aging and about people's experiences
in aging to perhaps "map the territory."

COMPUTER BASED MEMORY TRAINING AND

MEMORY PROSTHESIS IN OLDER ADULTS

Karl Syndulko, Valerie Crooks,
Robert Wang, and Wallace W. Tourtellotte

Neurology and Research Services
Geriatric Research Education and Clinical Center
VA West Los Angeles Medical Center and
University of California at Los Angeles

Human aging is associated with a fundamental slowing of central nervous system processes. The evidence is based on changes, often decrements, in the function of a wide range of perceptual, motor, cognitive and memory processes. Two of the most vulnerable processes are the short-term storage of information and the transition of new information to and from long-term memory storage. It has been estimated that about 10-20 percent of older adults show additional, often disease related, impairments in these processes. Behavioral and neuropharmacologic interventions have been proposed and to some extent tested for the remediation of memory deficits. The application of computer technology to these and related problems of the elderly offers considerable promise, but thus far has generally lagged behind applications in younger populations.

The computer is the most recent and direct technological tool for the extension of human information processing functions. Although still primitive in comparison to the total processing capabilities of the brain, human computer systems have evolved to emulate and extend the range of specific innate human processing capabilities. A particularly noteworthy extension has been in our ability to dynamically store, manage, and retrieve enormous quantities of digitally encoded numerical and text information. Computer augmentation of our short and long-term memory systems follows directly from similar use of other external, but relatively static forms of information storage such as carvings, paper, film, and tape. In the near future the computer will be integrated fully with systems capable of storing and retrieving complex audio-visual material (e.g., writable optical disks). Computers with optical storage will provide substantial augmentation of our ability to store and process detailed information about nearly all aspects of daily existence.

Two primary applications of computer systems are especially relevant to the aging human. The first application is the remediation or training of basic cognitive processes, particularly those involving memory functioning. The new technology involving interactive videodisk systems could be utilized as the foundation of an ecologically valid memory training, evaluation and maintenance program. The primary emphasis will be on: 1) memory challenges that are realistic and meaningful; 2) individually tailored programs that incorporate unique aspects of the person's life; 3) progressive training modules that dynamically assess memory function and its changes during the intervention process; and 4) intervention strategies that are based on our best current empirical and theoretical formulations of cognitive and memory processes in humans, especially as related to human aging.

Such computerized memory systems could extend into other aspects of daily life also. The use of modems, mixed audio-video-digital communications links, and computer network services could provide a much wider, more stimulating range of social and professional interactions, especially for those with limited physical mobility. For example, specific testing and monitoring of progress, question and answer sessions, problem solving, group therapy sessions, and social interactions could all be conducted via communications links to supplement live individual and group sessions. Data from biomedical monitoring devices hooked to a home computer and transmitted upon request over the communications link could be used to more thoroughly integrate medical care with psycho-social care.

The second major computer application involves the development of a portable microcomputer system as a real-time memory prosthesis. This function is a direct extension of currently available notebook-calendar systems directed primarily at the "busy" professional. These systems currently have software for some or all of applications such as word processing, electronic mail, teletext terminal for tapping remote knowledge bases, filing or data base systems for creating a personal data base, scheduling or calendar functions, calculator and spreadsheet capability, and simple graphics production. These programs could be tailored for general personal use and would support such uses as maintaining travel directions including maps, prompting for important dates and deadlines, filing names and related personal information, storing shopping lists and activity lists, inputting and organizing isolated facts, observations and ideas for later review and transfer to a larger home based system. Minimally, a portable memory prosthesis could provide direct prompting of scheduled events such as

medications and appointments. Additional portable uses include
games, cognitive training modules, education modules, self-help
skill acquisition or habit modification aids, diet monitoring,
exercise monitoring, and music or other artistic production. These
latter uses are also available currently in at least rudimentary
forms on home based microcomputer systems, but are well within the
capabilities of a portable system.

In its most comprehensive form the memory prosthesis would be
within audio range of the user day and night. It could be
programmed and monitored either by the user alone, or with the help
of others, or via a modem. It would utilize visual displays and
possibly synthesized speech for output, and keyboard or voice
recognition as input. Pre-scheduled reminders could occur until
recognition was indicated by active entry of a completion code. The
device could notify a remote site of the lack of response using a
built-in telecommunication link.

A personal computer memory aid could also be combined with home
station units and increased storage of complex audio-visual
information. Small digital cameras with alpha-numeric annotation
capability could augment a memory prosthesis by recording detailed
representations of individually important persons, places, and
events. These data, in turn, could be converted into computer
controlled media for long-term storage. Through use of
sophisticated cue-word retrieval systems, the complex data could be
reviewed periodically, e.g., prior to social encounters, and could
be incorporated into a personalized memory training or memory
maintenance program.

The primary concept in these proposals is the utilization of
technology to enhance the richness of daily life by expanding and
extending the current level of cognitive functioning, regardless of
what that level is. For impaired older adults a computerized memory
prosthesis could fully or partially compensate age-related
processing losses. For others it would augment normal processing
abilities. Age is a somewhat arbitrary distinction for use of a
memory prosthesis, as it is for use of visual, auditory, language or
biomechanical prosthetic devices. However, ergonomic factors, as
well as economic, social, ethical and personal issues specific to
older adults require special consideration. In addition, the
computer as a direct extension of human information processing
capacities has special potential for significantly increasing the
quality of life in older adults, a population with virtually no
direct exposure to the new technology.

AGING AND TECHNOLOGICAL ADVANCES:

HUMAN FACTORS

David B. D. Smith and Arnold M. Small

Human Factors Department
Andrus Gerontology Center
University of Southern California

The purpose of this report is to provide an overview of some of the major topics addressed by the human factors study group at the NATO Symposium on "Aging and Technological Advances." The authors have used as a framework the individual presentations and discussion of that study group, but the content represents our own assimilation of the group's deliberations, plus other relevant material that arose in the context of the overall symposium.

Consistent with the scope of the study group's activities was the following definition of human factors: Human factors is concerned with increasing and applying knowledge about people and their interaction with technology, the environment and with one another to the design, operation, and management of person-machine and organizational systems. The purpose of this effort is to enhance reliability, efficiency, safety, convenience, and the overall quality of life. As regards technology, human factors deals with both (1) hardware technology, e.g., the design of the computer's keyboard and visual display terminal to assure accuracy and reduce fatigue and eye strain, and (2) process technology, e.g., the application of behavioral science techniques to organizational design and to areas such as training, personnel decisions, decision support systems, and job design.

A life-span developmental view of aging was implicit in the deliberations of the study group. This led to considering the impact of technology in the middle years of life as well as later years. In addition, it emphasizes that since age is the convergence of both biological and developmental processes, adult age today may not adequately model a comparable adult age segment ten or even five years hence. This makes it important to be aware of the demographic, health, education and social changes that are likely to affect the future old.

Presentations and discussion in the study group centered around six major topic areas. These were assistive devices, environmental design, computer technology, training systems, personnel systems and management/decision making.

Assistive Devices

Miniaturization achieved through multifunctioning chip technology has increased technological applications for individual users. Potential applications for older persons include job aids, sensory and memory protheses and assistive devices for home and medical care facilities. Progress in electronic developments can make deficiencies in hearing, seeing and memory less restrictive for older persons at home and in the community. This is conducive to more personal interactions and active participation judged both satisfying and vital for older individuals.

Assistive devices that aid mobility may serve to maintain the independence of the older person and his or her access to available options, thus reducing the loneliness and depression common to the very old. On the other hand, such aids as computer shopping could act to decrease levels of stimulation, activity, and skill use. It has been reported that practice and/or appropriate task strategies maintain or restore the older person's capability for a specific task at a level comparable to younger ages. These issues suggest that a general criterion in human factors design for assistive devices should be the maintenance of appropriate activity, stimulation and skill levels of the user.

The importance of the occupational role in determining the quality of life is widely recognized. An important question is how future technological change will influence older people's options to work. On the one hand, there is evidence that technology, such as robotics, by taking over established functions will threaten work opportunities for many people, and the older worker is likely the most vulnerable. On the other hand, technology offers the potential of extending the productive work life by providing assistive support or by taking over those tasks elements for which the older person is disadvantaged. Research is needed on how and where human factors can use existing and new technology to preserve older people's work opportunities.

Environmental Design

For the older person there is an increased dependency upon social institutions and services and a physical environment likely designed, operated, or appropriate only for younger ages. Physical

pathology and needs for medical care are major contributors to the
dependency that accompanies aging. A high proportion of elderly
persons (65+) report a chronic medical condition and almost half
have a disability that limits activity. For many of the physically
impaired, access to transportation is limited by environmental
barriers, both physical and operational. Also, for the elderly,
housing is often old, inadequately heated and poorly maintained. A
significant trend in the United States, in recent years, has been
the growth in the number of elderly living alone. The high
incidence among the elderly of accidental hypothermia and
poisioning, and falls and accidents with antiquated equipment is no
doubt closely linked to the inadequacies of this environment.
Finally, the elderly's mobility in the community is affected by
their perception of crime, and we know that environmental design
plays an important role in both the perception and the incidence of
crime. Associated with these inadequacies of the environment are
the problems of loneliness, depression, social detachment and the
high incidence of alcoholism, homicide and suicide among the
elderly.

It is easy to conceive of ways that technology, through
environmental interventions, could ameliorate these problems. For
example, auditory and visual supports, continuous monitoring and
emergency signalling, memory aids, self-maintaining facilities and
access to services, such as, educational and recreational
activities, in-home shopping, and medical appointments come to
mind.

For human factors a central issue is how the person-
technology-environmental interface should effectively accommodate
the elderly person. Unfortunately, human factors design information
concerning the capabilities and performance limits of the elderly
has not been assembled in anything approaching a usable form. There
may be doubts whether the largely laboratory results in gerontology
are adequate as practical criteria for real world applications. In
any case, an important goal for future technological applications
will be the development of human factors criteria usable by the
practitioner in the design of housing, transportation, medical care
facilities, leisure environments and work situations.

Computer Technology

Hardware technology today is largely the application of
computers to almost every job and activity that people do. One
consequence of this has been to change the nature of tasks both at
work and in daily activities. An example is industrial robotics
which is said to result in the lowering of skill requirements, the
simplification and routinization of tasks, a reduction in individual
judgment and a loss of social contacts. In other highly automated
work environments, e.g., nuclear and conventional power plants,

aviation and military weapon systems, many jobs have become of the command control variety. These are characterized by monitoring and information processing tasks and transient, high demand levels. The wide-spread use of computers may also significantly increase the information processing demands of common daily activities, such as in the use of transportation and shopping. An important question for human factors to address is how older ages are affected by such changes in task structure. How capable, ready or willing are older persons to adapt to these changes? Finally, do such tasks affect overload stress, and is the older person advantaged or disadvantaged in this regard?

Training Systems

The introduction of new technology and growth in specialized bodies of knowledge has brought to the forefront the problem of skill and technical obsolescence. Not only is knowledge doubling every few years, but the nature of jobs is changing. Demographic data point to this obsolescence problem being progressively associated with the older employee. As an example, the percentage of persons in the United States in highly technical and professional occupations has more than doubled since 1950. These highly educated and technically trained persons began in the 1970's to constitute the middle-aged and older population.

Learning new skills is a major means by which people adapt to the obsolescence caused by technological change. In a rapidly changing environment, a lack of opportunity, inability or an unwillingness to train can seriously limit a person's options, economic future and quality of life. Older persons were once thought to be less able to learn. We can now be quite sure that with appropriate opportunities and training methods they can learn to perform tasks comparable to younger persons. Also, the traditional "youth" orientation of training is undergoing change because of legal sanctions, an older work force, and the growing importance of human resources in organizations. This leaves, as the critical problem in retraining older persons, a general reluctance to seek training. The nature of this issue points to the importance of an employee's perception of self worth, the potential for job progression, past training success, and the organization's design, incentives and supervisors attitudes. Methods need to be explored in how to change organizational environments and individual perceptions that will motivate the older person to learn new skills and update others.

Personnel Systems

An important way that organizations adapt to change is through their personnel decisions, policies, and practices. This in part may explain why in our volatile, economic and technological society human resource planning and its behavioral technology methods have come to play a significant and vital role. This, combined with the certainty of an aging work force, technological changes, and legal sanctions in job discrimination and fixed-age retirement, should serve to enhance the significance of personnel attitudes and decisions in general, and age-related ones in particular.

These trends would suggest a greater role in organizations for career planning as corporate policy, for skill updating and retraining as integral to the job description, for alternate career paths with pay progression less closely tied to job progression, and for behaviorally based assessment and job appraisal methods.

Over the recent decades a variety of behavioral technology methods have been developed to both facilitate change and to aid in personnel decisions. These include behavior modification, task analysis, behavioral role modeling, assessment centers, behavioral anchored rating scales, goal setting, career pathing, and job sampling techniques. Unfortunately age-related data for any of these methods are hard to find. Research is needed to determine the fairness and value of these methods across the working life span.

Management and Decision Making

A greying population, an aging work force, concerns over productivity and chronic unemployment, all problems of western industrialized nations, are reasons to ask questions about how age and the aging process are impacted by managerial decisions, and conversely how they might influence management practice.

Some of the questions to be asked are: How do attitudes and perceptions about the older person's productivity and adaptation to change affect managerial decisions? How and where can the issue of age and age/technology be incorporated into management education? Can managers be better trained to manage technological change with respect to older persons? How does age affect the competency of the older manager or professional, if it does? Do human values suffer when decision technologies, such as, systems analysis, modeling, simulation and decision theory are used to aid the decision maker? Finally, will new decision technologies, such as artificial intelligence and decision support systems limit or eliminate the older person's traditional advantage in experience and wisdom?

These are fundamental questions since managers are critical in the decisions of how, when, and where technology will impact people. At the present time there is limited evidence with which to address these questions because research provides us very little age-specific information on the management role.

Resolution

As a result of earlier papers and the activity of the NATO study group on human factors the following resolution was passed unanimously:

WHEREAS, The deliberations of the Study Groups of the NATO Symposium on "Aging and Technological Advances" have included presentations and discussion on many aspects of human factors and technological impacts on older adults both potentially positive and negative,

and

WHEREAS, Many problems were identified and current trends noted indicating too early applications of technology to permit adequate assessment

and

WHEREAS, Consideration of applications of technology must involve human factors among the primary factors in making decisions on specific applications,

Therefore,

Be it resolved that the Human Factors Study Group recommends to the Human Factors/ Ergonomics Societies in the NATO countries and the International Ergonomics Association that they assume leadership in (1) developing goals, criteria recommendations and guidelines for planning future balanced applications of technology as they affect older persons' lives and societies as a whole and (2) identifying and doing basic and applied human factors research in areas relevant to and supportive of the developments in (1) above.

Implementation

It is suggested that the first step in following up this resolution be mutual sharing of an initial plan from each Human Factors/Ergonomics organization indicating what is contemplated in connection with items (1) and (2) of the resolution and the method for accomplishing it. The Technical Group on Aging of the U.S. Human Factors Society offers to serve as a collection center for receiving and collating these, with their subsequent distribution to all. A target date for reporting progress is suggested as the 1985 International Ergonomics Association meeting in England.

THE OLDER PERSON IN THE RESIDENTIAL ENVIRONMENT

M. Powell Lawton

Philadelphia Geriatric Center
5301 Old York Road,
Philadelphia, Pennsylvania 19141

This Symposium on Aging and Technology is devoted to the
transactions between people and environment, technology being a
subset of all that is included in the more general environmental
category. This chapter will first suggest a structure into which
our problem, technology and aging, can fit. It will then consider
at some length the older person as the focus of the transaction, on
the assumption that the technology itself is better dealt with by
those with more expertise than I possess. More specifically I shall
concentrate on some intrapersonal factors that mediate the
relationship between the person and the technological environment.

A STRUCTURE FOR UNDERSTANDING THE PERSON IN THE TECHNOLOGICAL ENVIRONMENT

The characteristics of the person are, of course, where our
effort begins. The knowledge base in gerontology is well-provided
with information on older people, which emphasizes the diversity
among individuals and the consequent impossibility of designing
environmental resources for "older people" as a class.

The Person

I should like to define at the outset a framework for viewing
the person within the environmental context. Following the lead of
White (1959) and many others since then, I have suggested that
people be viewed partly in terms of their "competence," that is,
their abilities in the basic domains of biological health,
cognition, and psychological integrity, or "ego strength." These
domains theoretically represent intrapersonal characteristics, or

"capacities" in the sense that upper limits or bounded ranges of
functioning are hypothesized to exist for a given individual. The
typical indicators of competence are various laboratory and clinical
measures of cellular, organ, and physiological health and scores on
tests of intelligence and related functions; ego strength has thus
far defied measurement. This concept of capacity is primarily of
heuristic value, however, since functioning in even such basic
domains as these is to some extent dependent on the context in which
it is exhibited, thus introducing extrapersonal factors.
Nonetheless, these functions are relatively stable in the absence of
pathology and are "basic" in the sense that they condition strongly
one's performance in all domains, including those that depend
substantially on the environmental context.

    Although decline in the basic competences has not been shown to
be an inevitable concomitant of the aging process, the probability
is very high that decrements in some forms of biological and
cognitive functioning will occur as chronological age increases.
Thus we have a basis for defining several potential subgroups of
users of technological aids:

    1. The majority, that is, those without notable impairment.
Technology developed for the use of people in general will be useful
to the normal older person. Conversely, as we have already seen in
the application of behavioral principles to environmental design for
special user groups, we shall probably do better for the population
at large as we learn better how to design for these special users.

    2. People with sensory and perceptual deficits (Hiatt, 1981;
Pastalan, 1979).

    3. People with motor deficits and limitations in ambulation
(Koncelik, 1976; 1982; Raschko, 1982).

    4. People with problems in memory, information processing,
judgment, and other cognitive functions (Lawton, 1981).

The Environment

    The total context in which the person behaves has concerned
many scientists as a general focus for their research. How
environment has figured in our research and conceptual frameworks
will not be repeated at this time. It is important, however, to
make explicit three general points:

    First, as more human systems become involved, the ensuing
behavior becomes more highly related to environmental factors.
Physiology varies more with the environmental context than cellular
function; ambulation more than muscle strength; problem solving more
than sensory reception; or social behavior more than solitary

behavior. Since interventions of all kinds attempt to upgrade the functional quality of relatively complex behaviors, the environmental context of the functioning is often a very productive point for the intervention, as contrasted with the person (Lawton and Nahemow, 1973).

Second, "environment" must always be specified in terms of whether a physical or subjective definition is intended. Each has its place. The objective environment, which can be defined in physical terms independent of any person's judgment of it, is the focus for design and planning. The subjective environment represents the portion of the objective environment that is registered, comprehended, used, and interpreted by the person. In general, the more of the objective environment one's subjective environment includes, the greater the likelihood of a favorable outcome, in the sense to be described below. The effect of the objective environment on the person's behavior and affect is usually indirect, through the subjective environment.

Third, technological aids represent a special class of members of the objective environment. Viewed as systems, technology clearly demands the human presence as designer, user, observer, recipient of impact, and so on. Technological aids are clearly objective, yet in order to have an effect on the person, they must become elements of the subjective environment, the point that will be developed as the major focus of this chapter.

Outcome

I have suggested elsewhere that the outcomes of the transactions between a person with a given set of competences and environment of given objective and subjective qualities may be viewed in evaluative terms externally as adaptive behavior or subjectively in terms of psychological wellbeing or perceived quality of life (Lawton, 1982; in press). The search for environmental prostheses and other technological assistance is thus based on the idea that the quality of behavior or subjective experience can be elevated by attention to the environmental component. "Pure" manifestations of biological health, cognition, or ego strength rarely occur. These basic competences are, instead, put into action in situations with differing demands, barriers, facilitating factors and with differing degrees of involvement with other people and social systems. Thus, in the heirarchy of behavioral competence that I have proposed (Lawton, 1972; in press), most behavioral examples of competence are in the transactional, rather than the intrapersonal realm: functional health, most cognitive behavior, time use, and social behavior.

Figure 1 illustrates this person-environment problem. One may define physical indicators of housing quality, for example, and

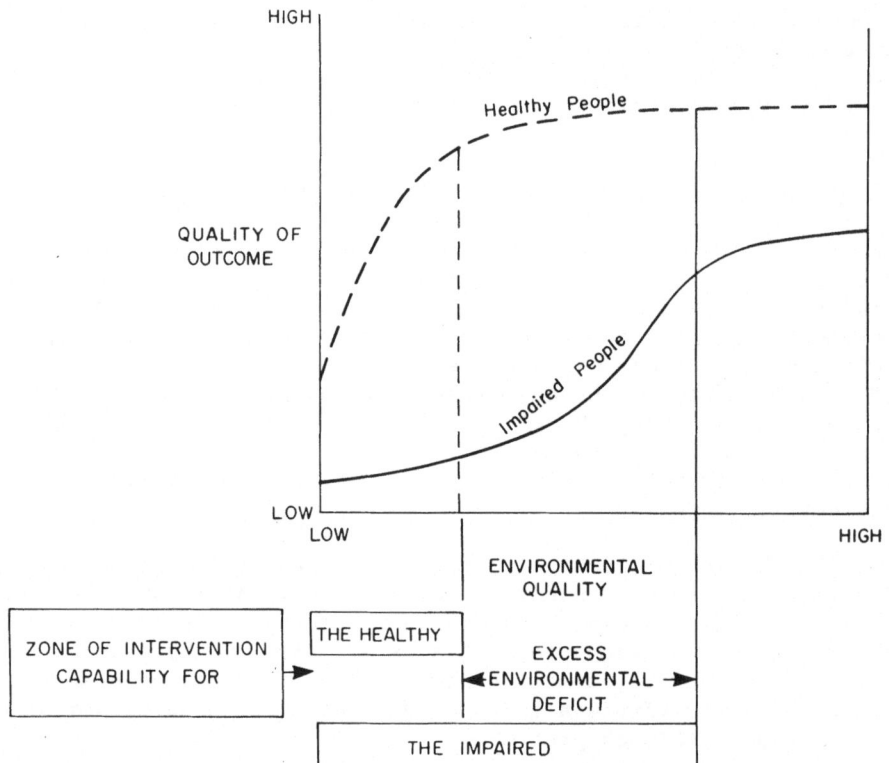

Figure 1.   Environmental Quality and Outcome Functions For Healthy
           and Impaired People.

measure this objective environmental characteristic, as indicated on
the abscissa. However, many objective deficits may have little
effect on the behavioral or affective outcomes for healthy people.
Competent behavior compensates effortlessly for minor environmental
deficits. By contrast, the same housing deficits may affect grossly
the quality of life of the impaired person. Thus for planning
purposes, whether a given level of housing quality is read as a
favorable or unfavorable social indicator depends very much on the
target population for whom the housing is intended.

     Thus far I have presented in only slightly altered form some of
the conceptions of the ecological model articulated by Lucille
Nahemow and myself some years ago (Lawton and Nahemow, 1973). It is
worth noting, however, the similarity of this conception to the
human-factors model that sees performance as the outcome of the
operator's manipulating the machine, performance quality varying as
a function of the operator's skill and the machine design; like
competence, skill is seen as a relatively stable characteristic of

the person, while performance is far more variable, a transactional characteristic.

Transactional Characteristics

A transaction by definition involves an element of the person and an element of the environment acting together in such a way that distinguishing between what is person and what is environment becomes difficult, even, to some, irrelevant (Ittelson, 1976). The subjective environment, for example, environmental cognition, cognitive maps, or perceived quality of life, is one variety of such transactional phenomena. Another is cognitive style, a characteristic manner in which a person processes external stimuli, for example, introversion-extraversion or repression-sensitization (Byrne, 1964; Lawton, 1982).

All too often, the traditional design process bypasses this member of the behavioral system. The user-oblivious approach assumes that all users have the same needs and characteristics; it designs a single product to suit the majority. The user-informed approach attempts to incorporate knowledge of user characteristics into the design of the product, resulting in diversity of products. This transactional approach thus incorporates knowledge that may be unique to a particular combination of user characteristic and product characteristic in the design process.

There are many specific classes of such transactional characteristics. The one on which this presentation will focus may be called the "utilization gap," that is, reactions of the user to particular environmental resources, whose outcome is behavior that may vary from complete failure to utilize the resources to full utilization.

THE UTILIZATION GAP IN A MODEL OF PERSON ENVIRONMENT TRANSACTION

Figure 2 schematizes the conceptual structure that has been proposed. The cross-hatched area where person and objective environment overlap represents the portion of the objective environment that is registered, comprehended, used, and interpreted by the person and therefore is available for the transactional process. The direct paths from person to outcome and from objective environment to outcome acknowledge that some effects of behavior may occur because of relatively pure intrapersonal factors or environmental factors that are not necessarily processed psychologically by the person, for example, industrial pollution. The feedback paths from outcome to the person and the environment depict the dynamic nature of the processes: An outcome that is positive (either as more adaptive behavior or as increased subjective wellbeing) is thus seen to be able to increase

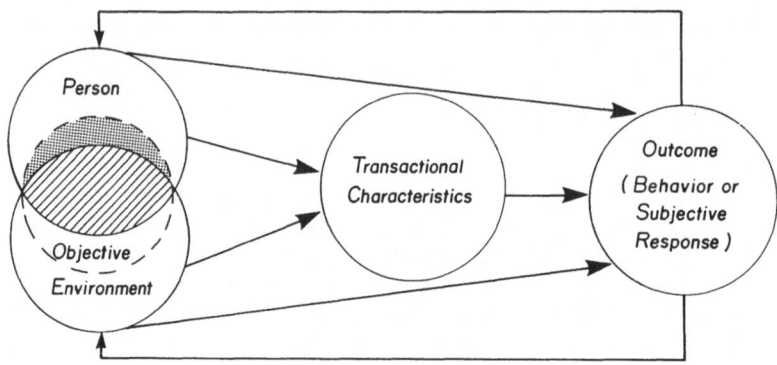

Figure 2.   A Model of Person-Environment Transaction.

competence.   Such as outcome is also able to alter the objective
environment: People choose and shape their environments to their
own purposes. Such an achievement brings some greater portion of the
objective environment under the control of the person (see the
shaded portion representing an increase in the overlapping portions
of P and E).   Adaptive behavior or positive affect increases the
extent of the usable environment, which should in turn lead to
increased adaptive behavior, and so on.   Thus we have an excellent
rationale for focusing some attention on the transactional processes
that facilitate adaptive behavior and thereby enhance the person's
control over the environment.

        I shall discuss environmental knowledge, familiarity and
acceptability.   Each of these terms represents an aspect of the
person-environment transaction that may be easily overlooked in the
total technological change process. Narrowing the utilization gap in
each of these instances may assist the goal of enhancing the
independent and productive life of the person.

        It is a matter of some interest to note the relative infrequency
of concern for such transactional factors in the Conclusions and
Recommendations that emerged from the 1981 Wingspread Conference on
Technology and Aging (Cohen, 1982).   While the needs of the elderly
as a special user group were prominent in the report, the factors of
consumer knowledge, product acceptability, and training for use are
mentioned only in passing, rather than as topics for in-depth consideration.   I suggest that awareness of such transactional factors
should constitute an essential element of the process that leads to
technological design, diffusion, and adoption.

Environmental Knowledge

        There is a science of knowledge diffusion of which we need to
be aware (Pelz, 1981).   However, relatively little is known about

older people as specific targets for spreading knowledge of environmental resources. Therefore we must rely considerably on a priori reasoning and marginally relevant research findings for insights into the responses of older consumers in this regard.

We must begin by acknowledging the structural factors that limit the diffusion of knowledge about new environmental opportunities for older people. Informal associations at work are one major source of knowledge that is curtailed for the elderly in general. As social networks diminish through the death of friends and physical impairments reduce spatial mobility, this source of information becomes further limited.

Some information about the history of utilization patterns for services for the aging is probably relevant to technological utilization. In the early days of senior centers there was a tendency for utilization to be greater among older people with more education and higher incomes. As senior centers became more numerous, use by the less-privileged elderly grew. In the area of housing, the same story was repeated in relation to the property-tax exemption program for older people in the state of Washington (Buczko, 1981); the relatively more priviliged were heavier users.

The recently-completed Experimental Housing Allowance Program (EHAP) offered a test of knowledge diffusion by age (Zais, Struyk, and Thibodeau, 1982). Older eligible recipients lagged behind younger people in their knowledge of the program where information was widely disseminated.

It thus seems clear that extra effort will be required to bring knowledge of new technologies to potential older users. The first natural point for such dissemination is the public media, with television the likely first choice. Older people in general report (using the time-budget method) watching television an average of about 18 hours per week (Hill, 1981), while Nielson estimates based on television viewers only range from 25-35 hours for total on-time (quoted in Kubey, 1980, p.17). Even the lower estimate shows the potential power in this educational medium.

The aging services network is another potentially powerful point of entry into the information dissemination chain. A chronic problem with this network, however, is its relative lack of concern for and expertise in housing problems, the result of the split at the national level between housing and social services. For many reasons, a head-on effort to train social agency personnel in some of the basics of housing and dwelling-unit design problems is necessary. While primary expertise in the design of technological aids is not appropriate to demand of social agency personnel, they may well be those on whom the burden falls for affirmative marketing

to the older consumer. Attention to their training for this purpose
is therefore in order.

The likelihood of involvement by the aging services network is
increased by the probability that specialized assistance will be
required selectively by the less competent, who will be also less
able to use ordinary sources of information of product
availability. Thus professionals dealing with the impaired will
require continuing exposure to channels of new information.

The families of older people represent still another entry
point to knowledge. The vigilance of this constituency was well
demonstrated recently in response to a brief letter to the editor of
a medical journal about some promising results from a drug given to
11 dementia patients; this physician's phone lines were clogged for
days with inquiries from hopeful relatives. Organizations of
relatives of impaired older people offer the opportunity for focused
dissemination of knowledge.

For residential opportunities in particular, research has shown
that friends are the favored channel of communications (Longino,
1982). Some attention might thus be given to identifying key people
who as "cosmopolitans" or innovators might be able to function as
information sources for their peers.

The first point in the transactional chain thus requires the
use of all possible actors in transmitting the knowledge. Those who
are most impaired may be in most need of the knowledge of product
availability but least able to obtain such information on their
own.

Environmental Familiarity

Almost by definition, a technological innovation will have some
element of unfamiliarity. Research on the cognitive functioning of
older people underlines the importance of crystallized intelligence,
that is, the ability to use past learning in the present. Where
fluid intelligence, which involves new learning or novel
applications, appears to be adversely affected by the biological
aging process, crystallized intelligence shows virtually no age
decline (Botwinick, 1978). A habit or reliance on previous learning
thus may be one force leading some older people to prefer what is
familiar. Another line of research shows that when given a choice of
responding or withholding a response to a stimulus, older people are
more cautious than younger people (Botwinick, 1969). "Behavioral
rigidity" (Schaie and Parham, 1975), more common in older than in
younger people, may be yet another behavior illustrating that the
familiar is easier to respond to than the unfamiliar.

Several studies of the effect of familiarity on cognitive
performance have shown that a reduction in the gap between old and

young subjects' performance was achieved when "age-appropriate"
stimuli were used, such as objects used at the turn of the century
(Howell, 1972), or "dated" words in a verbal learning paradigm
(Barrett and Wright, 1981).

Even for younger people it was shown that geometric figures
exposed for such short times that they were not consciously
perceived by subjects were rated later more positively than
nonexposed figures when inspected at length (Zajonc, 1968). If
aging increases people's cognitive comfort with the familiar, then
it is more even understandable for them to prefer the familiar in an
affective sense.

A special case of preference for the familiar is seen in
attachment to personal possessions, residence, or neighborhood.
Relatively little research has been done in this area, but it has
demonstrated that the majority of older people have a favorite
possession. Those people who reported having such a possession were
higher in life satisfaction than those who did not (Sherman and
Newman, 1977-78). People who wish to move report less attachment to
their homes (Howell, 1982); as yet not tested is the question of
whether the unattached are in fact more likely to move. Residential
attachment as a construct has been found to be multi-dimensional;
factors including competence in a familiar environment, traditional
family orientation and memories, and status value of home ownership
emerged from O'Bryant's (1982) analysis of attitudes toward one's
home. We have no direct age-comparative data on these kinds of
attachment. However we know that older people move much less
frequently than the young and that they have lived longer in their
present homes (Struyk and Soldo, 1980). The hypothesis that the
elderly may be more attached to their residential environments than
are younger people is thus at least consistent with some of these
observations.

Such thoughts on the cognitive and affective effects of
familiarity are relevant to our topic because they underline the
importance of recognizing familiarity as a factor in introducing to
older people new products, new forms of living environments, or new
organizations of goods, services, and accesses to them. We thus
need to be cautious lest our zeal for new forms begins to demand
such new operator requirements that the older consumer becomes
motivated to reject the innovation out of hand. If there is a
choice, a familiar form is preferable. If a new form can appear in
a familiar guise, the product can be more acceptable. For example,
the television screen and controls are no doubt thoroughly familiar
to most older people; this fact will probably enhance the
acceptability of the home computer.

If a sense of familiarity grows at a slower rate for the older
person, a correspondingly longer period may need to be given for the

introduction of the product, dissemination of knowledge regarding
its availability, and learning to use it.

Environmental Acceptability

Despite the fact that inadequate reception of knowledge and
resistance to the unfamiliar can constitute barriers to adoption,
new environmental forms can become acceptable to the consumer. For
evidence one need only look at the 2 million older people who live
in planned housing or the 1.2 million who live in mobile homes,
forms of residence virtually unknown at the time the cohort now
residing in them was in its adult working years.

An environmental option may become acceptable because it is
perceived as congruent with one's needs or when the potential user
learns to use it so that its congruence becomes apparent.

The perception of congruence is a transactional process whereby
the person assesses her needs and competences and the ability of the
environmental option to satisfy them. As competence changes, so
will the assessment of the acceptability of particular environmental
options. A tradeoff is made between the degree of need and the
desirability of the option. For example, few people wish to enter a
nursing home, but declining ability to care for one's own needs
leads some to accept the institution, however reluctantly. Almost
everyone prefers a private residence, but either limited income or
failing health may make acceptable shared housing and its
compromises with privacy, territoriality and control.

Learning how to use and value a new environment is an
achievement whose possibility we are very likely to deny older
people. If we press the earlier thoughts about the desirability of
familiarity too far we end up exactly with the ageistic view that
"older people can't learn." In fact, research in cognition provides
many examples of how the opportunity to learn a task, especially at
one's own pace and under other favorable conditions, benefits older
subjects selectively (Botwinick, 1978, Chapter 15). A recent report
documented that institutional residents can learn and enjoy some
computer games (Weisman, 1983). Thus in pondering the problems of
introducing new technologies to older people we must tread a fine
line between capitalizing on the familiar wherever possible, while
recognizing that new learning is something we continue to do and
often enjoy doing, regardless of age.

The problem, then, becomes first to recognize the ability of
older people to learn to use technologies and second to provide
appropriate training opportunities and experimentation with
different methods for such training. Whether a useful environmental
form will arouse anxiety and rejection or challenge and acceptance
will often depend on the person's perception of her ability to learn

to master its use. Corso (1977), for example, reported an estimate that 40% of older people who could be benefited by wearing a hearing aid find them personally unacceptable. Fields such as rehabilitation and audiology have built up bodies of knowledge on training to use prostheses (Alpiner, 1978; Maurer and Rupp, 1979). This knowledge will be extremely useful in facilitating the acceptability of new technologies to the aged.

It is clear that we shall have to concern ourselves much more with such matters as the phrasing, type size, and layout of instruction sheets or program announcements. The composition of cathode-ray tube screens, where considerble environmental learning will occur in the future, is another topic deserving of attention. A beginning technology is emerging for helping people to comprehend the large-scale environment, stemming from early efforts by Pastalan (1982) to prepare people for forced institutional relocation. More recent research (Walsh, Krauss and Regnier, 1981; Weisman, 1981) has been able to identify some of the critical aspects of environmental learning that may occur through representations such as photographs or scale models. Such techniques should ultimately find their way into the housing counseling process as methods for learning about opportunities and giving vicarious experience in comprehending new residential forms. Similarly, since we know that intimate knowledge of one's neighborhood and neighbors is likely to reduce the fear of crime (Liang, Sengstock,and Hwalek, in press) it is clearly worth the effort to use new methods to enhance such knowledge.

Recent data relevant to this topic is seen in the 1978 Annual Housing Survey, the only major Census Bureau survey to date that has combined extensive information on both health and housing. Struyk (1982) analyzed data from householders age 65+ who reported one or more members with a health problem or activity limitation; such households constituted about 57% of all elderly headed households. Among other interesting findings, Struyk reported the number of housing units that had special features designed to enhance their livability for one or more disabled members. Table 1 shows the numbers of households who reported each feature, out of the total of 8.6 million "disabled" households. It is of interest to note that by far the most ubiquitous feature is the grab bar or handrail, aids whose presence has become familiar because of their widespread use for people of all ages. One of the two closest competitors for frequency, special wall sockets or light switches, is likewise very likely to have been so frequent because its characteristics were favorable to all users--outlets at chair-rail height, lighted switches and so on.

Extra vigilance on someone's part was required to know that these products were available and, for some features, to follow up knowledge with an effort to retrofit the dwelling: Some special sink features, the special telephone, wide doorway, flashing light,

Table 1. Number of 65+ Households with an Impaired Member,
          which Contain Specific Adaptive Features

ADAPTIVE FEATURE	NUMBER
HANDRAIL OR GRAB BARS	568,000
SPECIAL SINKS, FAUCETS OR CABINETS	103,000
SPECIAL WALL SOCKET FEATURES OR LIGHT SWITCHES	103,000
ELEVATORS OR LIFT CHAIRS	69,000
SPECIALLY EQUIPPED TELEPHONE	69,000
RAMPS	60,000
EXTRA WIDE DOORS OR HALLS	60,000
BATHROOM FOR WHEELCHAIR USE	26,000
DOOR HANDLES INSTEAD OF KNOBS	26,000
FLASHING LIGHTS TO ANNOUNCE VISITORS	26,000
RAISED LETTERING OR BRAILLE	(9,000)
PUSH BARS ON DOORS	(9,000)
OTHER FEATURES	164,000

Source: Adapted from Table 3, page 15 of R.J. Struyk, The demand
         for specially adapted housing by elderly-headed
         households. Washington, D.C. Urban Institute, 1982.

door handles. Even rather expensive additions like ramps and
elevators did manage to become incorporated in some homes.

     Nonetheless, the total number of special features is relatively
low for even the most frequent (6.6% had handrails); all other aids
are so scarce as to be insignificant. We do not know which factors
may have led people to incorporate these features. Interestingly
enough, Struyk did not find that household income was associated
with their presence. When we consider that all of these
technological features have been around for a couple of decades, at
least, we get an idea of how even more slowly many fantasy
innovations are likely to become known, accepted, and adopted.

     These are only a few examples of how the utilization gap may be
reduced. For the most part, it is plain that the effort to enhance
familiarity, knowledge, and acceptability will benefit all users.
Even when the relevant measures appear to be age-specific, it is
very likely that the age-specific solution will illustrate the
payoff that results from user-sensitive design. Applying the same
principle to designing for other user groups thus leads to creative
diversity.

CONCLUSION

     Finally, the major point should be restated. The ultimate
purpose of user-friendly technology is to bring some increased

proportion of the external environment within the realm of the
person's control. One aspect of the person-environment transaction
is the person's subjective response to environmental innovations.
We must take account of pre-existing attitudes and experiences,
present channels of information transmission, the environmental
feature's acceptability,and the user's perceived competence as
necessary factors in adoption of a helpful environmental
innovation.

REFERENCES

Alpiner, J. G., 1978, "Handbook of Adult Rehabilitative Audiology,"
     Williams and Wilkins, Baltimore.
Barrett, T. R., and Wright, M., 1981, Age-related facilitation in
     recall following semantic processing, J. Geront., 36:194.
Botwinick, J., 1969, Disinclination to venture responses vs.
     cautiousness in responding: Age differences, J. Genet.
     Psychol., 115:55.
Botwinick, J. 1978, "Aging and Behavior," Second edition, Springer,
     New York.
Buczko, W., November, 1981, Utilization of property tax relief
     provision by the elderly. Paper presented at the annual
     meeting of the Gerontological Society of America, Toronto.
Byrne, D., 1964, Repression-sensitization as a dimension of
     personality, in "Progress in Experimental Personality
     Research," 1:169.
Cohen, E. S., 1982, Conclusions and recommendations from the First
     National Research Conference on Technology and Aging,
     Gerontological Society of America, Washington, DC.
Hiatt, L. G., July, 1981, Technology and chronically impaired
     elderly. Paper presented at the National Research Conference
     on Technology and Aging of the Gerontological Society of
     America, Racine, WI.
Hill, M. S., 1981, Patterns of time use, in: "Essays in the Use of
     Time Among American Households," Draft reports of work in
     progress, F. T. Juster and F. P. Stafford, eds., Institute for
     Social Research, University of Michigan.
Howell, S. C., 1972, Familiarity and complexity in perceptual
     recognition, J. Geront., 27:364.
Howell, S. C. and associates, 1982, Determinants of housing choice
     among elderly: Policy implications. Final Report, Grant
     90-AR-2116, Architecture Department, Massachusetts Institute of
     Technology, Cambridge, MA.
Ittelson, W. H., 1976, Some issues facing a theory of environment
     and behavior, in: "Environmental Psychology," second edition,
     H. M. Proshansky, W. H. Ittelson, and L. G. Rivlin, eds., Holt,
     Rinehart, and Winston, New York.
Koncelik, J., 1976, "Designing the Open Nursing Home," Dowden,
     Hutchinson and Ross, Stroudsburg, PA.

Koncelik, J. A., 1982, "Aging and the Product Environment,"
    Hutchinson and Ross, Stroudsburg, PA.
Kubey, R. W., 1980, Television and aging: Past, present and future,
    Gerontologist, 20:16.
Lawton, M. P., 1972, Assessing the competence of older people, in:
    "Research, Planning and Action for the Elderly," D. Kent, R.
    Kastenbaum and S. Sherwood, eds., Behavioral Publications, New
    York.
Lawton, M. P., 1981, Sensory deprivation and the effect of the
    environment on management of the senile dementia patient, in:
    "Clinical Studies of Alzheimers Disease and Senile Dementia,"
    N. Miller and G. Cohen, eds., Raven Press, New York.
Lawton, M. P., 1982, Competence, environmental press, and the
    adaptation of older people, in: Aging and the Environment:
    Theoretical Approaches," M. P. Lawton, P. G. Windley, and T.
    O. Byerts, eds., Springer, New York.
Lawton, M. P., in press, The dimensions of wellbeing, Exp. Aging
    Res.
Liang, J., Sengstock, M. C., and Hwalek, M. A., in press,
    Environment and criminal victimization of the aged, in:
    "Housing an Aging Society," R. J. Newcomer, M. P. Lawton, and
    T. O. Byerts, eds., Hutchinson and Ross, Stroudsburg, PA.
Longino, C. F., 1982, American retirement communities and
    residential relocation, in: "Geographical Perspectives on the
    Elderly," A. M. Warnes, ed., John Wiley, London.
Maurer, J. F. and Rupp, R. R., 1979, "Hearing and Aging: Tactics
    for Intervention, " Grune and Stratton, New York.
O'Bryant, S. L., 1982, The value of home to older persons, Res. on
    Aging, 4:349.
Pastalan, L. A., 1979, Sensory changes and environmental behavior,
    in: "The Environmental Context of Aging," T. O. Byerts, S. C.
    Howell and L. A. Pastalan, eds., Garland Publishing, New
    York.
Pastalan, L. A., 1982, Research in environment and aging: An
    alternative to theory, in: "Aging and the Environment:
    Theoretical Approaches," M. P. Lawton, P. G. Windley and T. O.
    Byerts, eds., Springer, New York.
Pelz, D. C., July, 1981, Technology transfer for the elderly:
    Building linkages between technology and gerontology. Paper
    presented at the Wingspread Conference on Technology and Aging,
    Racine, WI.
Raschko, B. B., 1982, "Housing Interiors for the Disabled and
    Elderly," Van Nostrand Reinhold, New York.
Schaie, K. W. and Parham, I. A., 1975, "Manual for the Test of
    Behavioral Rigidity," (2nd revised edition), Consulting
    Psychologists Press, Palo Alto, CA.
Sherman, E. and Newman, E. S., 1977-78, The meaning of cherished
    personal possessions for the elderly, Int. J. Aging Hum.
    Devel., 8:121.

Struyk, R. J., 1982, "The Demand for Specially Adapted Housing by Elderly-headed Households," Urban Institute, Washington.

Struyk, R. J. and Soldo, B. J., 1980, "Improving the Elderly's Housing," Ballinger, Cambridge, MA.

Walsh, D. A., Krauss, I. K., and Regnier, V. A., 1981, Spatial ability, environmental knowledge, and environmental use: The elderly, in: "Spatial Representation and Behavior Across the Life Span," A. H. Patterson, and N. Newcombe, eds., Academic Press, New York.

Weisman, J., 1981, Evaluating architectural legibility: Way-finding in the built environment, Envir. and Beh., 13, 189.

Weisman, S., 1983, Computer games for the frail elderly, The Gerontologist, 23:361.

White, R. W., 1959, Motivation reconsidered: The concept of competence, Psychol. Rev., 66:297.

Zais, J. P., Struyk, R. J., and Thibodeau, T., 1982, "Housing Policy for the Elderly Renter: Implications of the Reagan Program," Urban Institute, Washington.

Zajonc, R. B., 1968, Attitudinal effects of mere exposure, J. Pers. Soc. Psychol., Monograph Supplement, 9:1, (2, Part 2).

AGING AND TECHNOLOGICAL ADVANCES IN

TELECOMMUNICATIONS

Elias S. Cohen

WITF TV/FM, Harrisburg, Pennsylvania and
Division of Public Policy and Information Dissemination
Mid-Atlantic Long Term Care Gerontology Center
Philadelphia, Pennsylvania

There is both the temptation and the hazard for casting technology and its promise in either a heroic or a villainous role. Whether, for example, the home washing machine is a boon or a bane depends on whether you are a laundry man, laundress or homemaker. Today, more people are washing their own clothes than ever before in the twentieth century. The issue was put somewhat more elegantly in Ellul's account of the technological phenomenon in our twentieth century society (Ellul, 1964). He urged that modern technology is, or soon would be, completely autonomous, that it had its own imperatives which could only be reconciled with human values with difficulty and that it soon was, or soon would be out of effective human control. Men, he argued, had become slaves of the servant.

There is an assumption, too, born of what was a labor-intensive society where much work was physically hard, that labor-saving devices distributed more benefits that detriments, and that technology (which is a term loosely used to suggest substitution of mechanical or electrical force for human generated force) necessarily conferred benefits. This led to solutions seeking problems. And sometimes, even worse--solutions when applied, creating problems--the "technogenic" illness, as it were. And of course, newer technologies are inevitably better than old ones.

This brings me to one of the newer technologies and its applications: telecommunications, or more specifically, telecommunications applied to the service of the elderly in home and community.

But consider first John Naisbitt's caveat (1982). He
introduces an important notion under the general heading he calls
"High Tech/High Touch." Naisbitt suggests that "whenever new
technology is introduced into society, there must be a
counterbalancing human response, i.e., high touch, or the technology
is rejected. The more high tech, the more high touch." He cites
numerous examples suggesting that the high tech explosion of the
'60s was necessarily accompanied by the high touch human potential
movement (see Reich, 1970). High tech medical care, he points out,
has been accompanied by a revival of lying-in maternity units, the
development of the hospice and home care; the high tech of TV has
not replaced the high touch of movie theaters where people laugh,
cry and get scared together. Indeed, one might suggest that the
NATO Symposium on Aging and Technological Advances itself could have
been organized for world-wide participation by a satellite TV
transmission of video signals, audio return by telephone and
participation by many more people (although varying time zones might
have produced peculiar problems, no doubt). So, a word of caution
here: while the high tech of telecommunications offers the
potential of providing access to substantial data-stored systems, or
to shopping services or other links for communication, the "feel" of
it must be right as well as the logic.

A new technology, while flashy, may not add significantly to
existing technology. A video interactive scheme of placing an order
with the supermarket may not offer any significant improvement over
a straight forward telephone order system--one which I suggest
worked better 50 years ago than it does today (largely because there
were grocers and butchers who delivered foodstuffs then and there
are virtually none today). And in some instances, as I will
indicate below, reliance on technology may produce precisely the
opposite effect from that intended and may be more harmful than
helpful.

The General Scope of Telecommunications Technology as Applied to the Elderly

Telecommunications, as discussed in this paper, embraces five
general purposes or utilities using singly or in combination the
transmission of telephone signals on a point to point basis with a
high degree of selectivity by the user; radio, an audio transmission
to a broad audience of receivers (in general); and television, the
transmission of video and audio signals simultaneously. These
utilities are as follows:
1. Entertainment. Entertainment is the most common use of radio
    and television experienced by the general public.
2. General Purpose Information. Radio and television are used to
    provide news and information to consumers, citizens, members of
    groups, neighborhoods, general public and others. Frequently,

there is opportunity for conversation between points as noted
below.
3.   Entertainment/Engagement. This is a use of radio and
     television that is not common but does exist nonetheless. The
     first magazine format television program directed at an
     audience of the elderly was The Time of Our Lives produced in
     1967 in South Central Pennsylvania. Each segment of the
     program was designed to engage the viewer in some contact or
     activity beyond passive viewing. Thus, if information was
     provided about Medicare benefits, there was always an
     opportunity to write in for a Medicare booklet. If there was
     an exercise or singing segment, it was produced in a way that
     was designed to have the viewer exercise along with the
     exercise instructor or sing along with the conductor. If
     hobbies were displayed, opportunity was provided to secure
     information. But in no instance was the viewer left merely
     with the sole option of listening and nothing more.
4.   Community Participation/Community Organization. Radio and
     television have been used from time to time to bring groups
     together, to mobilize groups into activity, and urge some kind
     of community action or participation. In 1983, public
     television aired a series of programs entitled The Chemical
     People designed to produce community action in the fight
     against substance abuse by juveniles. To do this, the
     producing station, WQED (Pittsburgh) made available to the
     Public Broadcasting System materials to be utilized by
     discussion leaders where groups were organized to watch the
     program and take action. The national effort was based upon a
     local effort in Pittsburgh undertaken a year earlier. Similar
     activities have been undertaken around such simple issues as
     collecting money for UNICEF and, in some instances, to deal
     with crises such as flood, racial disturbance, or the Three
     Mile Island accident.
5.   Information/Training. Radio and (more particularly) television
     have been utilized to produce teleconferences for professional
     groups, notably attorneys, and industrial organizations. This
     is accomplished via scrambled signals to discrete locations, or
     through some combination of broadcast and telephone
     communication.

     The following are some definitions which may be useful in
understanding the discussion below:

     "Broadcasting" is the sending of a signal (radio and/or video)
over the air waves to receivers (radio and television sets) equipped
to receive the signals being broadcast.

     "Narrow Casting" is a term which encompasses a variety of
methods and systems for transmitting a signal to a limited group of
potential receivers. The most common form of narrow casting is
cable television, where signals are transmitted only to receivers

connected by a cable to a special transmitter.

"Low Power Television" is a form of broadcasting using low
wattage transmitters which send out a signal over 10 or 15 miles
depending upon topography. Low power television can be used to
transmit locally produced programs of strictly local interest to
receivers within its range, or to bring to a rural area not served
by cable, and/or blocked by physical interference, signals from a
distance received on the low powered station's satellite receiver
and then sent out over its antenna.

"Cable Television" is a narrow cast system of distributing one
or more television signals over a wired system to individual
receivers in a given geographical area. If a cable system carries
two cables, so that signals can flow to and from receivers, it is
possible to secure reactions and communications with individuals who
are hooked up to the cable system. An example of this system,
currently in use in Columbus, Ohio and some other communities, is
called Qube.

"Microwave" transmission provides a means of delivering a
highly directional signal between two points on a line-of-sight
basis. This technique aims a signal from a transmitter at a
receiving antenna. It can be used to connect a remote site to a
central location for rebroadcast by another means (e.g., broadcast,
ITFS, cable), and to connect cable company antenna sites.

"Translators" are devices that receive a signal on one
frequency and convert it for transmission on another frequency.
Thus, one station broadcasts on a particular channel and a
translator in another city receives the signal and treats it so as
to broadcast it in its area on another channel.

"Interactive Telecommunications" is a term to describe two-way
electronic communications. This can mean transmission of a video
signal to a receiver with a return electron "yes-no" impulse (the
QUBE system); a video transmission to a receive site with an audio
return (via a separate cable or radio link, or via leased telephone
lines); or a two-way video system where those at remote receivers
can also be seen by those at the transmitting source.

Most familiar to the general public are the variety of uses of
radio. These include not only one-way broadcasting, but the ever
popular radio broadcast (down stream) and the telephone return (up
stream) for the call-in show. Here, listeners get to hear both
sides of the conversation between the announcer or commentator and
the person calling in. Citizen band radio involves broadcast on low
wattage "stations" where the conversation may be heard by anyone
tuned in to the proper frequency.

The combination of video/audio possibilities is less well known. The variations may include video down-stream with audio (via telephone) upstream (the television call-in show); video interconnect (e.g., two studios that have visual and audio two-way communication, with or without audio (i.e., call-in) capability, upstream; video down-stream and data transmission upstream via a digital system with limited bit capacity (e.g., QUBE); and video downstream to a live trainer who interacts with trainees (a mode contemplated for ITFS).

Some Experience to Date

The variety of options noted above suggests that it is possible to tailor the use of telecommunications to the particular needs and desires of the elderly. To some extent, that has been done. Some adaptations, like call-in radio shows, are ubiquitous.

Apart from the telephone, the advent and development of cable television is the single most important innovation in placing telecommunications at the service of the elderly in recent years. Requirements imposed upon cable operators to provide public access channels, (Real, Anderson and Harrington, 1980) and the recognition of the potential impact of cable TV in the late 1960s and early 1970s led to the National Science Foundation experiments and to a variety of community efforts to serve the elderly (Housing the Elderly Report, 1983).

A thorough review and history of the early background and literature on social services and cable TV may be found in the 1976 report of the Cable Television Information Center published by the National Science Foundation (Social Services, 1976). That report lists 48 different projects under the general rubric of "social services" including 4 specifically targeted on the elderly, 19 on the general public and 5 on other groups which might include the elderly such as the deaf, divorcees and undereducated adults. Of the total group of 48 projects, 19 were discontinued or suspended, including 2 of the 4 addressed to the elderly. The report concluded that cable operators are unlikely to develop these services in significant ways for a number of reasons.

First, there is a lack of incentive to do so. There is little evidence that provision of such services attracts or retains subscribers, builds audience or in other ways justifies the heavy investment required. Furthermore, such service competes with demands for investment, expansion, marketing, programming and equipment.

Second, municipalities are hindered by economic factors. There is little hard evidence that resources allocated to cable services for service delivery are better used, more efficient, or more

effective than resources applied to conventional means of service
delivery. Some examples cited were the failure of TV monitoring of
public places as opposed to monitoring by live persons.

Third, experimentation has been haphazard, and therefore
frequently not fruitful. Some experiments have simply grown out of
existing systems, often relying on volunteers or a single interested
person. Such projects are not evaluated. Some projects are almost
random out-growths of a government or foundation grant without lead
time to design evaluation materials, and so forth. A small number
of experiments, funded by The National Science Foundation, were
carefully planned and executed, and those are discussed below.

Fourth, the entire past history and current effort continue to
be infected with the unrealistic expectations that the technology,
because of its ability to distribute information and data, because
of its potential availability to all and because of its potentials
for memory and ultimately recording through hook-ups to computers,
would be able to resolve a myriad of social problems even if
technology could not solve such simple problems of laundering
clothes. By 1975, it was becoming apparent that the predicted
"wired city," "checkless society" and the "interactive community"
simply were not going to happen.

It was in the mid-seventies that the National Science
Foundation undertook to finance a number of experiments in
interactive television, one of which was the New York
University/Berks Community TV consortium experiment in Reading,
Pennsylvania. That experiment is generally acknowledged as a
success and much can be learned from it.

The project had four objectives: (1) establish three
neighborhood communication centers which would be interconnected by
cable TV; (2) train a staff of citizens to operate the system; (3)
involve senior citizens and public agencies in the system; and (4)
provide two-way public service programming through which senior
citizens could communicate with each other and with service delivery
organization.

Reading is a city of almost 90,000 persons, 60 miles northwest
of Philadelphia; Berks County, in which it is situated, has almost
300,000 people. The population 65 and over is 16 percent of the
total of the county. Initially, only 117 elderly could receive the
programs. Subsequently, the programs were carried to all 35,000
subscribers to the local cable system. City hall and some schools
were also hooked into the inter-connect. Programming was determined
by senior citizens and was carried two hours a day. The centers
could interact with other locations, and senior citizens could call
in to the broadcast centers by telephone.

The senior centers were established in different settings; a three story senior citizen center, a 150-unit high rise senior citizen housing project, and another housing authority facility serving two housing projects.

The experimental service was evaluated (Moss, 1978) and found to have some effect in three general areas: (1) knowledge of public services; (2) involvement in political processes; (3) participation in social and community activities.

Keep in mind that senior citizens themselves set programming priorities. Thus, when gaps in public services were identified, the TV resource became a catalyst in producing service from non-public sources. For example, the expressed need for having help with writing wills produced pro bono service from the Berks County Bar Association when Community Legal Services was prohibited from offering such service.

Similarly, information about nursing homes and funeral parlors was publicized. Peer group counseling emerged on issues such as · sex, insomnia, and when to stop driving a car. Thus, the diversity of interests diverted attention from the question of whether the two-way cable system increased use of specific public services, and there were no conclusive results on this score. Also, while awareness of public services may change, utilization may not. Programs on food stamps, for example, produce significant increase in the numbers considering the use of food stamps but very little change in actual utilization.

Political activity produced perhaps the most exciting results. The two-way discussion between City Hall and the three communication centers began with a monthly "Meet The Mayor" program and led to a weekly program with a member of City Council appearing each time. This then expanded to a television call-in program through which individuals receiving the program in their homes over the cable system could query council members in the studio. Without citing the evaluative measures utilized, suffice it to say that participating seniors felt a greater sense of involvement in the political process. Finally, measures of participation by senior citizens in community activities, education programs, and senior center activities showed increases among viewers as opposed to non-viewers.

An overall assessment of the experiment may be seen from the response of participants in the experimental phase of the project to the question whether anything had been done to improve their lives during the experimental year. Approximately 25 percent of attendees at the wired community centers and 80 percent of the home viewers identified the interactive cable TV as the major improvement in their lives during the year.

The system has expanded and is now community supported. It has both daytime and prime time programs. A microwave interconnect permits participation from two towns, Kutztown and Hamburg, about 25 miles away. Penn State University uses the system for adult education.

Programming between young people in schools and seniors occurs regularly and origination sites are rotated through use of a mobile unit. The communication between young and old people has been most effective. The use of the interconnect where the young people can see the older individuals who are appearing on the screen and vice versa (a split screen technique) produces the right mix of direct communication and remoteness to provide the feeling of immediate conversation without the potential embarrassment of proximity when questions like sex are being discussed, or other issues arise that are sometimes deemed relatively personal.

The lessons from Reading suggest the following elements in developing later active television and the elderly: (1) the key role of senior citizens as initiators of the program; (2) use of neighborhood facilities as origination sites; (3) aggregation of organizations to generate a diversity of programs; and (4) programming to serve diverse groups and subgroups.

Having said that, let me offer an anecdote which suggests a caveat. Among the programs offered by Berks Community Television is a "sing-along" which connects one of the centers with the Wernersville State Mental Hospital. The people participating at Wernersville can see the conductor and the conductor can see the patients at Wernersville simply by looking at the screen. Viewers see both, either by virtue of a split screen image or alternately. The conductor can be chatty and recognize individual participants at the hospital who are singing along with the attendees at the communication center. This is intended to bring the patients at Wernersville Hospital into the community orbit. In fact, it is reasonable to suggest that it does the opposite. Bringing the patients from Wernersville physically to the communications center where they could be directly in contact with people in the community in the sing-along would be the way of producing a real connection to the community. The use of the interactive connection permits the appearance of connection to the community while maintaining the mental patients in a remote location where one need have little contact with them. I suggest that here interactive television produces precisely the opposite result from that intended and, in fact, may assist in maintaining the mentally ill as pariahs distant from the community rather than bringing them closer in every sense of the word.

What does one conclude about applying the uses of interactive cable TV to the benefit of the elderly? First, interactive cable

TV, using a digital return, has little marginal utility over the telephone (for ordering goods or services) or print material (e.g., catalogs), little utility over and above one-way television, and none over conventional classes for teaching or training (Lucas, Heald and Bazemore, 1979).

As for interactive video with either audio or audio-video return, I suggest there is probably high utility in the following arenas: (1) political "meetings" or "programs"; (2) intergroup meetings; (3) activities for senior citizen groups for producing local programming, i.e., using the gimmickry and access channel to produce both one-way and two-way TV; (4) producing interesting call-in shows; and (5) interconnecting nursing homes, hospitals and other institutions to join patients and residents and staff among facilities for a wide variety of purposes. However, whether this is more appropriate for cable or for ITFS is difficult to say.

Note that every one of these instances suggests the utility of the technology when supplemented by a person--an interlocutor, if you will--so as to produce the possibility of discussion or something happening beyond the program.

That brings me to a discussion of applications of telecommunications technology in terms that include systems technology and the technology of the social sciences as well as the technology of electronic communications. The remainder of this paper is addressed to what some believe is the third wave of public broadcasting--the development of the community network.

Telecommunications' Third Wave

The new technology has launched public television's "Third Wave." It is the logical outcome of public television's evolution. The first wave of educational programs was heavily weighted toward classroom presentations beamed out over the airwaves with teachers standing at lecterns and writing on blackboards, or panels discussing interesting topics.

The second wave is the one with which we are most familiar: the fine documentaries, the superb presentations of drama; music; dance; the special, informative, in-depth news reporting and weekly financial, political and legislative news updates; and the intelligent interview shows that are so enriching. The third wave that lies ahead is, like Toffler's "Third Wave," grounded in technology.

The "Third Wave" puts telecommunications technology at the disposal of traditional and nontraditional community groups and organizations, educational institutions of every level, health care institutions and agencies, governmental agencies at state and local levels, and cultural groups, so that they can communicate between and among each other, conduct training and continuing education programs effectively and efficiently, hold conferences, develop cultural events and affairs, and enhance community organization and democratic processes, all through the new devices of narrowcasting (as opposed to broadcasting), low power television, hub configurations, microwave relays, and a host of other telecommunications techniques and devices. The new approach combines telecommunications and community organization in ways that permit the first significant change in methods of organizing groups in the community for the benefit of the community that have occurred in the last fifty years.

The implications for the aging, the aged, planners and practitioners are striking--particularly when considering links with health care facilities, senior centers, senior housing, and adult education centers. While the equipment and hardware are now common in telecommunications (if not commonplace), the combination of technology, community organization and social, health and cultural organizations joined in a common enterprise is not.

The Corporation for Public Broadcasting has provided funds to the South Central Pennsylvania Educational Broadcasting Council[1] to undertake the first stage of planning for the third wave. The first step by WITF staff was to assemble a consortium of community health, educational, welfare, social service, legislative, public and private agencies to ascertain uses to which the new technology might be put, the technological requirements and new opportunities the technology might present.

Staff of Senate and House Committees on Public Health and Welfare from the Pennsylvania General Assembly suggested the conduct of open committee meetings on television, with audio response to committees through leased telephone lines, from legislators, administration officials or others viewing the program from home or special viewing locations. Interactive interconnects between studio locations would enhance exchange of views. Recognizing the complexity of much of today's legislation and rulemaking they also suggested narrowcast applications for both text and training program transmission in social service and Medicaid programming. The applications to and for the elderly are extensive. But more importantly, they will demonstrate where interests of older people and others coincide.

Educational institutions at both public instruction and higher education levels foresaw in-service training applications, shared coursework, shared concerts, plays and public lectures made available via ITFS, cable or other narrowcast means, teleconferencing, preparation and distribution of videotaped material, unified library catalog systems, utilization of existing

cable systems, and interactive programming similar to the Berks
Community TV experience described above. For the elderly, the
applications for late life learning and utilization of community
services are apparent.

The Hospital Council of Central Pennsylvania suggested the use
of televised in-service training and other teleconferences for
staffs of approximatly 40-member hospitals. Those needs were
projected as great enough to consume 24 hours per day of some form
of narrowcast service. In-service training could be provided by a
single trainer to various kinds of staffs of several hospitals
simultaneously. Possibilities also exist for improving patient
education information through the production of patient education
tapes which could be broadcast within hospitals or called up on
demand over a closed circuit hospital channel. Given hospital
censuses that average 40 percent of patients 65 and over, such
telecommunication service for providers clearly has impact on the
elderly. Similar applications have been suggested for long term
care facilities.

The development of a system that might undertake the activities
suggested above requires innovative organizational development and
staffing; enhanced technological capacity, i.e., the provision of
necessary hardware; marketing the device and training potential
users; and community organization and community development
activities with special interest groups which currently do not have
adequate capacity for developing organizational power.

For an 11-county area like South Central Pennsylvania
comprising approximately 1,850 square miles, the technological
requirements include the following: (1) ITFS transmitter origination
equipment for four channels, with repeaters to permit hub
configurations at major urban centers in the area with microwave
return capability (to permit interactivity); (2) low power broadcast
transmitters at three urban locations; (3) microwave interconnects
with cable systems in major cable service areas; and (4) programming
capability from all sites.

Provision of full four-channel capability is indicated since
the assessment of potential market indicates strong utilization
potential with a high probability of rapid acceleration.
Utilization for hospitals and educational institutions alone have
been estimated at about two full-time channels leaving two and
one-third channels for other institutional, commercial, and national
users. Given the estimates of the use by and interest of elderly,
supported by the experience of Berks Community TV, the importance to
the elderly of new approaches to technological applications of
telecommunications is enormous. Provision of four channels with
return interactive capability reflects the institutional assessment
finding that there is a high degree of interest in teleconferencing

for program and service delivery. The fact that institutions from
the entire region could participate in teleconferences from
conveniently located sites would increase the likelihood of use by
institutions, local and state governments, and commercial users.
This, in turn, would expand the revenue-generation capability of the
system.

This projection is well within reach of telecommunication
planners and providers. It requires relatively little investment of
dollars by potential organizational users. Conceptually, it
requires considerable shifts in notions about the uses of audio and
video, the nature of community organization (in its social work
sense), the techniques of community organization, and the
relationship between the new capabilities of telecommunications and
the needs of organizations, institutions and agencies at the
community level. Applied to the elderly, the possibilities are
especially exciting. But perhaps most importantly it requires some
understanding of the limitations of the new technology lest we
repeat the errors of excessive promise that accompanied the advent
of cable TV.

Conclusion

This paper ends as it began: with a plea to balance the
promise, dazzle and glitz of the new technology with an
understanding of both the old technology and the needs, demands and
capacities of users. While I do not necessarily adopt an Ellulian
approach, I have considerable respect for the knowledge, experience
and wisdom of the elderly, operating both individually and as a
"market" to identify and select out those technologies which work
and those which do not.

The new telecommunications technology offers enormous promise.
The challenge is to place that promise at the service of the elderly
in accordance with the needs and demands of the elderly rather than
in response to technological experimentation.

NOTE

[1] The corporate parent of WITF-TV and WITF-FM, located in
Harrisburg, Pennsylvania.

REFERENCES

Ellul, Jacques, The Technological Society, Knopf, 1964.
Housing the Elderly Report, "2-way cable TV won't be for years but
     elderly are already exploiting it," CD Publications, August,
     1983.

Lucas, W.A., Heald, K.A., and Bazemore, J. The Spartanburg
    Interactive, Cable Experiments in Home Education, Rand Corp.,
    Santa Monica, CA, 1979.

Moss, M. "Two-Way Cable Television: An evaluation of community uses
    in Reading, PA." Final Report to NSF, Summary, NYU-Reading
    Consortium NYU, Alternate Medical Center/School of Arts,
    Graduate School of Public Administration, 1978.

Naisbitt, John, Megatrends: Ten New Directions Transforming Our
    Lives, Warner Pub., 1982.

Social Services and Cable TV, Report Prepared for: National Science
    Foundation, Government Printing Office, Washington, D.C.,
    1976.

Real, M.R., Anderson, H.L., and Harrington, M.H., "Television Access
    for Older Adults," J. of Communication, V. 30, No. 1, Winter,
    1980.

Reich, Charles, The Greening of America. Random House, 1970.

TECHNOLOGY IN THE SERVICE

OF THE AGING VETERAN

Paul A. L. Haber

Western Region, U.S. Veterans Administration
Palo Alto, California

INTRODUCTION

In this paper I share a view of the promise of technology in the service of the aging veteran which, although it may be parochial, has had the virtue of some testing and some actual deployment. I will further discuss a basic division of effort in technology in the service of the aging into two major components: health care technology and quality of life technology. It is important to recognize this distinction because the two different subgroups of technology tend to develop in isolated fashion when interchange between them might be mutually beneficial.

In order to show how the Veterans Administration is involved in this program I will sketch out the demography of the VA. There are something less than 30 million living veterans in this country and the one phenomenon that is most arresting is that this group is rapidly aging. In a very real sense we are hip deep in what the rest of the country is ankle deep in. There are 12 million veterans of World War II whose average age is currently 63. By the year 1995 three out of four American males age 65 and over will be veterans.

To care for this growing need on the part of veterans we have a system which is budgeted at about $8 billion per year. We operate 172 hospitals, 102 nursing homes, 255 outpatient clinics, 16 domiciliaries and a number of other ancillary programs including hospice, adult day care, residential care and state home programs.

TECHNOLOGY

As I mentioned at the beginning I believe that it is convenient
to consider technology in two separate divisions and I will first
briefly discuss health care technology. Health care technology is
that division of technologic advances which relate specifically to
the provision of health care. This would encompass diagnostic
devices, diagnostic procedures, therapeutic devices, therapeutic
procedures, communications technology in health care, drugs, and
health care research. An example of diagnostic devices would be the
now widely accepted and used CT Scanner. Diagnostic procedures
would include such items as radio-isotope scanning. Therapeutic
devices might be exemplified by the use of cord stimulators to treat
intractable pain. Therapeutic procedures could be exemplified by
the use of percutaneous coronary angioplasty. Communications
technology would be the use of those devices and computers to
quantify and communicate large masses of analytic procedures. Drug
technology needs no further explanation. Health services research
is the search for more cost effective ways of caring for older
people, particularly the use of non-hospital environments.

The Veterans Administration has made an admirable beginning in
attempting to come to grips with the use of health care technology.
In a series of programs called Special Medical Programs we were able
to apprehend, test, validate and deploy with subsequent reevaluation
a whole host of new technologies. This methodology has served us
well and was used in the deployment of such new technologies as open
heart surgery, renal dialysis, and the CT scanner. In this
methodology we would justify a priori to the Office of Management
and Budget and the Congress the particular thrust of the new
technology and we would in sequential fashion acquire numbers of
these devices or procedures while monitoring their cost
effectiveness, their utility, their patient acceptance, the dangers
attendant upon their use, and the maintenance costs and
requirements.

The second division of technological implication is that
involving the quality of life. In this area my basic plea is for a
more concerted logical approach, understanding that the hazards of
the marketplace and the opportunities for capital return must
dictate the pace at which new developments are marketed. I feel
that there needs to be some effort made by the gerontological
community to bring some order into what is an otherwise chaotic
situation.

I would like to consider two separate sections under
application of technology in the section dealing with the quality of
life. These two areas are first, application to interior design,
including technology development in the activities of daily living
and second, environmental concerns.

In dealing with these issues we must understand the problems of compensating for the disabilities of aging. Several approaches might be used. Obviously there are physiological decrements in every organ system even if the process goes on at different rates. In terms of implications for design for both institutional and non-institutional living only five systems call attention to themselves. They are the visual system, the auditory system, the proprioceptive system, the musculoskeletal system and the central nervous system. I will deal briefly with the description of the physiological declines in each of these systems and the importance in implications of design will flow from that. Although there are comparable decreases in function of the cardiovascular system, the gastrointestinal system, the genital urinary system, the hemotologic system, the endocrine system and so on for each of the component physiological systems, their impact on design is clearly less significant.

Data gathered by the U.S. Department of Health and Human Services indicate that the chronic condition causing limitation in those age 65 years and older that is greatest is arthritis and rheumatism accounting for 1.691 million persons; heart conditions were second accounting for 1.628 million, visual impairments accounting for .554 million and hypertension without heart involvement .510 million (USDHEW, 1976). Impairments of the lower extremities and hips came next accounting for .423 million persons. Other data on the most common conditions for which worker disability allowances were granted in 1973, by age, indicates that the incidence of chronic ischemic heart disease increased from 22.3% to 26.9% going from age groups 50 to 54 up to 60 years and over. The incidence of osteoarthritis increased from 4.9% of the total in the group under 54 years of age to 9.2% in the group age 60 years and older. All these data tell us what we already know: that the incidence of physical limitation increases with age and tends to be very prevalent and the most common cause of disability in older persons.

Hiatt (1980) has pointed out that all of these disabilities do have a pronounced effect on the interior design of the environment. The Veterans Administration has had a pool of experience now, having built over 100 nursing homes with these considerations in mind.

It is necessary for me to digress briefly to quickly cover material relating the perceptual and functional decreases in the elderly. The first of these are decrements in visual perception. The problem is complex because many of these visual difficulties have intricate interrelationships. These are discussed in the Handbook of the Psychology of Aging (Birren and Schaie, 1978). There are physiological changes which cause a loss of elasticity of the lens, decrease its transparency and produce yellowish coloration. The ciliary muscle and bodies lose some of their adaptive capability. The aqueous and vitreous humours are also changed by decrements in the refractive capability and the retina is heavily impacted by the process of aging. Not only are there physical and anatomical changes but there are also many adaptive changes so that the length of time necessary for information to be presented to the aging eye, in order for it to be correctly transmitted to the retina and interpreted by the brain, increases dramatically. It turns out that approximately twice as much time in millisecond is required for processing of visual information in the older person as against the younger person. We find that macular disease causes about 45% of the eye problems encountered in the aged patients. Cataracts, on the other hand, account for about 33% of such visual defects. Vision losses for all purposes including macular degeneration, diabetes melina, cataracts, glaucoma and retinal disease all increase rapidly after the age of 70. Peripheral vision tends to be lost in stepwise fashion by the elderly and when this occurs it results in tunnel vision. This tunnel vision presents a severe handicap when the individual is trying to orient himself in space and since side vision is critical in perceiving motion. The loss of peripheral vision represents a greater hazard for the older person because it makes it difficult for him to detect early warning signs of oncoming vehicles or other individuals. On the other hand, macular disease tends to affect the central visual area of the eye and this one loss also poses difficulty in discriminating details, fine print, faces and so on.

It thus appears that there are at least three sets of considerations which control visual perception and recognition. There are first those which relate to the actual physiological and physiopsychological effects of vision. Secondly there are a whole series of considerations which deal with orientation, pattern recognition and understanding of visual information and finally there are all the aesthetic considerations involved. To my mind no single unified theory has been developed which embraces all three sets of considerations. A general rule might be made that the more illumination provided the better, however we must recognize the susceptability of the older eye to glare and to the confusion and disorientation that glare creates. There seems to be a trade off between the amount of glare and the amount of illumination. Shiny surfaces ought to be avoided wherever possible. Single point sources of light or sources of light which cause strong shadows for long distances ought to be avoided. The usual practice in nursing homes seems to be to have a large picture window gracing the end of any long corridor. While this might seem to be aesthetically pleasing and certainly increases the amount of illumination it also produces a situation in which the elderly patient proceeding down the hall is immediately faced with a large glare surface.

A few general rules about color in residences and institutions housing older people are appropriate. Color can be very effective in helping to orient older people, marking junctions of different functional spaces in producing an appropriate mode and in providing interest and variety of an indeterminate nature. We have had disappointing experience with color coding, particularly when colors of the same intensity and brightness are involved. Blues, greens and blue-greens are difficult to distinguish. So also are darker colors, deep reds, brown, grays, and blacks. Color perception diminishes with age due in part to the yellowing of the lens. Very light colors, pinks, salmon, light yellows, pale greens, off whites, and light blues are also difficult to distinguish and should not be used with color coding as the objective. The use of darker colors for doors and door frames as opposed to the light colors for walls does serve to mark entrances to rooms and facilities. In the final analysis it is enormously individualized affairs that one deals with in choosing colors and only the general rules described above can be used without attention to individual preference.

The second sensory modality which is decreased with aging is that of hearing and a very brief overview of the physiological decrements is in order. As with other sensory modalities and indeed all functional processes of the aging, it is important to distinguish between pathological decreases due to disease and the decrements found with increasing age that are more "normal" in perspective. Hearing, as directed particularly towards speech, consists essentially of three steps: 1) conduction of the sound through the external and middle ear; 2) conversion from physical sound waves to bioelectrical signals by elements of the inner ear; and 3) transmission of these signals to the brain through the auditory nerve. Disorders of the first step are generally accorded to be conductive hearing defects and this is amenable to amplification through the use of hearing aids. Those defects caused by dysfunction in the inner ear of the auditory nerve are now termed sensorineural hearing losses. Disorders of the first step are generally accorded to be conductive hearing defects and these are amenable to amplification through the use of hearing aids. Conductive losses may be due to loss of air conduction or both air and bone conduction. The audiogram is helpful in distinguishing between these. Loss of flexibility of the typmanic membrane while common in the aging does not usually cause major losses of hearing. On the other hand, otosclerosis which causes fixation of the foot plate of the stapes in the oval window, while beginning in youth, frequently continues throughout middle and late life. When complicated by other hearing difficulties this represents a major problem. The most common affection of the inner ear are the sensorineural changes manifested in presbycusis. This is a progressive disorder which, although it may begin early in life, reaches a plateau and then may remain stable for long periods of this time. First to be affected are the hearing for high pitched tones; then middle frequencies are lost and finally the lowest frequencies are involved. Since human speech is composed of a range

of high and low frequencies the imbalance and the loss of hearing
for various frequencies results in distorted perception of speech.
When these natural decrements are further complicated by competing
sounds it requires intense concentration for the older person to
distinguish what is being said to him. Weston (1964) has reported
that some hearing loss is almost universal among elderly people
although this may not be always new to them.

It should be understood that hearing aids can sometimes be of
help in sensorineural deafness although the greatest use of such
hearing aids is in conductive deafness situations. It is important
to remember that a hearing aid cannot make correction for deficit
hearing in the same way that eye glasses do for deficits in vision,
but rather it can only serve to amplify sounds.

The design implications of hearing deficits are several. One
of the most important is that background noise, to the extent
possible, should be countered by the use of appropriate surfaces.
Acoustical tile does not always work as advertised, but frequently
the use of wall hangings, draperies and fabric surfaces will tend to
diminish the background noise. This is particularly important in
dining areas. Dining areas are usually designed with maximum
attention being paid to cleanliness and cheerfulness which often
comes to mean "hard" surfaces which are highly reflective of light
and sound. The difficulty in understanding spoken speech on the
part of the older person is complicated by the fact that several
conversations may be continuing at once. If one adds to this the
clatter of dishes and flatware in a dining area the situation may
well become impossible.

Of more than passing importance is the role of closed caption
hearing and television programs, particularly news programs. These
are on the increase and in the nursing home environment, or even in
the individual's own home, the caption hearing programs do serve as
a stress free communications device for the hearing impaired
elderly. Innovative ways should be found to use these as subjects
for group discussion in group living situations and as an
opportunity for the hearing impaired elderly to participate in
community activities to a much greater extent.

The third disability of importance to us in aging is the
influence of musculoskeletal disorders, arthritides, and muscle,
bone, and joint problems. We can get some idea of the relative
frequency of musculoskeletal impairments from the general population
by consideration of data presented by Kelsey, Pastides, and Bisbee
in their monograph "Musculoskeletal Disorders - Their Frequency of
Occurance and Their Impact on the Population of the United States."
Musculoskeletal disorders heads the list of disabilities among
person referred for vocational rehabilitation, representing 41% of
all such referrals.

If we look at data on the average number of persons with
limitation of activity by selected chronic conditions causing
limitation by ecologic and economic impacts of disabilities in the
aging, it should be noted that as of 1978 about 20 million people in
the United States had musculoskeletal impairments and that the total
annual economic costs attributable to these conditions was estimated
to be about $20 billion per year.

The majority of the population in the United States is at some
time affected by back pain which is the most frequent cause of
activity limitation in younger persons. Arthritis is almost
universal in the elderly and, when all age groups are combined, is
the leading cause of mobility limitation and the second leading
cause of activity limitation in the United States. Osteoporosis
occurs to varying degrees in most elderly women and substantially
increases the risk for fractures in the elderly. Among women of 75
years of age and older almost 90% have x-ray evidence of
osteoporosis. It is a truism that certain types of musculoskeletal
trauma such as hip fractures in the elderly are associated with long
recovery periods. Over half the women age 55 years and older with
hip fractures have not regained their ability to carry on their
usual activities within six months after the injury has taken
place. It is estimated that musculoskeletal conditions account for
about 20% of Medicare hospitalization costs of over $1 billion.

If we look at a list of the estimated number of persons in the
United States in 1971 with limitations of activity attributable to
specific conditions, musculoskeletal system disabilities heads the
list; about 10 1/2 million persons are so afflicted. The design
implications of accomodating physically handicapped patients with
musculoskeletal disabilities is basically one of access. The design
and the furniture of the institution and private home must permit
ready access of the patient to the area. There are obvious
inferences with respect to the door and hallway widths and surfaces
which must be dealt with.

A codicil note should be added to the descriptions of the
disabilities resulting from musculoskeletal deficits and this is the
loss of sense of balance in older patients. The loss of the sense
of balance is very complex but can be due to the loss of
proprioceptive sensory nerves. The extremely long axons which
convey proprioceptive impulses from the soles of the feet to the
brain are nourished by cell bodies in the central nervous system and
they appear to lose some of their ability to provide sustenance to
these very long neurons.

It is estimated that virtually 80% of people over the age of 70
have diminished proprioceptive sense. This then means that they are
more likely to have impaired input for balance and therefore are
more likely to fall. When one combines this with the previously

mentioned musculoskeletal disabilities, it is clear that patients may sustain serious falls and fractures due to a combination of these disabilities. One way of compensating for the proprioceptive defect is to enhance other sensory modalities particularly visual ones, and again bright illumination of a non-glare type is a useful antidote to the loss of balance.

The fourth disability which we must mention is that of dementia. The incidence of dementia of Alzheimer's Disease has received a great deal of popular attention. It is estimated that 5% of the population of 65 suffers from some degree of dementia, mild, moderate or severe but that the incidence goes up precipitously so that by the age of 80 approximately 20% of the population is so afflicted. We are talking now about surveys of patients or people living in the community. If one looks at the institutionalized population the percentage of demented patients is much higher, running as high as 50-60% in some surveys. It is clear from various kinds of surveys that most of the patients with dementia do not end up in facilities of an institutional nature but are kept at home. Therefore it behooves us to try to care for the demented patient. The symptoms of the dementia vary with the course of the disease but almost invariably they include, before the final stage, characteristic losses in cognitive ability, disorientation for time, space and person and emotional liability including irascibility, temper tantrums and violent outbursts. Some patients exhibit docile, almost submissive attitudes at one stage of the disease and can be led or placed in position where they are not likely to move for hours at a time.

The design implications are pretty obvious. In the first place the design should be such that patients can familiarize themselves with a particular locus so as to prevent confusion. Again the use of color may be advantageous in trying to help orient the patient. Such obvious clues as permanent displays of date, time, place, etc. as part of the reality orientation can be a very important part in helping reassure the patient as to his orientation.

Application of Technology to Environmental Design

When the VA Nursing Home Program was first begun in 1964, we were able to afford the luxury of determining the kinds of patients we could care for. This is unique because no other VA program has the leeway in determining the mix of patients. We, therefore, tended to get an even bell shaped distribution of patients so as to provide a maximum of care with the staffing that we were permitted. The VA categorizes patients according to need for nursing care, one through four; one is most in need and four least in need. Since the VA staffing was roughly 100 staff for one hundred patients (an obviously 1 to 1 ratio), and since 85 percent of this was in nursing staff, we felt that we could afford an even distribution of patients through the four categories. Thus it was that categories one and four each had 5 percent, and categories two and three had 45%.[1] This ideal distribution obviously didn't last very long because we found that as time went on we were accumulating more and more seriously ill patients requiring a lot of nursing home care.

We therefore felt that it was important to establish certain parameters of the physical environment based upon the characteristics of our system. One of the most important of these was a climate of expectancy. We made a conscious endeavor to create in the VA a variant of the prevailing doctrine about nursing homes in 1964 that no one left a nursing home alive, and this was highlighted by our development of the idea of a climate of expectancy. This meant that some patients could look forward to discharge to their homes or to other levels of care which were less intensive than nursing home care. In order to achieve this climate of expectancy, certain conditions were required.

First of all, we were concerned with the element of privacy. This became so important to us that we had to depart from the usual four to eight to twelve ward bedrooms of the hospital in order to create a mix of beds, and at first we started out with a mix of 50% multiple bedroom, that is to say three or four beds in a room and 50% singles and doubles. Although it is now commonly accepted, it was very unusual for the VA at that time and was deemed a great luxury. Not only were we interested in the privacy of the bedrooms, but there was some attempt to install individual bathrooms in some of these bedrooms.

A second principle under the climate of expectancy was to preserve the individual's identity. In order to do this we felt that the individual should retain as much as possible of his former life in the nursing home. One specific objective was to keep the individual in a position where he could use his own clothes rather than the anonymous hospital gowns and bathrobes that are so frequently found in VA settings. In order to do this, we had to build in more room for the storing of personal clothing and valuables, but it was important that we keep the patients in their own clothes. This, of course, presented a problem of cleanliness and providing laundries, and in some of our VA nursing homes, as a matter of fact, we have washing machines and dryers so that patients, at least the more ambulatory ones, can do their own laundry.

A third principle was to preserve the life-style of the individual. This meant the insistence upon more flexible dining hours than is generally true in the hospital, although even here we were constrained by staffing problems, etc. It also meant that there should be opportunities for recreation which came as close as possible to the individual's prior life-style.

A second requirement of our nursing home was that, to the greatest extent possible we want our patients to continue to be ambulatory. One of the ways in which this was accomplished was by the creation of dining rooms. The tendency in institutions to feed patients in bed may be a desirable one in general in that it tends to prevent unnecessary movement of patients and because of the saving of space in the form of communal dining rooms. In the nursing home, however, we felt that if we had dining rooms, patients would be more likely to ambulate to them or to be brought to them in wheelchairs than if it were not the case. It was important for us, therefore, to have the dining room located immediately within the nursing home care unit. This was extremely difficult for us because in many VA hospitals with a multistructured physical plant, there are already several dining rooms, as for instance in some psychiatric hospitals where the Hahn-type hospital was prevalent in the 50's and 60's. For us to require an additional dining room for the nursing home, therefore, represented an enormous expenditure. We, of course, use the dining room for other purposes -- particularly for recreation and group meetings when it is not in use for meals. Yet another characteristic of ambulation was that we wanted patients to have access to the out-of-doors. Although many of our hospitals are located in severe climates, it was felt important to permit the patients to have easy egress from the building and therefore we wanted to have either one-story construction wherever possible, or to have the lowest floor if it were a multi-story structure.

However, the requirements for the nursing home care unit we have developed are subject to change. Two forces are at work requiring the change. In the first place we see that a much larger proportion of patients are disabled than we had originally expected to care for. While fifty percent of the patients were in categories one and two, that is to say requiring a great deal of nursing care and just slightly less, the press of time has forced us to accept a higher percentage of patients in these two most disabled categories so that now there are seventy percent of these patients. This means that there are more nursing time needs and the patients are more medically ill than was formerly the case. Greater thought has to be given to further treatment of a medical nature in the nursing home care unit.

The second effect has been that, as geriatric medicine has become more popular, increasing numbers of trained physicians in geriatrics are available. This was not the case when the VA Nursing Home Care Program was first begun. Now we are engaged in training Geriatric Fellows. Indeed, such Fellowships exist at twelve VA medical centers and we have eight Geriatric Research, Education and Clinical Centers which are also producing trained clinicians in geriatrics. These people will want to work in nursing homes and should be encouraged to do so. If they are to be encouraged, then we must, perforce, recognize that they need additional aids. While in the original nursing home design no thought was given to the presence of a physician's office since these nursing homes were designed in approximation of the hospitals, now, with increasing numbers of trained geriatricians, we will want to consider building physician offices in the nursing home. In one hospital, for instance, rather sophisticated electrocardiographic monitoring equipment was asked for in the nursing home. Under previous rules that would not have been permitted because the patient was to be transferred immediatly to the acute hospital if detailed electrocardiography was indicated. With the presence of geriatric fellows and geriatric clinicians in the nursing home care unit at one of our medical centers in Florida we have had to reverse that ruling and sophisticated EKG monitoring equipment is now available in at least a few of our nursing homes. What the future holds is still uncertain, but clearly the excess of trained internists, and particularly of geriatric specialists, will make available a large pool of trained and interested manpower which we should certainly put to work in the nursing home care unit -- this will undoubtedly have an effect upon the design. Our experience has been very satisfactory, but much remains to be learned about this important aspect of long term care.

REFERENCES

Birren, James E. and Schaie, K.W. "Handbook of the Psychology of Aging," 1978, Van Nostrand Reinhold, New York.

Hiatt, L.G. "Care and Design," Nursing Homes, Vol. 29, No. 3, May-June 1980, pp. 34-40.

Kelsey, J.L., Pastides, H. and Bisbee, G.E. "Musculoskeletal Disorders," Prodist, New York, 1978.

US Department of Health, Education and Welfare, "Public Policy and Chronic Disease," Washington, 1976.

Weston, T.E.T., 1964, Presbycusis, A Clinical Study, J. Larygn. Otol., 78, 273.

EVALUATION OF A PERSONAL EMERGENCY RESPONSE SYSTEM

Margaret Gatz, John Eiler, Cynthia Pearson,
Michael Gilewski, Max Fuentes, Mary Zemansky,
Charles Emery, and Linda Dougherty

Institute of Policy and Program Development
Andrus Gerontology Center, and
Department of Psychology
University of Southern California, Los Angeles

INTRODUCTION

Among the services thought to be essential to maintaining frail
and vulnerable older adults in their own homes is emergency
assistance when needed. A variety of electronic technological
systems are now available to provide such a service. This paper
reports some initial results from an evaluation of one program,
Emergency Alert Response System (or EARS). The EARS program uses
equipment from Lifeline Systems, Inc.

Lifeline is a technological system for connecting a frail older
person's telephone to a central emergency operator at a hospital.
The subscriber may summon emergency help actively by pressing a
button on the Lifeline unit or on a portable trigger, or passively
via a monitoring device which notifies the central operator if a
preset timer is allowed to run out. Upon receiving a signal, the
emergency operator contacts predetermined responders (a neighbor
with a key or a nearby relative) and/or paramedics, police, or other
services as needed. The service afforded by such a system is, in
fact, twofold: first, the actual use of the unit in case of an
emergency; and second, peace of mind from having the system
constantly available.

Sherwood and Morris (1981) previously conducted a demonstration
project with Lifeline. Subscribers, who lived in public housing,
were identified to meet the criteria of three screening groups: (1)
those who were severely functionally impaired and socially isolated,

(2) those who were severely functionally impaired but not socially
isolated, and (3) those who were not severely functionally impaired
but medically vulnerable and social isolated. Sherwood and Morris
found that, for the group of users who were severely functionally
disabled but not socially isolated (group 2), Lifeline resulted in
more comfort and confidence in the ability to live independently and
less use of nursing home care than matched controls who did not have
the unit. For groups who were socially isolated and either severely
functionally impaired or medically vulnerable, there were slight
increases in anxiety. Sherwood and Morris subsequently developed a
screening instrument for classifying subscribers into the three
screening groups.

    Writing about the same demonstration project, Dibner, Lowy, and
Morris (1982) reported an average of .44 emergencies per Lifeline
subscriber per year. Physical illness or accidents accounted for
73%, while environmental emergencies (assaults, maintenance
problems) accounted for 27% of the calls. While emergencies
typically were signalled by pressing the button, 22% of the time the
emergency operator was reached by means of a telephone call placed
by the subscriber or by a friend or relative.

    There have been other studies, primarily in sheltered housing
in the U.S. and Great Britain, which maintained records of the use
of emergency alarm systems (e.g., Brenner, 1981; Butler, 1981;
Garrow, 1976). The nature of the emergencies in all of the studies
tended to be similar. The most common physical problem was falling;
psychologically-related incidents tended to involve disorientation;
environmental problems involved such things as vandals, kitchen
fires, or inadvertently locking oneself out of the house.

    Dibner (1982) surveyed the program coordinators at 72 of the
more than 700 hospitals and agencies in the U.S. that have purchased
Lifeline. These data are the first to describe naturally developing
programs. There was an average of .84 emergencies per person per
year, with quite a number of subscribers experiencing multiple
emergencies. Of the incidents, 90% were physical health-related
(again, falls were most frequent, followed by heart attacks) and 10%
were environmental. Program coordinators felt that the system
served to delay institutional placement for one-sixth of the
subscribers.

    In our study we were particularly interested in a number of
issues related to the effect of this technological program on the
elderly subscribers and their families: First, we wanted to look at
the informal support system of the EARS subscriber. For instance,
what is the role of the family in the decision to install emergency
response equipment, and what is the effect of the program on the
relationship between the family and the subscriber? In particular,
does the family feel less burdened? Furthermore, how does the

neighbor who is participating as a responder feel about EARS?
Second, we wanted to follow up on the differences attributable to
EARS being a naturally developing program rather than a
demonstration project. For example, in the current study, some
units were given to low-income elderly, and others were leased to
people who requested them. Not all of these individuals fell into
one of the screening groups identified by Sherwood and Morris.

METHODS

The four hospitals selected by the Area Agency on Aging to
participate in the EARS program each agreed to provide the research
team with the names of all subscribers whom the hospitals had
approved to receive a unit. Subscribers were telephoned to request
their participation in an evaluation study. Those who agreed were
interviewed either before their unit was installed or within a short
time of installation. They were also asked whether we might contact
a member of their family and a neighbor who was serving as an
emergency responder. If they agreed, we phoned these other people
to request interviews. Post-test data were collected one year after
pretesting. In addition, subscribers were phoned by a member of the
research team every three months to ask about emergencies and other
use of health-related services. Finally, with the subscriber's
permission, we obtained copies from the hospital of all emergency
incident reports.

The battery of measures for subscribers was designed to assess
the constructs of interest to us. For comparison purposes, several
measures similar to those used by Sherwood and Morris were
included. The subscriber battery encompassed: (a) client
descriptive variables: demographic information, the Lifeline
screening instrument (Sherwood and Morris, 1981), mental status
(Kahn, Pollack & Goldfarb, 1961), self-rated health (USHEW, 1978),
physical illnesses (Pfeiffer, 1975), and activities of daily living
(Pfeiffer, 1975); (b) mental health outcome variables: happiness as
assessed by the Affect-Balance Scale (Bradburn, 1969), sense of
mastery (Pearlin & Schooler, 1978), and psychiatric symptoms as
assessed by the Brief Symptom Inventory (Derogatis, 1977); (c)
outcome variables related to sense of security: general anxieties
about living independently (Sherwood and Morris, 1981), specific
worries related to being a frail older person (an original scale),
opinions about institutionalization (Zarit, 1982), and fear of crime
(Patterson, 1978); (d) outcome variables related to social contact:
frequency and purpose of social interaction (an original scale).

The family interview encompassed the perceived condition of the
EARS subscriber (health, activities of daily living, happiness),
opinions about institutionalization, frequency and purpose of social
contacts with the subscriber, and sense of burden (Zarit, Reever, &

Bach-Peterson, 1980). The neighbor interview encompassed the
perceived health of the EARS subscriber, and reactions to the role
as emergency responder.

RESULTS

Background Information

   While 60 pre-test interviews with subscribers were completed,
at this time we have posttested and performed preliminary analyses
of data from only the first 28 subscribers and a smaller number of
family members and neighbors. Ten of the 60 subscribers are now
deceased, and the remaining subscribers have not yet had their unit
for one year. Consequently, these results are offered as an initial
glance at our findings.

   The age range of the sample was 54 to 99, with a mean of 78.
Over three-quarters were women; over three-quarters were Caucasian;
exactly 75% were widowed. The mean number of years of education was
11; about half of the respondents had an income of $4000-7000 per
year. Over half lived in houses, and most of the rest in
apartments; 15% lived with a family member, the rest lived alone.
On the mental status exam, 73% scored in the unimpaired range, while
the others were mildly impaired; 63% had noticed changes in their
memory in the past year. Not surprisingly, their self-rated health
status was poorer than national data for those 65 and older
published by USHEW (1978), and 55% had fallen down in the past
year.

   On the Lifeline screening instrument, 47% met the criteria of
group 2 (severely functionally impaired but not socially isolated)
and 13% were distributed across the other groups. The remaining 40%
did not fall into any of the screening groups; many of them were
medically vulnerable but not socially isolated.

   Slightly more of the family members were female than male, and
their mean age was 54, with a range from 28 to 75. Among the
neighbors, 70% were female, and their mean age was 62. Family
members tended to live quite nearby (an average driving distance of
18 1/2 minutes), while most of the neighbors lived next door.

Expected Benefits

   Before the units were installed, we asked subscribers and their
families about the benefits they expected from EARS; a year later we
asked what benefits had been obtained. At pretest, the two greatest
benefits expected by the subscriber were an increased sense of
security and the ability to obtain emergency help if needed. Others

stressed that EARS alleviated their families' concerns about their
living situation, and a handful mentioned the possibility that EARS
might increase their independence and self-reliance and enable them
to get out more. Two other variables that figured in subscribers'
responses were the fact that they lived alone or were in poor
health, especially having a history of falling. For instance, the
unit was described as a "mechanical dog." At posttest the greatest
benefit reported was the sense of security, followed again by
emergency help if needed, frequently combined with comments about
the value of EARS for someone living alone. However, fully a third
did not discern that having the unit made any changes in their
life. Twenty-five percent of the subscribers attributed their
getting the unit to their family's hearing about it, and at posttest
78% saw the unit as providing more peace of mind and independence
for their family.

Families' perceptions of the benefits of having EARS also
centered on an increased sense of security, both their own and the
subscribers', and the availability of emergency help if needed. The
second most often noted change was increased independence both for
the family and the subscriber. The family felt able to get out
more, and to see or telephone their aged parent less frequently.
Over a third of the families in turn saw the subscriber as able to
live alone more confidently and to get out (e.g., into the yard) and
do more things.

Lifeline Usage

Over the year there were an average of .28 emergencies per
subscriber, which is somewhat lower than previous reports. However,
if we consider only those subscribers who met the criteria of
Sherwood and Morris' screening groups, the average number of
emergencies per subscriber per year was .43, which is comparable to
the figure reported by the demonstration project (Dibner et al.,
1982; Sherwood and Morris, 1981). The emergencies predominately
involved physical illness or accidents--50% or more entailed
falling; chest pains were second. One call involved feeling
confused. Interestingly, in 25-30% of the emergencies, someone
other than the subscriber (more often a family member than a
neighbor) pushed the button to signal an emergency.

A partial compilation of just those false alarms that were
recorded by the hospital on emergency incident sheets indicates 1.10
per subscriber per year. Most of these entailed the subscriber's
failing to reset the unit, sometimes because of neglecting to turn
off the timer when leaving town. In these instances, neighbors
often were asked to respond; sometimes the hospital checked with a
relative; and, rarely, the paramedics were called and broke into the
house. However, one failure to reset indicated a genuine emergency
in which the subscriber was very ill.

We also asked subscribers about emergencies they had had when
they did not use the Lifeline unit. The rate per year was some 40%
greater than the rate of actually using the unit for emergencies.
The nature of the emergencies for which EARS was not used was
similar to those for which it was used--falling, heart problems,
panic attack. The main reasons given for not using EARS were (a) a
family member happened to call or the subscriber phoned the family
member directly instead of using EARS to summon help, (b) the
subscriber wasn't wearing the trigger, (c) the subscriber didn't
want to bother the neighbor, or (d) the subscriber didn't know
whether to regard the problem as an emergency or whether to handle
it herself.

Pre-post Changes on Dependent Variable for Subscribers

On correlated t-tests comparing the subscribers' pre and post
score levels, there were some statistical trends but basically
little change. Mean scores on selected variables are shown in Table
1. There was a slight improvement in self-rated health status. In
addition, there were trends toward less concern about having a
medical emergency such as fainting or a heart attack and less
anxiety about what to do in the case of an emergency, although there
was no decrease in general anxiety. On other measures, there were
trends in both directions, e.g., less fear of violence but slightly
more negative affect and slightly reduced estimate of the likelihood
of remaining in one's present living situation over the next two
years.

Comparisons of extent of change between socially isolated and
non-isolated subscribers were made using independent groups t-tests
between the two difference scores (post minus pre). As shown in
Table 2, socially isolated subscribers decreased significantly in
sense of mastery, while tending on the other hand to endorse fewer
specific worries. These findings parallel Sherwood and Morris'
reports of some paradoxical results for socially isolated
participants, as if having the unit may increase the social
isolate's sense of vulnerability.

Those who were included in one of the Lifeline screening groups
were significantly worse at pretest on at least five of the
dependent measures than those subscribers who did not meet the
criteria for any of the screening categories. There was little
difference in extent of change from pre to posttest, however; only a
statistical trend suggested that those who met screening group
criteria decreased in anxiety, while those who were less impaired
may have increased anxiety.

Finally, there were virtually no differences in extent of
change between subscribers who used their Lifeline unit in an
emergency and those who experienced emergencies but chose to rely on

Table 1. Pre and Posttest Mean Scores on Dependent Measures for
          Subscribers

	Pre	Post	$\underline{t}$(N=28)
Self-rated health status[a]	2.75	2.54	1.80[#]
Extent of health concern[a]	2.18	2.04	0.58
Positive affect[b]	3.38	2.92	1.63
Negative affect[b] (Bradburn)	1.30	1.41	-0.43
Affect-Balance[b]	22.08	21.51	1.93[#]
Sense of Mastery (Pearlin)[a]	14.35	15.24	-0.95
General anxieties about living independently[a]	24.02	23.23	0.82
Specific worries about frailty[a]	21.17	19.53	0.98
Probability of institutionalization[a]	1.19	1.23	-0.30
Likelihood of remaining in present living situation[b]	2.00	1.88	1.81[#]
Feeling about EARS[b]	4.42	4.57	-0.56

[a] low scores indicate a more positive response (e.g., less concern,
more sense of mastery)

[b] high scores indicate a more positive response (e.g., more
positive affect, more pleased)

[#] $p < .10$

resources other than their unit. This result suggests that the
potential service of having the unit available may be as important a
benefit of the system as its function in providing emergency
assistance.

Pre-post Changes on Dependent Variables for Neighbors

Neighbors are an important link in the Lifeline system because
they are called first in an emergency. Taken as a whole, they did
not appear to be affected positively or negatively by fulfilling the
responder role. At posttest they expressed the view that

Table 2.   Pre-Post Difference Scores for Socially Isolated and
           Not Isolated Subscribers

	not isolated	socially isolated	t
	(N=23)	(N=5)	
Self-rated health status	-0.26	0.00	-0.83
Affect-Balance	-0.65	0.62	-1.42
Sense of Mastery	0.52	3.33	-2.26*
General anxieties about living independently	-0.54	-2.22	0.62
Specific worries about frailty	-0.45	-8.50	1.77#
Feeling about EARS	0.17	0.00	0.26

#p<.10

*p<.05

subscribers felt more positively about the EARS program and were
less concerned about their health than at pretest (Table 3).

Pre-post Changes in Dependent Variables for Family Members

When asked directly whether they perceived change as a function
of the EARS program, families saw more change than did subscribers.
While only slightly over 50% of subscribers reported having
experienced changes because of the program, 80% of the families said
that they had seen change in the subscriber. Yet, on the various
scales that measure families' perceptions of the subscribers'
happiness, health, and probability of institutionalization, there
was little actual difference from pre to posttest (Table 4), only a
trend toward seeing less negative affect in the subscriber.

A clue to understanding this pattern of results is found in
that the families described themselves as feeling less burdened:
There was an overall trend toward decreased burden on the burden
scale; the item showing the most impressive change was a significant decrease in feeling angry toward their aged relative. The
interviewers' ratings of the apparent extent of burden of the family
also showed a significant decrease. Moreover, those families whose
relatives used the unit showed a significantly greater decrease on

Table 3.   Pre and Posttest Mean Scores on Dependent Variables
for Neighbors

	Pretest	Posttest	$\underline{t}$(N=10)
Subscriber's health status[a]	2.80	3.20	-1.18
Extent of subscriber's health concern[ac]	3.11	2.56	3.16**
Likelihood of subscriber's remaining in present living situation[b]	1.89	2.00	-1.00
Pleased with present living situation[b]	2.00	2.00	0.00
Subscriber's feeling about EARS[bc]	3.90	4.80	-2.59*
Neighbor's feeling about being a responder[b]	4.40	4.10	0.90
Interviewer's rating of neighbor's sense of burden[a]	1.80	1.90	-0.26
Interviewer's rating of neighbor's extent of involvement with subscriber[b]	2.80	3.30	-1.63

[a] low scores indicate a more positive response
(e.g., less concern, less burden)

[b] high scores indicate a more positive response
(e.g., more pleased, more involved)

[c] neighbors' indication of their impressions of the subscribers'
feelings

* p<.05

** p<.01

Table 4.   Pre and Posttest Mean Scores on Dependent Variables for
            Family Members

	Pre	Post	t(N=20)
Subscriber's health status[ac]	2.75	2.90	-0.77
Extent of subscriber's health concern[ac]	3.20	3.00	1.29
Subscriber's positive affect[bc]	3.26	3.05	0.72
Subscriber's negative affect[ac]	2.61	2.05	1.78[#]
Subscriber's Affect-Balance[bc]	20.66	21.00	-0.71
Probability of institutionalization[a]	1.26	1.37	-0.49
Likelihood of subscriber's remaining in present living situation[b]	1.89	1.89	0.00
Pleased with present living situation[bc]	1.90	2.00	-1.45
Subscriber's feeling about EARS[bc]	4.25	4.35	-0.34
Family member's feeling about EARS[b]	4.90	4.75	0.77
Sense of Burden (Zarit)[a]	47.82	44.88	1.99[#]
Interviewer's rating of family's sense of burden[a]	3.55	2.55	3.45[**]
Interviewer's rating of family's extent of involvement in caring for subscriber[b]	1.80	2.05	-1.00

[a] low scores indicate a more positive response
    (e.g., less concern, less burden)

[b] high scores indicate a more positive response
    (e.g., more pleased, more involved)

[c] family members' indication of their impressions of the subscribers'
    feelings

[#] p<.10

[**] p<.01

the burden scale ($\overline{X}$ = -7.52) than those families whose relatives did not use the unit ($\overline{X}$ = -0.27), $\underline{t}$ (17) = -2.77, $\underline{p}$ = .01. In sum, several analyses converge to support the hypothesis that a major benefit of the EARS program and Lifeline technology is in decreased burden for the families of frail older adults.

We asked families as well under what conditions they would consider institutionalizing their older relative (Table 5). At pretest they were equally divided among saying that institutionalization would occur when the subscriber couldn't care for him or herself, when they were unable to obtain the necessary level of in-home care, or when the family was unable to do enough for their older relative. At posttest a new category emerged: fully 25% now asserted that under no circumstances would they consider nursing home placement.

DISCUSSION AND CONCLUSIONS

In summary, while we are presenting preliminary data and drawing occasional inferences from statistical trends, these results bear interestingly on the effect of a technological innovation on older persons and their families. For subscribers, in general there was little change, and benefit did not seem to depend on whether or not the device was used. Thus, the potential service of having the unit always available would seem to be as powerful an intervention psychologically as the actual use of the device in the case of an emergency.

The types of emergencies for which EARS was used were similar to reports from alarm systems elsewhere (e.g., Brenner, 1981; Dibner et al., 1982). Consistent also with previous reports (e.g., Brenner, 1981; Garrow, 1976), subscribers rarely used EARS for matters that did not require immediate attention. What was more often the case was that subscribers did not use EARS even though the situation could have warranted it. The reasons for not using EARS again are similar to reasons that others (e.g., Butler, 1981) have reported for not using an alarm in an emergency--the person called directly for help or did not think the need was urgent enough to involve emergency responders. Although these findings indicate that the system was not being misused, another observation of Butler's may pertain: He suggested that people sometimes used the alarm system when, although the emergency was genuine, other ways of coping with it were available. In our study, for at least a quarter of the calls, the emergency was signalled by someone other than the subscriber pushing the button. This fact may be an instance of the phenomenon identified by Butler, because presumably these individuals could have used the telephone directly to call for help. Alternatively, the family member or neighbor may have been demonstrating to the frail older person that the situation was

Table 5.   Conditions under which Subscriber Might Enter Nursing Home

	subscriber		family	
	pretest	posttest	pretest	posttest
never, under no circumstances	25%	15%	0%	28%
when became a burden on the family; family not able to do enough; family decided it was best	15%	15%	35%	17%
when helpless, impaired, unable to care for self; last resort	40%	52%	30%	33%
when can't obtain necessary help (unable to obtain home help, can't cook own meals, etc.)	8%	7%	35%	17%
when it's a better choice; when can enter a desirable facility	5%	7%	0%	6%
never have thought about it	7%	4%	0%	0%

sufficiently serious to warrant using EARS, or the individual may actually not have known where else to turn for help.

The most striking effect of EARS was found in the families of subscribers, who indicated feeling more peace of mind, more freedom, less sense of burden, and more commitment to maintaining their relative outside of an institution. We can suppose that the greater dismissal of nursing homes as a possibility for their older relatives is reflective of their reduced feeling of burden. Given that the decision to institutionalize an older person often involves the family (e.g., Kutza, 1980; Linn & Gurel, 1972), these changes on the part of families become quite important. Consequently, families should be included in future studies of the cost-benefit of emergency alert technological systems.

A further finding was that the screening procedure proposed by Lifeline received support in predicting which subscribers were more apt to use the system. However, benefit did not seem to depend on whether or not the subscriber met the criteria of one of the screening groups. The distinction which emerged as being most salient was whether or not the subscriber was socially isolated. Consistent with Sherwood and Morris (1981), we found that socially isolated subscribers showed some effects opposite of those hypothesized--in particular, their sense of mastery declined appreciably.

In conclusion, EARS offers an example of how advances in technology can be employed to help older persons. Some (e.g., Sewel, 1983) have raised concerns that alarm systems represent a "technological fix" in the face of economic restraints and cuts in service provision. Along similar lines to our preliminary findings for socially isolated subscribers, Butler (1983) has expressed concern that alarm systems may at times unintentionally serve to undermine the independence of an older person. Moreover, he has suggested that the psychological support afforded by alarm systems may be more for the social providers than for the subscribers. Others (e.g., Dibner, 1982) have seen the technology as offering the older person another option, complementing other services in a long term care system. Our results suggest that effects for the family were possibly greater than for the subscriber, but that the psychological support afforded to the family may indirectly benefit the subscriber. Clearly discretion is called for in determining how much support to offer to the frail older person and in setting up a personal emergency response system such that it widens choice for older persons and their families.

NOTE

We would like to thank the following: UCLA/USC Long Term Care Gerontology Center, USC Human Relations Center, AoA Grant No. 09-AT-33/01, Los Angeles City Area Agency on Aging, and the participating hospitals.

REFERENCES

Bradburn, N. M. (1969). The structure of psychological well-being. Chicago: Aldine.

Brenner, D. (1981). The Southwark community alarm partnership scheme. In A. Butler & C. Oldman (Eds.), Alarm systems for the elderly. Workshop held at University of Leeds, England.

Butler, A. (1981). The efficacy of alarm systems for the elderly. Paper presented at the XII International Congress of Gerontology, Hamburg, Germany.

Butler, A. (1983). Towards an effective evaluation of alarm systems for older people. In M. McGarry (Ed.), Community alarm systems for older people. Conference held by Age Concern Scotland, Glasgow.

Derogatis, L. R. (1977). Symptom check list, revised version (SCL-90), manual. Towson, Maryland: Clinical Psychometric Research.

Dibner, A. S. (1982). A national survey of lifeline programs. (Available from Lifeline Systems, Inc., 1 Arsenal Market Pl., Watertown, MA 02172.)

Dibner, A. S., Lowy, L., & Morris, J. N. (1982). Usage and acceptance of an emergency alarm system by the frail elderly. Gerontologist, 22, 538-539.

Garrow, W. C. (1976). The planning and implementation of a night emergency service for the elderly living in congregate housing. Gerontologist, 16, 410-414.

Kahn, R. L., Pollack, M., & Goldfarb, A. I. (1961). Factors related to individual differences in mental status of institutionalized aged. In P. Hock & J. Zubin (Eds.), Psychopathology of Aging. New York: Grune & Stratton.

Kutza, E. A. (1980). Allocating long term care services: The policy puzzle of who should be served. Long Term Care Symposium, Williamsburg, Virginia.

Linn, M. and Gurel, L. (1972). Family attitudes in nursing home placement. Gerontologist, 12, 220-224.

Patterson, A. H. (1978). Territorial behavior and fear of crime in the elderly. Environmental Psychology and Non-Verbal Behavior, 2, 131-144.

Pearlin, L. I. and Schooler, C. (1978). The structure of coping. Journal of Health and Social Behavior, 19, 2-21.

Pfeiffer, E. (Ed.). (1975). Multidimensional functional assessment: The OARS methodology. Center for the Study of Aging and Human Development, Duke University, Durham, N.C.

Sewel, J. (1983). Highland helpcall--A case study of a dispersed alarm system. In M. McGarry (Ed.), Community alarm systems for older people. Conference by Age Concern Scotland, Glasgow.

Sherwood, S. and Morris, J. N. (1981). A study of the effects of an emergency alarm and response system for the aged: a final report. Grant # HS01788, NCHSR. (Executive summary available from Lifeline Systems, Inc., 1 Arsenal Market Pl., Watertown, MA 02172.)

U.S. Department of Health, Education, and Welfare. (1978). Health--United States 1978 (DHEW Publication No. 78-1232). Washington, D. C.: Author.

Zarit, J. M. (1982). Predictors of burden and distress for caregivers of senile dementia patients. Unpublished doctoral dissertation, University of Southern California, Los Angeles.

Zarit, S. H., Reever, K. E., & Bach-Peterson, J. (1980). Relatives of the impaired elderly: correlates of feelings of burden. Gerontologist, 20, 649-655.

COMPUTER AND TELECOMMUNICATIONS APPLICATIONS TO ENHANCE

THE QUALITY OF LIFE OF OUR ELDERLY CITIZENS

Marvin Kornbluh

Congressional Research Service, Library of Congress
Washington, D.C.

INTRODUCTION

Over the next few decades, the number of elderly persons in the
United States is expected to increase at a rate faster than the
general population. The number of persons over 65 is projected to
increase by nearly 50 percent before the end of this century; the
percentage of those 80 years old or more will also increase as
medicine becomes more adept at prolonging life. These individuals
will differ from their parents and grandparents. They are likely to
be better educated, wealthier, healthier, and possess more social
and political awareness. They will want more than a decent place to
live and enough to eat; they will want more independence, continued
productivity, and be less tolerant of custodial care. In other
words, they will want an enhanced quality of life in their twilight
years.

There also appears to be an increasing prevalence of chronic
degenerative disease among older citizens coupled with rising costs
of health care, social services, and income maintenance.
Traditional patterns of medical care may be traced to a model of
illness called the "Medical Model," where the human body is viewed
as a machine with illness and disease regarded as malfunctions of
its operation. The tasks of physicians and medicines are to
diagnose and repair such malfunctions. By attending to what has
occurred, it is reactive in nature. A newer, proactive approach is
a model of wellness referred to as a "Holistic Model." It focuses on
identifying potential health problems and preventing them from
happening. Holistic health care is more than just preventing
disease; it concerns the whole person--his social, spiritual,
physical and emotional well-being. It emphasizes that an

425

individual's health is largely his own responsibility and that he
must be a participant in maintaining wellness. It emphasizes health
education, health appraisal, behavior modification, development of
self-care skills, and home care rather than institutional care.

As elderly persons experience major life changes--increasing
debilitation, changes in daily routine, modifications to the
external environment, and new socio-cultural demands--wellness
becomes more difficult to maintain and illness more difficult to
prevent. However, information technology--personal computers,
telecommunications networks, and microprocessor-based devices--can
significantly assist our elderly citizens in reaching their goals
with the possibility of reducing the high expenditures our society
makes on behalf of its elderly as well. Indeed, there may be no
real choice for the aged members of our society. They must begin to
"fit into" the emerging information age characterized by a vast
array of information-based products, services, and applications.

APPLICATIONS OF INFORMATION TECHNOLOGY TO AID ELDERLY PERSONS

There are a vast number of current and potential applications
of computer and telecommunications technologies to help us towards
an improved quality of life. There are also a few that are
especially useful to our elderly population. These are highlighted
below.

Information Retrieval

By attaching a device called a "modem" (modular-demodulator) to
a computer, an individual can tap into large databases (collections
of organized information in electronic form) to obtain information
on virtually any topic. A modem holds the telephone receiver and
changes computer signals into signals that can be sent and received
over telephone lines. Different types of databases are available
online (immediate access); they differ by subject, scope, geographic
and chronological coverage, and updating frequency. Some contain
reference information, that is, they refer users elsewhere
(documents, organizations, and individuals) for needed information;
others are source databases, that is, they contain the needed
information. The fall 1983 issue of "Directory of Online Databases"
published by Cuadra Associates, Inc. of Santa Monica, California,
describes 1,878 databases. Some actual or possible databases to
assist elderly persons include:

   a) Filing medical insurance forms;
   b) Regulating diet and weight control;
   c) Body care, conditioning and relaxation exercises;
   d) Services for the aging;
   e) Safe and wise use of drugs and medications;

f) Recognizing signs of infection, coma, and shock;
g) Vocational counseling and placement organizations;
h) Handling diabetes, arthritis, and high blood pressure;
i) Sleeping techniques and materials;
j) Nursing homes and senior citizen centers;
k) Improving working efficiency in the home;
l) Handling stress, pain, constipation, and depression;
m) Dealing with and preventing personal crime; and
n) Preparing wills.

Such databases are or could be quite comprehensive and organized for easy access. For example, safe and wise use of drugs and medications could contain:

a) Drug names (brand and generic);
b) Prescription and non-prescription drugs;
c) Colors, types and sizes of medications;
d) Possible side effects and sensitive reactions;
e) Times to take drugs--frequency and length;
f) Mixing drugs per se and with ordinary foods and liquids;
g) Possibility of physical dependence;
h) Alternatives to taking drugs;
i) Special diets, vitamins and mineral supplements; and
j) Stimulants and depressants.

An elderly user could create and maintain his own personal databases such as reminders (birthdays, appointments, and medications) and lists (possessions, friends, and recipes).

Electronic Networking

A network can be regarded as a collection of individuals, small groups, and/or organizations informally and voluntarily linked together for mutually beneficial exchanges of ideas, opinions, and facts. Elderly participants in networks can selectively and simultaneously dialog with each other with varying degrees of interaction at different times. The elimination of occupational interactions tends to weaken ties of friendship for older persons. Networks can replace friends left on the job. They can provide stimulation through sharing of values, experiences, aspirations, and increased self-awareness. Accurate, personalized, messages can be exchanged with new and old friends and family by turning the computer into a word processsor. Standard phrases can be obtained from the computer's memory; electronic bulletin boards can be established among participants and messages can be picked up at one's convenience.

Computer connected mutual support groups could be formed to provide elderly persons having common physical, social, or emotional problems a way to share feelings, frustrations, and successes.

These would be similar to such current groups as Weight Watchers,
Alcoholics Anonymous, and Handicapped. Other possible support
networks inlude those who have recently lost their spouses, new
arrivals into an area, and those with common illnesses.

Networks could also be focused on political action. Elderly
persons could express their political views on local and national
issues from their homes and participate more fully in democratic
government; this may reduce some feelings of hopelessness with
respect to governmental awareness of their problems.

Education

The aged are a group in the United States whose potential usage
of educational services has been largely untapped--perhaps because
they perceive education as less critical than social, income, and
health services. However, there appear to be major advantages to
elderly persons to continue their education with a personal
computer. Some prevalent ones are:

- a) Self-paced--time to absorb and comprehend the material
    without inconveniencing other persons;
- b) Little embarrassment--when mistakes are made there is
    privacy and only the computer knows;
- c) Immediate feedback--the user can obtain immediate knowledge
    concerning his responses in learning programs;
- d) Objective evaluation--the computer bases its evaluation of
    the student solely on performance; personal characteristics
    of the student are not considered; and
- e) Drill and practice--the computer never gets tired of
    teaching the student.

An organization entitled "Telelearning Systems, Inc." has
established an "electronic university" to make education accessible
to those lacking the time, money, or ability to attend traditional
universities. Using specialized modems and software, with their
computers, elderly students can progress at their own rates, ask
questions, get answers, and obtain test results immediately after
electronic grading. Electronic mail would enable students and
teachers to exchange messages.

Four types of learning could enhance older persons'
capabilities to deal with their major problems:

- a) Learning for economic self-sufficiency; this would include
    job-hunting and money management skills;
- b) Learning of practical life skills; these would help
    restructure coping skills developed over a lifetime to the
    changed environment of an older person;
- c) Learning for community participation; this would enable

older Americans to help government cope with its mounting
social and economic problems;

d) Learning for personal satisfaction; this would include
exposure to liberal and fine arts and physical and social
sciences.

Electronics-based Services

There are a number of electronics-based services that could
lessen stress upon elderly persons. They currently can or will be
able to receive services such as the following without leaving their
homes:

a) Information services. These are services which provide
information subject to rapid changes such as news, sports,
travel schedules, stock quotations, hotel room availabili-
ties weather, and obituaries;

b) Financial management services. These include budget and
cash flow analysis, investment portfolio analysis, tax
computation, and general investment analysis;

c) Purchasing services. Participants can request displays of
items for sale or rent on their computer screens--clothing,
food, cars, and apartments--as well as details about prices,
styles, terms, dates and even consumer reports. Orders can
be given via computer-originated forms;

d) Bill payment services. By inserting credit card information into the computer or by requesting savings or checking
account transfers, elderly persons can pay bills.

e) Reporting services. These services could include bank
statements, various transaction histories, summaries of
service charges, and credit changes; and

f) Administrative services. These would provide listings and
descriptions of locations, services, products, hours, etc.
of business and professional firms.

Elderly Physically Handicapped

Life for elderly physically handicapped persons often means
barriers, high energy expenditures, and frustrations as they try to
function in everyday life--regardless whether the impairment
involves speech, mobility, hearing, mental acuity, or sight.
Elderly handicapped have to satisfy their self-preservation needs
before they can expend energy for social and political tasks.
Almost any aspect of human activity that has been impaired could be
improved through use of personal computers, thus improving the
quality of an elder's life.

The personal computer can be turned into a communication and
control prosthesis for mobility, visual, hearing, and speech
impaired elderly persons. Two approaches are taken to do this. One

is through software whereby special computer programs are written to
enable physically handicapped persons to use the computer as a
"paper and pencil." The second is through hardware where "adaptive
devices" are linked to the computer. There are hardware and
software that make audio and visual information more easily
understood. For example, visual displays can be expanded or
converted into auditory displays, and printed text can be translated
into speech and braille. In addition, the computer has been
employed to aid motion impaired individuals improve their speed of
communication by increasing the amount of information released with
a given number of signals or switches; there are special switches
which can be operated by the tongue, eyes, head, mouth, voice, and
by light. They permit handicapped persons to take maximum advantage
of their residual skills.

Alerting and Monitoring

The personal computer can perform the "guardian" function; it
can monitor aspects of an individual's health, detect unsafe
conditions in an area, control expenditures of electrical energy,
and sound alerts.

Computer-controlled energy monitoring of residences can
optimize the distribution of energy for heating, cooling,
ventilation, purification, lighting and electrical appliances; the
computer can also give notice of needed maintenance of energy
operated systems. Thus, there is potential for large reductions in
utility bills of the elderly.

A personal computer can give older persons property and
personal security not previously possible without high expense.
Sensing devices linked to a computer can monitor an area for sounds
caused by intruders, detect fire and smoke on the premises, and
check an individual's mental and physical alertness and breathing.
Hazardous conditions can be reported to predesignated authorities
(police and fire departments, rescue squads, hospitals, or family
members). Further, fire control equipment and cameras can be
automatically activated and occupants of a residence warned of
emergencies.

Emergencies of elderly persons who have special medical
conditions such as heart trouble, diabetes, serious allergies,
epilepsy, implanted pacemaker, glaucoma, or are taking
anticoagulants can be quickly detected by a computer. An elderly
person can perform standard tests at home to reduce the number of
doctor visits. He could test breathing patterns, anemia levels,
blood pressure values, blood sugar counts and cholesterol levels and
be informed when to contact a physician.

Medicine has developed questionnaires to detect conditions that a physical examination could overlook. Health hazard appraisals based on nutritional surveys, personality inventories, and stress profiles can "zero-in" on life styles that could contribute to future illness. Risks associated with particular life styles can be assessed considering social, environmental, psychological, and genetic factors.

Leisure Activities

Retired older persons have more time that is relatively uncommitted. This leisure time can be consumed in many ways--by contemplation, scholarship, self-expression, reading, hobbies, travel, games, relaxation, painting, music composition or appreciation, movies, television, spectator sports, participative sports, exercise, and other diversions. The leisure pursuits of networking and education were discussed. Another possibility is playing computer games--of chance, strategy, and skill. Computer games can be beneficial to aged individuals for a number of reasons; to stimulate brain activity, coordinate the use of hands and eyes, increase personal interaction, and add a measure of self-esteem. The elderly could play checkers, scrabble, poker, chess, and a video and non-video games with their peers, grandchildren, and family members--even if remotely located. Games can be modified for adoption by older cohorts; play can be slowed down, images enlarged, rules simplified, and number and complexity of video images decreased.

There are computer-based "electronic palettes" which can be used to foster artistic creativity--without messy paint. Computers can form images--realistic and surrealistic--change colors and sizes of and rotate images, and even animate them. Electronic music can also be composed by a computer. Musical notes can be created, displayed and edited on a screen and played in various synthesized modes--from a single instrument to a full orchestra.
With respect to hobbies, lifetime collections--stamps, coins, antiques, recordings, matchcovers, whatever--can be easily listed, cross-indexed, and accessed.

The computer can become a word processor through software. A retiree can write memoirs and letters accurately and quickly; errors can be corrected and punctuation, spelling, and some grammatical mistakes can be flagged automatically; words, sentences, and paragraphs can be added, deleted, and moved without retyping. Margins, page numbers, headings, bold faces, and underlines can be automated.

Work and Retirement

Retirement can be one of the most stressful times in one's life. Some are able to handle it in mature, constructive ways; others go through long periods of denial or adopt overcompensatory activities. Some form of retirement preparation appears needed to cushion the transition and formulate new life goals. Personal computers, with their access to databases, educational facilities, varied services, social networks, and leisure activities, can play important roles for retirees. They could search databases for descriptions of places to live and get advice on retirement income and taxes, second careers, insurance coverages, and travel possibilities.

There is a need for potential retirees to become "computer literate" if they desire second careers, multiple careers, continuation of disrupted careers, or late careers (particularly for women after child-rearing is completed or divorced after many years of marriage). Pre-retirement counseling could be offered remotely with computers. Subjects could cover psychological aspects of aging, emotional adjustments to retirement, housing, money management, and legal matters.

Careers for older persons could employ "telecommuting" or working at home--part or full-time. The home computer would communicate over telecommunications networks to computers located in business firms and elsewhere. Careers could entail clerical work, writing, specialized consulting or marketing and other opportunities. The computer is an excellent repository for "canned information" which the retiree could use in conducting business at home. One example is storing different kinds of business letters. There are advantages to retirees working at home. Tensions and the physical efforts of commuting are reduced, out-of pocket expenses for commuting are decreased, and work scheduling is flexible.

One way of easing retirement and imparting computer literacy to a potential retiree is mentoring. A younger person guided by an older person gradually develops and assumes increasing responsibilities. The older "mentor" can feel that he is leaving a legacy behind--the occupational identity of a capable younger employee. In return, that employee teaches the near-term retiree use of computers.

The previous list of computer applications is far from exhaustive. Applications of information technology to elderly persons is limited only by human imagination, motivation to experiment and available resources. Personal computers with telecommunications capability may be located in different places. They can be placed directly in the homes of elderly persons or in homes of family members. Elderly persons could also share computers in nursing homes, community and senior citizen centers, libraries, associations catering to the needs of the aged, and congregate housing for the elderly. Health providers--physicians, social workers, counsellors, therapists--could install computers in their offices and conduct experiments on behalf of and with elderly individuals.

BARRIERS IN APPLYING INFORMATION TECHNOLOGY TO AID ELDERLY PERSONS

Because a number of significant barriers are present, applying computer and telecommunications capabilities to meet the needs of our older citizens and adequately implement the applications will require great patience and strong support from government, business, and key trade, professional, and public interest groups. Four kinds of barriers deserve recognition.

Social Barriers

Society appears to have negative images of older people: Some gerontologists declare that aged persons are judged more on what they cannot do anymore than by what they still can do. These images have limited opportunities for the elderly to use computers. Employers may be reluctant to invest in training mature workers in computers--fearing poor returns on their efforts. Increased use of computers has also aroused concerns about unauthorized access to private information by unscrupulus individuals and investigatory organizations. Demands for information privacy and security can easily conflict with demands to collect and use collected information. Elderly citizens may be especially vulnerable since they would likely find it more difficult to recover losses stemming from unauthorized access to information.

Psychological Barriers

In the minds of many elderly, family members, and service providers with little exposure to a computer, its use raises negative reactions. It suggests something incompatible with judgment, compassion, and other human qualities and a threat to their competence and autonomy. The jargon of computer professionals can be confusing to them. Gaining mastery of a computer culture takes time and can be difficult. An elderly person may be unwilling or afraid to learn and apply computer instructions, machine operating procedures, and error correction techniques.

Financial Barriers

Despite the decreasing cost of personal computers over the last few years, those with the capability to perform the applications discussed will likely cost a few thousand dollars. This initial

investment cost is compounded by operating costs for data
transmission, database access, maintenance, software, and training.
Such costs can be significant.

Physiological Barriers

Older persons may experience psysiological restrictions
including declines in sensory acuity, immunity system failures,
slower reaction times, memory blockages, and degenerative diseases.
However, learning and productivity need not be adversely impacted
with sufficient motivation, persistance, and discipline by older
persons.

PUBLIC POLICY ISSUES

Increasing purchases of personal computers and increasing
numbers of aged persons present an opportunity for their symbiosis.
The degree to which this interdependency will be realized depends on
society recognizing and resolving many public policy issues. A
sample of such issues are:

Balancing the Focus on Wellness and Illness

Some health experts believe the health of our elderly depends
primarily on the promptness and adequacy of medical treatment
received; others believe maintenance of wellness and prevention of
illness is more critical. The contribution of computers to medical
treatment of the elderly is well-known and accepted; the
contribution of computers to the maintenance of wellness by the
elderly is less known and accepted. How should government and
business allocate their resources to adequately treat illness of the
elderly while trying to maintain their wellness?

Equitable Access to Information Technology

If our society is to receive all the benefits stemming from
applications of information technology, its less privileged
segments--which are likely to include many elderly citizens--need to
be guaranteed the same opportunities to employ the technology as the
affluent and organized groups. How can the public and private
sectors of our society interact to ensure equitable access by
elderly persons to information technology and to the concomitant
benefits such as computer literacy?

Cost Containment--Service Balance

A major focus of government and private enterprise has been to
contain rising health and medical costs. The use of computers by the
elderly and health providers offers the promise of reducing such

costs. How can government encourage the use of personal computers by the elderly and health providers without excessive outlays of public funds?

Balancing Human Judgment and Electronic Advice

As elderly persons and providers apply information technology they may become overdependent on it. This risk would be manifested by older persons procrastinating in seeking human assistance and failing to exercise sufficient judgment in applying "canned" electronic advice. How can our society assure that older persons exercise sufficient judgment in using information technology?

OPTIONS FOR RESOLUTION OF PUBLIC POLICY ISSUES

Roles of the Government

Options for resolving public policy issues would entail major roles by government, at all levels, and by the private sector--including business and trade and professional associations. The Federal Government could review its regulatory, subsidization, and taxation roles with respect to our aging population and assess whether market forces and local interests are encouraging the use of information technology by and for the elderly. State and local governments could consider subsidizing the acquisition of computers for use by elderly persons and providers in the expectation that other subsidies for older persons may be reduced as a consequence. Federal, state, and local governments could also provide incentives for manufacturers of information technology and gerontological centers to undertake research into the use of computers by the elderly and to develop appropriate hardware and software. Such research could have "positive spillover effects" on the "mainline efforts" of the manufacturers and broaden the boundaries of gerontology centers.

Roles of the Private Sector

The private sector could find it financially beneficial to respond to the demands of the elderly for enhanced quality of life. The market of maturing individuals, their families and service providers, is a significant one. Manufacturers of information technology could orient their products and services to the needs of the elderly. The media could portray elderly persons in more positive ways; the entire business community could provide them with more opportunities to learn and use information technology.

NOTE: The opinions expressed in this article are solely those of the author and do not reflect those of the Congressional Research Service of the Library of Congress.

TECHNOLOGY AND AGING: IDENTIFICATION AND EVALUATION

OF PRODUCTS TO ENHANCE AN INDEPENDENT LIFESTYLE

Cheri K. Krauser
George T. Baker, III and
Robert G. Lynch
University of Maryland

Frank J. Carmone, Jr.
Drexel University

INTRODUCTION

The process of aging is a complex biological phenomenon which may be broadly characterized by the progressive decline in the capacity of an organism to withstand environmental stresses. The time-dependent changes that occur with advancing age manifest themselves as anthropometric, physiological and biochemical alterations in the organism (Shock, 1981). Although these changes in physiological adaptability render aging individuals more susceptible to the stresses of everyday living, the mature population can exhibit a high level of physiological and cognitive performance given environments which would accommodate the changes accompanying biological aging (Baker and Andrews, 1981). Much of the built environment in our society is a reflection of products and design that accommodate the physiological and psychosocial needs of a young adult. The development and application of technology-- broadly defined as adaptations and/or alterations to the environment that would enhance the independence of older individuals--could ameliorate some physiological changes and enhance a mature individual's fully integrated role in society.

The trend towards an older population in the United States has been much reported, and at the risk of belaboring the commonplace we might mention a few statistics here. According to the 1980 census, 21% of the population of the United States is now 55 years of age and over. Several European countries have higher proportions--West Germany has 25%, as does Denmark; France has 24%--but the United States figure is higher than the world average. The median age will

rise sharply over the next three decades: in 1980 it was 30 years;
in 2000 it is predicted to be 36 years; and in the year 2010, 38.4
years.   In addition, as more and more Americans grow older and are
in relatively good health as medical technology advances, this
nation will be something it has never been before, namely old.
Furthermore, it is estimated that households headed by individuals
55 and older (ca. 30 million--one in three households) have at least
28% of the discretionary income in the country today.   This amount
is nearly double that of households headed by persons 35 and under;
households headed by those 25 and under have only 1% of all
discretionary income (Allan, 1981; Baker et al., 1982).   This is not
to dismiss the fact that approximately one fourth of the elderly
live at or about the poverty level.   A significant proportion of
these are women living alone.

     Given the demographic trend toward an older society and the
need for an environment responsible to the needs of all ages,
existing products that address physiological limitations imposed by
the normal aging process and personal needs of mature individuals
were identified (Krauser and Baker, 1982).   Products were selected
from items in manufacturers' catalogs, store catalogs, various
retail outlets, contacts with businesses, focus group interviews
with mature consumers, and articles and advertisements in newspapers
and magazines.   Products were classified via three schemata:
1) technological area, generated from analysis of focus group
interviews held with mature consumers throughout the United States
in 1981-1982; 2) physiological limitations categories identified by
biomedical researchers; and 3) personal need areas, identified by
gerontological research in needs assessments of the elderly.   Each
product was given a preliminary score on each of six assessment
criteria.   The items identified by this data have been published in
a Catalog of Products and Services to Enhance the Independence of
the Elderly (Baker and Krauser, 1982).   Classification schemata and
assessment criteria for the products in this data base are discussed
in the following sections.

CLASSIFICATION AND ASSESSMENTS

     In this section we briefly discuss the schemata for classifying
products by technological group, physiological limitation category,
personal needs requirements and, finally, selection and utilization
of the assessment criteria.   For a fuller explanation of the
classification schemata and assessment criteria see Krauser, et al.
(1983).

Technological Groups

     The basis of this classification schema was generated from
analysis of focus group interviews held with mature consumers

throughout the United States in 1981-1982. Analysis of the focus groups, not suggestions by the researchers, resulted in following groups: 1) appliances, 2) clothing, 3) communications, 4) food, 5) health & safety, 6) home building/design, 7) recreation, 8) transportation, and 9) catalogs and manuals.

Groups one through six and eight were originally defined in the focus groups; groups seven and nine were added during product identification to improve ease of classification. It was our experience that products fell easily into one of these categories. For example, electric teakettles obviously fit into group one, while modular housing goes into group six. A few products, such as scissors or foam tubing, satisfy multiple objectives and were not readily placed into a unique group. Therefore, placement of products was a multistep process; all assignments were reviewed by a multidisciplinary project staff before finalization.

Physiological Limitation Categories

These categories relate to physiological limitations imposed by the "normal" aging process and/or the more common health problems which occur with age. It should be emphasized that the process of aging is not a disease but rather a phenomenon which results in a decreased overall physiological capacity of an organism to withstand the stress of everyday living. From the biomedical literature, including Finch and Hayflick (1977), Strehler (1977), Shock (1981), and Baker (1981), the following categories were enumerated for the purposes of this project: 1) vision, 2) hearing, 3)speech weakness, 4) dietary needs, 5) manual dexterity, 6) upper extremity weakness, 7) lower extremity weakness, 8) mobility, and 9) physical/psychological comfort. Although a few products could be placed into multiple categories, each product was classified into only one category.

Personal Need Areas

These areas relate to chores or tasks that a person performs daily and activities/feelings that add to the quality of life. Capacity for self care decreases with age; by age 80 more than 1 of every 10 persons has trouble dressing, putting on shoes, bathing, cutting toenails and getting about the house (Shanas, 1980; Neugarten, 1978). Independence is maintained and enhanced by a person's ability to manage basic, everyday tasks that pertain to personal care. Additionally, other personal and social activities that are one step beyond basic care can add to the quality of life. Considering these two parameters, the following personal need areas were enumerated: 1) communication, 2) food and food preparation, 3) dressing and undressing, 4) personal hygiene, 5) home maintenance, 6) recreation, and 7) comfort and security. As in the physiological limitation category, each item was placed into one area even though a few items could fit into more than one area.

Assessment Criteria

Assessment criteria for evaluation of products were distilled
from a larger list used by the project staff. During the course of
project discussion, these criteria were decided upon as being most
relevant to the evaluation of products for potential inclusion in
the data base; that is, products had to be either currently designed
for and/or used by the aged or, in the best estimates of the project
staff, adaptable for use by the aged. Each product received a score
of 1-6 on each assessment criterion (scales omitted): 1) compensates
for physiological limitations, 2) substitutes or reduces social
support services, 3) enhances psychological/sociological independence, 4) availabilty, 5) benefit/cost ratio to consumer (cost =
retail price), and 6) market potential (percent of the population
affected by the limitation the item addresses).

A product was ultimately included (excluded) in this data base
by its overall assessment score, which was developed in the
following way: each criteria was scored 1 to 6, with 1 being the
least effective/desirable level and 6 being the most
effective/desirable level. The assessment score was the simple sum
of the project staff's evaluation of the product on each of the six
attributes. This is referred to as a simple compensatory model;
that is, low evaluations on some assessment criteria can be
compensated for by a larger value on other criteria. The total
score was then compared to an arbitrarily selected minimum of 18
(one-half of the maximum score of 36). Generally, products scoring
higher than 18 were in- cluded in the data base and those scoring
less were not.

The project staff was assisted by the Pennsylvania Department
of Aging in defining and scoring criterion two, substitutes/reduces
social support services. Criterion four, availability, was able to
be defined and scored more objectively than the others. The project
staff felt the least comfortable with criterion three, enhances
psychological/sociological independence, because this one, more than
any other, pertains to a person, not a product. Criterion five,
benefit/cost ratio, was treated as a benefit/price ratio and was
scored as follows: low cost was less than $25.00; moderate,
$25.00-$50.00; and high, above $50.00. Benefit related to the
relative scores on criteria one to three: high benefit was a score
of 5 or 6 on criteria one to three; moderate, a score of 3 or 4; and
low, a score of 1 or 2.

In summary, products were classified by technological group,
physiological limitation and personal need area, with an evaluation
given on each of the six assessment criteria; there are 158 products
included in this study.

Analysis Procedure

In checking for differences in assessment level for each of the products, stepwise discriminant analysis was used. Three separate runs using the BMDP 7M program for stepwise discriminant function analysis were made. Each run used one of the three schemata (technological group, physiological limitation, personal need area) for the dependent variable; in all runs the predictor variables were the six assessment criteria. (For the purposes of this analysis, the scale of measurement for the assessment criteria was assumed to be an interval scale.) This enabled us to identify assessment criteria whose levels were statistically different across the groups as defined by each schema.

RESULTS

If we can assume that the assessment criteria can be used as a measure of how well a product meets the needs of the elderly, then the results can imply that certain technological groups are closer to meeting these needs than others. Mean values on the assessment criteria for all products and the highest and lowest scoring groups are shown in Figure 1. The communications group scored average or the highest on all criteria except for number one, compensates for physiological limitation, where the recreation group scored slightly higher. Although the recreation group scored the highest on criterion one and above average on criterion five (benefit/cost ratio), this group's overall scores were below average or the lowest. Also, this group's above average score on criterion five may be explained in part by the fact that the average price in this group is slightly more then $5.00, while the average price in the communications group is approximately $85.00.

For the analysis of the physiological limitation categories, the eight products in technological group 9 were dropped because catalogs and manuals address not one area but all or a majority of areas in this schema. In addition, categories 1-3 were combined into one group (VHS) because they address similar physiological functions--vision, hearing, speech--all pertaining to communication. Category four (dietary needs) was dropped from the discriminant analysis because it contained five very similar products which scored the same on the assessment criteria, and it could not be reasonably combined into another group. Also, categories upper extremity weakness and lower extremity weakness and mobility were combined into one category, called mobility. The mean values and highest and lowest group scores are shown in Figure 2. The VHS group scored the highest on all assessment criteria. The mobility category, although it scored average or slightly above on some criteria, scored lowest overall.

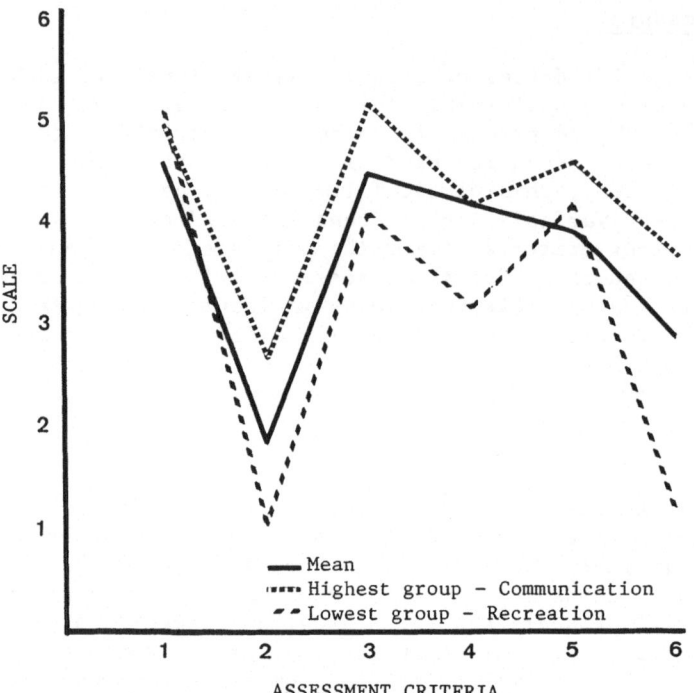

Figure 1. Decline in special senses and muscle strength with age.

Also for the analysis of the personal need areas, the eight
products in technological group 9 (catalogs and manuals) were
dropped because they address all or a majority of the personal need
areas. Areas two to four (food and food preparation, dressing and
undressing, personal hygiene) were combined into one area, "daily
tasks," because they all deal with tasks necessary for daily
functioning. Areas five and six, home maintenance and recreation,
were combined because both deal with activities that add to the
quality of life but are not necessary for daily functioning,
although a certain amount of home maintenance is necessary to
maintain a safe environment. The mean values and highest and lowest
scoring groups are shown in Figure 3. The communication area scored
the highest or above average on all assessment criteria. On the
other hand, home maintenance/recreation tended to have the lowest
values on all the assessment criteria except criterion one,
compensates for physiological limitation.

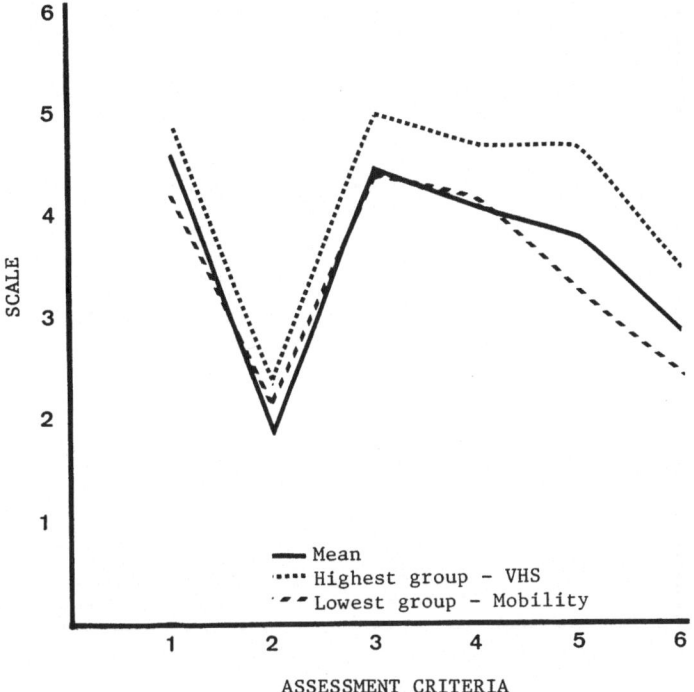

Figure 2. Decline in seleted physiological functions with age.

The above sections have discussed the classification and analysis of the products that, by means of an assessment score, were included in the data base. These "included" products account for approximately one-third of the products that were examined. The remainder of "excluded" products fell into two broad categories. The majority addressed a personal need but were difficult or cumbersome to use, considering physiological limitations that accompany normal aging (e.g., small buttons, small print or digital displays, weight and balance problems). Other excluded products, according to the assessment criteria, did not address a physiological limitation or a personal need. These items could be broadly categorized as gadgets or gimmicks--items which might be interesting or "fun" to use but did not aid in daily functioning or attendant activities. (Some excluded items might be of interest to a mature individual, but it was the purpose of this project to include items which address a physiological limitation or personal need.)

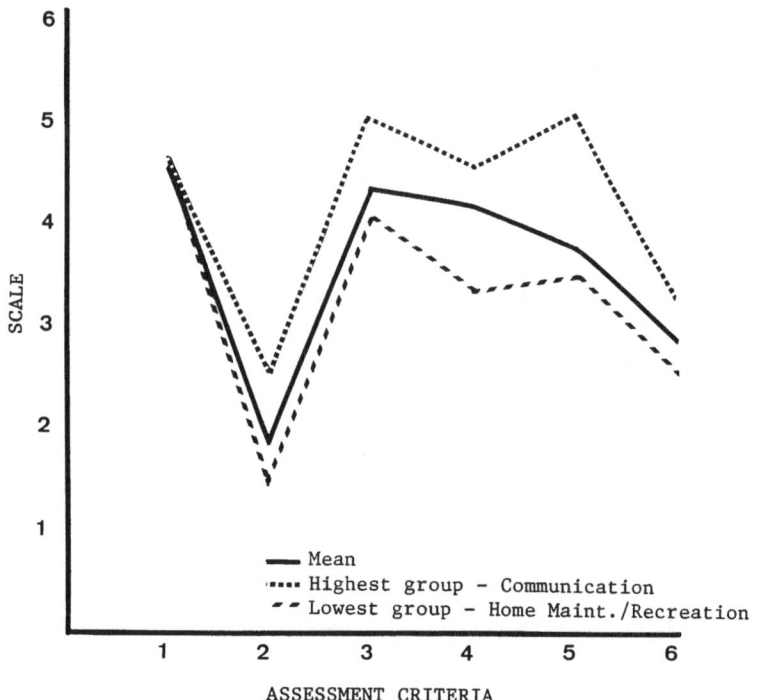

Figure 3. Mean, highest and lowest scores for Technological
           Groups.

CONCLUSIONS

     Although there are some limitations to the data--non-exhaustive,
exhaustive, highly selective inclusion of products in the data base;
arbitrary selection of assessment criteria; and limited controls on
product evaluations vis-a-vis the assessment criteria--some conclusions can be drawn from this analysis. As Figures 1 through 3 show,
the range of values on the assessment criteria tend to follow a
pattern no matter how the products are classified. That is, physiological limitations and psychological/sociological independence
parameters seem to be addressed better than other criteria. These
criteria are followed by benefit/cost ratio, availability, market
potential and, lastly, substitutes/reduces social support services.
If the assessment criteria can be used as a relative measure of how
well needs of older individuals are being met, then a common thread
appears throughout the results. That is, no matter how products are
classified--by technological group or physiological limitations or
personal needs addressed--the communication industry seems to be
meeting the personal communication needs of the elderly, and the
physiological limitations of a vision/hearing/speech (all dealing
with communication) seem to be better addressed than other

limitations. Conversely, these data would suggest that items that address decreased mobility and items that address home maintenance and recreation needs of the elderly are not readily available and that these needs are not being met.

Although many products were excluded from the data base on the basis of their assessment scores, in some technological areas, e.g., food and clothing, there seemed to be a dearth of readily identifiable products to assess. In talking to various industries and distributors in these two areas, the project team found the responses to coincide with this lack of products. In other words, there was an apparent reluctance on the part of food and clothing manufacturers/distributors to have their products associated with the elderly or older consumer. The clothing industry in particular seemed somewhat apprehensive about associating their products with the elderly. The hesitancy on the part of the food industries is somewhat more complex to interpret or explain. Aside from the implied, as well as stated, comments that this market is too small, or they were already doing what they should be doing in this area, there seemed to be a genuine concern that the public image of their products might suffer if associated with an older consumer. It is our interpretation that the major reason for this reluctance was based on an erroneous stereotypic view of the elderly as a nonviable market segment. This view is substantiated by Bartos (1980) who stated that industries, when considering the older consumer, tend to think of the market as monolithic, based on age alone, and do not consider the resources, attitudes and needs of this growing consumer group.

These conclusions are not definitive; further research is needed in the following areas: 1) a more exhaustive search for products in order to have a better representation of what is available; 2) a re-evaluation of the assessment criteria to refine the scales, definitions and to translate the psychological/socio-logical independence parameter into a product oriented criterion; and 3) a re-evaluation of criterion two, substitutes/reduces social support services, to define whether the products' consistently low scores on this criterion are an artifact of the scoring procedures or, indeed, suggest that products that address this criterion are not readily available. Further research into the "excluded" product areas is also indicated, specifically to determine if these stereotypic views are representative of industrial perceptions, if these views are changing and if industry, as a whole, is cognizant of the needs of mature individuals for consumer products.

REFERENCES

Allan, C. B., 1981, Over 55: Growth Market of the 80's, Nation's Business, 4: 23-32.

Baker, G. T., III, 1981, "Synopsis of Presentations on the Productivity, Profitability and Pride of the Mature Worker," Silver Spring, MD, George Meany Center for Labor Studies.

Baker, G. T., III, and Andrews, D.C., 1981, "Aging and the Built Environment," Racine, WI, First National Research Conference on Technology and Aging.

Baker. G. T., III, Griffith, B. C., Carmone, F. J., Jr., and Krauser, C. K., 1982, "Report on Products and Services to Enhance the Independence of the Elderly," College Park, MD, University of Maryland Press.

Baker, G. T., III, and Krauser, C. K., 1982, "A Catalog of Products and Services to Enhance the Independence of the Elderly," Drexel University, Philadelphia, PA.

Bartos, R., 1980, Over 49: The Invisible Consumer Market, Harvard Business Review, 140-148.

Finch, C. E., and Hayflick, L., 1977, "Handbook of the Biology of Aging," New York, Van Nostrand Reinhold.

Krauser, C. K., and Baker, G. T., III, 1982, Technology and Aging: A Catalog of Products and Services to Enhance the Independence of the Elderly, Gerontologist, 22: 209.

Krauser, C. K., Carmone, F. J., Baker, G. T., III, and Kriner, R. E., 1983, Technology and the Mature Consumer, in "Advances in Health Care Research," S. Smith and M. Venkatesan, eds., 2: 61-65.

Neugarten, B. L., 1978, The Future and the Young-Old, in "Aging into the 21st Century: Middle-Agers Today," L. F. Jarvik, ed., New York, Gardner Press.

Shanas, E., 1980, Self-Assessment of Physical Functions: White and Black Elderly in the United States, in "Second Conference on the Epidemiology of Aging," S. Haynes and M. Feinleib, eds., Bethesda, MD, NIH Publications.

Shock, N. W., 1981, Indices of Functional Age, in "Aging: A Challenge to Science and Society," D. Danon, N. W. Shock and M. Marios, eds., New York, Oxford University Press.

Strehler, B. L., 1977, "Time, Cells and Aging," New York, Academic Press.

ACKNOWLEDGEMENTS

This project was supported in part by the Pennsylvania Department of Aging. Bell of Pennsylvania, Colonial Penn Insurance Group, GTE Service Corporation, Alcoa Foundation, General Electric Company and the American Association of Retired Persons provided the intellectual and financial support. Special thanks are due to Mr. J. Eicher, Center on Aging, University of Maryland, for invaluable assistance. This data was, in part, presented at the Health Care Conference, Snowbird, Utah, April, 1983. Inquiries should be addressed to G. T. Baker, Center on Aging, University of Maryland, College Park, MD, 20742.

COMMUNITY ALARM SYSTEMS FOR OLDER PEOPLE -

AGE CONCERN SCOTLAND

Jan Killeen

Age Concern Scotland
Edinburgh, Scotland

There are at present three existing community alarm schemes in Scotland: in Lothian, Central and Highland regions, with several other authorities in the planning stages. Age Concern Scotland realised it was vital to hold a conference to examine the philosophy behind this trend, the reliability of the technology, the implications for other services and, most importantly, what old people felt about this "crisis response." Studies were presented by John Sewel, Institute for the Study of Sparsely Populated Areas, and Alan Butler, University of Leeds.

A Framework for Evaluation

Tilda Goldberg's[1] recently published book The Effectiveness of Social Care for the Elderly suggests that in order to mount an effective evaluation of a social provision the following dimensions need to be addressed: Aims, Needs, Means, Outcomes. Butler analysed each dimension in relation to alarm systems and it is to the analysis of needs that I shall turn now.

Needs

The claim is often made that older people "need" alarm systems and yet the term "need," as Bradshaw[2] has so usefully demonstrated, may itself be viewed from at least four perspectives:
a) Normative Need: Expert or professional assessment of need.
b) Felt Need: What people say they want when asked. c) Expressed Need: Demand, felt need turned into action. d) Comparative Need: Gap between service provision in one area and another.

At the moment, normative need would appear to be the most powerful in making a case for alarm systems. But is an alarm system the appropriate response to make, or are there alternatives which fit the bill just as well, or even better? For example, if one wants to provide an alarm in order to reduce anxiety, then a number of other devices and methods might be more appropriate: telephones, entry locks on doors, burglar alarms, fire detectors, etc. If an alarm system is seen as a way of minimising falls and their effects, then a number of other prosthetic devices,--barrier free housing, walking frames, etc. may need to be investigated. If an alarm system is to provide a means of contact, then a telephone or a visiting warden may fit the bill. There is a danger that because the new technology is available individuals' needs are ignored and an alarm offered as a cure-all. Having once created a system, it demands to be fed, and other local services, lacking perhaps the glamour of alarms, may become drained of resources.

The following points are listed by Age Concern Scotland as a brief summary of the main issues raised which need further investigation: 1. more research is necessary on the use made of alarms particularly with regard to the rapidly developing community or dispersed alarm systems; 2. up-to-date, independent, assessment of the reliability and practicability of alarms is required; 3. trials need to be conducted comparing the utility of alarm systems with such alternatives as telephones, emergency buzzers, etc.; 4. there is a need to examine any secondary effects, e.g., on patterns of interaction; and 5. far greater efforts should be made, when introducing a system, to ensure that the elderly recipients understand how it works, and feel comfortable about using it.

NOTE: Copies of the report should be requested from: Age Concern Scotland, 33 Castle Street, EDINBURGH, EH2 3DN.

REFERENCES

1. E. Goldberg and N. Connolly, "The Effectiveness of Social Care for the Elderly," Heinemann and P.S.I, (1982).
2. J. Bradshaw, The taxonomy of social need, in: "Problems and Progress in Medical Care," G. McLachlan, ed., OUP, 1972.

CHANGES IN THE HOME AND COMMUNITY ENVIRONMENT:
CHANGING ENVIRONMENTAL REQUIREMENTS AND SAFETY IN
RESIDENTIAL SETTINGS

Arthur Stern

Frankfurt Association for Aid to the Elderly
and Handicapped

The city of Frankfurt is generally known as the business
metropolis of the Federal Republic of Germany--the city of the
bankers--and for its international airport. Frankfurt is also a
city which has always featured prominently in the development of
social policy. An important part of Frankfurt's social policy is
based on the charitable welfare organizations, who provide social
services jointly with the city of Frankfurt. In particular, the
social policy for the senior citizens is primarily the task of the
Frankfurt Association for Aid to the Elderly and Handicapped
(Frankfurter Verband fur Alten- und Behindertenhilfe e.V.), which is
practically a part of the social service department of the city of
Frankfurt. Our association was founded 65 years ago, towards the
end of the first world war in 1917/1918, with the aim to organize
and provide homes to shelter the elderly destitutes. After a period
of relative inactivity the association started, after the 2nd world
war, with new tasks, such as opening residential homes, day-centers
and clubs for the aged. In 1974 the association had 340 employees,
7 homes for the aged and just a few clubs for the elderly. With a
staff of 762 employees and 210 voluntary helpers, it now provides a
most comprehensive service for approximately 15,000 citizens.

The city of Frankfurt has a population of 630,000, of which
110,000 (18%) are over 65 years old; 42,000 are over 75 years.
According to projections there will be little change in the number
of older citizens in the next 8 to 10 years. In order to enable
older persons to remain in their familiar environments and thus
maintain their social contacts, priority will be given to the
modernization of existing rental units . Whenever essential, the
elderly and/or handicapped citizens will be provided with home-help,
meals on wheels, visiting services through voluntary helpers or

449

neighbors, a Home Care Alarm System and day-care centers.

In connection with these services, a HOME CARE ALARM SYSTEM (Telecare) was introduced in 1982. There are now more than 400 elderly and/or handicapped citizens connected to the system. The rapid advances made in recent years in the field of electronics, the latest and most sensational of which are the microprocessors, provided the new, hitherto undreamed of possibility to develop this system. The aim of this Home Care Alarm System is to improve the feeling of security of those in need of help within the environment to which they are accustomed. The system consists of a subscriber unit and the emergency center (exchange). The exchange has a microprocessor-based control and indicator panel with loudspeaker and microphone, as well as a minicomputer with keyboard and terminal. This system guarantees a fast access to all stored data on the person initiating the alarm. The exchange automatically receives the alarm and the individual data on the person calling (such as name, address, medical practitioner, next of kin, neighbors, voluntary helper, home-help etc.) is immediately displayed on the terminal. A printer and tape-recorder are connected to provide documentation of the whole procedure. The subscriber initiates an emergency call by pressing a button on the home unit or by the use of a compact battery-operated radio device, carried either in the pocket or on a loop hung around the neck. The subscriber's unit also has a built-in microphone and loudspeaker, thus enabling the exchange to establish a verbal connection without the need to lift the telephone receiver. This method of initiating an emergency call is based on the assumption that the subscriber seeking help is, in fact, in a position to operate the device. An additional feeling of security is provided for the subscriber by an activity check. By pressing a button on the home unit, one or more times daily, a monitor is activated. If there has been no activity within a predetermined length of time, on account of loss of consciousness or some other cause, the subscriber's unit automatically initiates an emergency call.

In conjunction with the commencement of operation of the Home Care Alarm System, the association extended the neighborly services and increased the voluntary helpers in order to provide more home help and care for the elderly and handicapped citizens in their homes. Since the introduction of this system, more than 10% of the subscribers would have needed institutional care, had they not been connected to the Home Care Alarm System. Our observations confirm our opinion that the persons connected to the Home Care Alarm System show a higher degree of satisfaction; are happier regarding their situation; have more frequent contacts with neighbors; and are more responsive and less troubled by pessimistic thoughts, than those persons who are not connected to this system. The conclusion is that wellbeing in old age requires an environment which meets the needs of the elderly.

HOME SAFETY CHECKUPS FOR SENIORS

Sam Zagoria

U.S. Consumer Product Safety Commission
Washington, D.C.

The Consumer Product Safety Commission is developing an idea
that could help save possibly two billion dollars in Federal, local
and family expenditures. It is an idea we are convinced will also
bring longer and happier lives to millions of our nation's senior
citizens. An estimate was recently developed by a noted independent
research institution, the Buffalo Organization for Social and
Technological Innovation, for the Administration on Aging. It
concludes, and CPSC studies support the estimate, that the overall
impact of accidental injury to those over 55 years of age costs our
economy something in the neighborhood of two billion dollars a
year. This is a staggering figure. For example, it represents
something like two-thirds the figure to be spent on all federal law
enforcement in fiscal year 1984.

Since many of the accidents generating this outlay of public
and private funds could be avoided, this two billion dollars
represents an incredible and unnecessary waste of hospital and
health resources, convalescent facilities and productivity, not to
mention its drain on federal medicaid and medicare programs and the
economic and emotional resources of friends, family and community.
We are convinced that this dollar cost and the human suffering
associated with it can be dramatically reduced and we believe the
answer lies in technology which has already been successfully
applied in such national programs as energy conservation and fire
prevention. What we propose is a one-on-one, voluntary, direct
intervention program to safeproof the residential environment of our
elderly. If public programs have encouraged people to insulate
their homes and install smoke detectors, why can't we also get them
to remove or reduce the home hazards that yearly are killing,

451

nationwide, approximately 15,000 of our elderly? Can we not
interest service clubs, local health and building inspection units,
consumer groups and senior citizen centers to participate in a home
inspection program for the elderly? Wouldn't such a program reduce
the 7,000 or so yearly deaths of those over sixty-five due to falls
resulting from or contributed to by such things as unsafe stairs and
tubs and floor coverings? Can we not reduce the maiming and deaths
due to burns involving flammable wearing apparel and hazardous
cooking practices, or those due to scalding from unsafe hot water
tap temperatures? Why can't we insure that new residential
construction incorporates our best thinking on a safe environment
for elder citizens? Why can't we alert elderly residents to the
hazards like overly steep stairs, uneven risers, slippery tubs and
out-of-reach cupboards, which were merely an inconvenience when the
young residents moved in a generation or more ago, but now are
potential death traps to these older residents? We are convinced
these things can be done and we are funding a priority project for
fiscal year 1984, beginning October 1, to test our conviction.

The Consumer Product Safety Commission is a federal independent
regulatory agency. The general mandate of the Commission is to
monitor the safety of most consumer products. Our particular
concern for the elderly stems in part from our statute that requires
us to pay particular attention to this population and from our
experience that the safety problems of the elderly are indeed
unique. Our proposal is conservative and yet we believe it could
lead to significant reductions in the demands on already hard
pressed medical and convalescent resources. We have allocated a
modest $75,000 and 30 staff months for the development of a pilot
program to test whether or not the national program described above
could work and, if so, how it could be implemented.

Briefly, the intended project is as follows: (1) Develop a
check list for home safety. This would include not only a search
for products that have been banned by the Commission or otherwise
have been found to constitute a safety hazard but also advice
concerning fall and fire hazards about which we have developed
significant expertise. (2) Select and train volunteers to conduct
the safety check. (3) Select a community with many elderly
residents that presents a representative cross-section in terms of
hazard potential. (4) Conduct the safety check, door-to-door, on a
voluntary basis in the selected community. (5) Do a series of
return visits to investigate changes in the safety environment
brought about by the safety check. (6) Evaluate the program in
terms of its replicability, its cost and efficacy.

AGING AND TECHNOLOGICAL ADVANCES:

HOME AND COMMUNITY

Victor A. Regnier
School of Architecture
Andrus Gerontology Center
University of Southern California

Elias Cohen
WITF Communications Center
Harrisburg, Pennsylvania

INTRODUCTION

This chapter summarizes the comments of the home and community environment study group. This panel was perhaps the most eclectic of the four study groups and thus the range of discussion was wide and the topics available for review were varied. The issues presented ranged from the development of computer aided robotic devices to the design of comfortable therapeutic furniture. The range of topical issues made for stimulating and interesting discussion. Originally it was thought that consensus could be developed around questions dealing with the effects of technology on residential and community settings. However, the hoped for indepth analysis and integrative commentary on important issues was hampered by the varied backgrounds of panel members and the lack of time available for discussion. What did occur, however, was a clear enumeration of concerns regarding how high technology monitoring and enabling equipment will be accepted and/or made acceptable to older people. Study group participants were quick to identify a number of technological influences that would likely have a major impact on the lifestyles of older residents. The one single interest the study group members shared was a concern about how technological advances would directly affect the everyday lives of older people. The group recognized these advances with cautious optimism. Many participants were skeptical of the numerous positive predictions which portray life with technological aides as a paradise or absolute panacea. Others viewed these advances as inevitable change agents in society that should be directed and orchestrated to bring

453

about the most positive outcomes for older people. The following
enumerates some of the issues raised during the discussion and
highlights the varied perspectives which were brought to bear on the
problem.

WHO IS GOING TO CONTROL THE NEW TECHNOLOGY?

The issue of paramount concern vividly illustrated in two
presentations was the question of how and for what purposes
technological advances would be employed? Bowersox presented an
architectural case study of a new high rise housing project for the
elderly in Minneapolis to illustrate how existing technology was
being employed in the newest, most advanced purpose built housing.
The typical high-technology features available in new construction
are security, communication and fire safety. These applications
primarily improve the function of efficient building management and
share the primary advantage of making the building easier to control
and protect. The major benefits of technology typically occur to
management, by allowing savings to take place from reducing
personnel or increasing the accuracy and predictability of the
system's function. With a few exceptions (e.g., color T.V. hook-ups
to a "broadcasting" room on the first floor) the communications
system technology seems to be employed to manage rather than benefit
the residents directly. The application of this technology to
stimulate self expression, social exchange and intellectual
development seems lost. These first applications may represent the
"seating" of this technology in a system where cost effective
reasoning seeks to first replace labor intensive management tasks.
However, it does not bode well for a future vision of technology as
liberator and saviour of the human condition. Liefer's presentation
describing recently designed robotic devices at the Stanford
Veterans Administration center underscored in a dramatic way the
pace of technological advancement. Slides of previous generations
of robotic arms were presented along with newer versions that could
perform an even greater number of highly intricate manipulations
driven by even smaller computer consoles.

Just as the silicon chip has allowed a miniaturization of
calculation hardware, so has this same advanced technology made it
possible to construct a compact version of a slave helper.
Experimentation with disabled veterans demonstrated through video
tape test trials reinforced the promise of this technology.
Demonstrations of technological devices that enable the disabled to
walk, the blind to see and the deaf to hear cannot be viewed without
an emotional response. However, the application of these same
devices to older persons, many of which have lost the cognitive
capacity to manage or direct the hardware, raises a number of
questions about the practicality of a "user centered" philosophy of
machine control. Liefer's philosophy and the philosophy of others

who are working to develop these innovations emphatically underscore
the requirement that human control over technology continue as a
necessary pre-condition of further machine development. The
discussion in the final wrap-up session of the conference about a
new demonstration project employing robotic arms as feeding devices
in nursing homes brought back concerns about how "management" may
adapt this new technology to increase efficiency and maintain
greater control. How technology will be used, whether to free or
more efficiently entrap the older person, is an issue that will
continue to concern technologists and gerontologists. This is of
particular concern because the severely mentally impaired patient
who requires the greatest amount of attention may be the one for
whom a new technological device like this would be considered most
cost beneficial.

CAN THE APPLICATION OF TECHNOLOGY BE DIRECTED TO IMPORTANT PROBLEMS?

The robotic devices introduced by Leifer demonstrated this
concern; however, the study group voiced some concern over the
seemingly random application of technological solutions to problems
that may not be as critical as others. For some people, the option
of having one hundred cable television stations from which to choose
may be an important factor influencing the quality of life; but many
panel members felt a number of other problems were more critical and
should perhaps be addressed first. Among the suggestions made by
the group was one which dealt with the use of technology in making
the person's dwelling unit safer. How can technology be creatively
harnessed to avoid accidents in the home, in particular the costly
ones that lead to broken hips and legs that tax the older person
through premature institutionalization as well as our publically
financed health care system? Another important problem identified
by the group dealt with the use of technology to bring people
together. A future of electronic image transfers may reduce the
number of social encounters made at the bank, the grocery store, or
the department store. How can we influence the development of
compensatory communications systems that will make up for these lost
social exchanges and contacts? Are there other critical attributes
of our society that may be unavoidably altered by this new
technology? If so, perhaps an investigation of compensatory
responses should be undertaken.

HOW CAN PUBLIC-PRIVATE SECTOR COMMUNICATION BE ENHANCED TO ENCOURAGE THE EXPLORATION AND DEVELOPMENT OF NEW TECHNOLOGIES?

Several presentations dealt with the general concern of
defining the market potential of new products. Gollub presented a
model of corporate innovation and development, in which he argued

that many corporations are limited by past experiences and by past
successes in exploring new avenues for product development. His
model suggested a "fit" must exist between the corporate sponsor's
experience and the product's potential for development. The
conservative nature of many corporate development policies may mean
that a new product might not be introduced until the market
potential is great enough to support sales. New technologies that
appeal to only a small group of people may never be sanctioned as
commercially feasible and thus may not be introduced until a market
demand can be demonstrated. Decreased public agency support for
research may also encourage the development of more public-private
consortiums. These organizations may take a lead in developing new
technologies as academic institutions and industry work together on
problems of mutual interest.

HOW CAN INTERNATIONAL COOPERATION LEAD TO THE DEVELOPMENT AND REFINEMENT OF NEW TECHNOLOGICAL ADVANCEMENTS?

Several presentations made by Europeans and Americans dealt
with emergency call systems that linked older people in their own
independent housing unit to an emergency response team in the
community. Gatz reported a recent evaluation of the popular "life
line" system operating in the Los Angeles area under the acronym
EARS (Emergency Alert Response System). Killian and Stern spoke
about similar European systems which had been operational for a
number of years. Typically the European systems were more
ubiquitous because of widespread public support for these systems.
Issues ranging from the cost/beneficial aspects of various system
designs to the detailed protocol involved in securing aid and
assistance were introduced and made for a lively discussion and
exchange of ideas. The real value of an international symposium was
aptly demonstrated by the discussion that ensued around this topic.
Each presenter described an emergency system with unique features
developed in different socio-political, historical and cultural
contexts. Exchanges of this sort can advance the state-of-the-art
in a way that no single empirical evaluation is able to do. Viewing
a similar problem from a quite different cultural perspective can
lead to creative insights about the problem and one's approach to
that problem.

WHAT KINDS OF "LOW TECH" SOLUTIONS CAN IMPROVE THE HOUSING ENVIRONMENT AND LIFESTYLES OF OLDER PEOPLE?

The description of "hi-tech" solutions was balanced by
presentations and commentary emphasizing the need to make
improvements that do not necessarily require new electronic
gadgetry. Haber's presentation began with an indictment of
"hi-tech" hospital based diagnostic and treatment equipment and led

to an emphasis on the design of the environment and the encouragement of simple but effective appliances for older persons. He argued that the quality of life would be more dramatically affected by simple changes in environmental design than by the acquisition of another multi-million dollar diagnostic device. Krauser's presentation closely followed this theme by presenting a framework for the use and application of simple as well as complex technologies.

Conclusions

The preceeding issues were among many the study group dealt with indirectly and directly during the two days it met. More questions were raised than answers provided during the session and all hopes of arriving at any agreement ended when it was discovered the group was represented by a dozen different disciplines. However, the discussion led to a strong consensus shared by group members that systematic evaluations of how new technological advances work for different people must be encouraged. Experiments should be designed in such a way as to allow the full inter-disciplinary, interactive and intergenerational effects to be measured and carefully considered. The future is above all else highly unpredictable, but, if we can enter it with as much knowledge as possible about how new technologies may affect society, perhaps we can exert influences that will lead us toward technologies that stimulate and liberate rather than those that would degrade or enslave.

INDEX